Water Treatment

PRINCIPLES AND PRACTICES OF WATER SUPPLY OPERATIONS SERIES

Series Editor, Harry Von Huben

Water Sources, Second Edition

Water Treatment, Second Edition

Water Transmission and Distribution, Second Edition

Water Quality, Second Edition

Basic Science Concepts and Applications, Second Edition

Water Treatment

Second Edition

American Water Works Association

Water Treatment, Second Edition
Principles and Practices of Water Supply Operations

Library of Congress Cataloging-in-Publication Data

Water treatment--2nd ed.
xvi, 523 p., 18x23 cm.--(Principles and practices of water supply operations)
Includes index.
Originally published: Introduction to water treatment.
ISBN 0-89867-789-0
1. Water--Purification I. American Water Works Association. II. Introduction to water treatment. III. Series: Principles and practices of water supply operations (Unnumbered)
TD430.157 1995
628.1'62--dc20 95-32337
CIP

Disclaimer
Many of the photographs and illustrative drawings that appear in this book have been furnished through the courtesy of various product distributors and manufacturers. Any mention of trade names, commercial products, or services does not constitute endorsement or recommendation for use by the American Water Works Association or the US Environmental Protection Agency.

ISBN 0-89867-789-0

Cover and book design by Susan DeSantis
Editor: Phillip Murray
Composition: Carrie Dubois

Printed in the United States of America

American Water Works Association
6666 West Quincy Avenue
Denver, CO 80235
(303) 794-7711

Printed on recycled paper.

CONTENTS

FOREWORD

Water Treatment is part two in a five-part series titled Principles and Practices of Water Supply Operations. It contains information on commonly used water treatment processes and an overview of treatment plant instrumentation and control.

The other books in the series are

Water Sources
Water Transmission and Distribution
Water Quality
Basic Science Concepts and Applications (a reference handbook)

References are made to other books in the series where appropriate in the text.

A student workbook is available for each of the five books. These may be used by classroom students for completing assignments and reviewing questions, and as a convenient method of keeping organized notes from class lectures. The workbooks, which provide a review of important points covered in each chapter, are also useful for self-study.

An instructor guide is also available for each of the books to assist teachers of classes in water supply operations. The guides provide the instructor with additional sources of information for the subject matter of each chapter, as well as an outline of the chapters, suggested visual aids, class demonstrations where applicable, and the answers to review questions provided in the student workbooks.

The reference handbook is a companion to all four books. It contains basic reviews of mathematics, hydraulics, chemistry, and electricity needed for the problems and computations required in water supply operations. The handbook also uses examples to explain and demonstrate many specific problems.

ACKNOWLEDGMENTS

This second edition of *Water Treatment* has been revised to include new technology and current water supply regulations. The material has also been reorganized for better coordination with the other books in the series. The author of the revision is Harry Von Huben.

Special thanks are extended to the following individuals who provided technical review of all or portions of the second-edition outline and manuscript:

William J. Daly, City of Concord, Department of Water Resources, Concord, N.H.
Tom Feeley, Red Rocks Community College, Lakewood, Colo.
Stephen E. Jones, Iowa State University, Ames, Iowa
Kenneth D. Kerri, California State University, Sacramento, Calif.
Gary B. Logsdon, Black & Veatch, Inc., Cincinnati, Ohio
Benjamin W. Lykins Jr., US Environmental Protection Agency, Drinking Water Research Division, Cincinnati, Ohio
William Moorhead, MacDonald–Stephens Engineering, Mission Viejo, Calif.
Thomas J. Sorg, US Environmental Protection Agency, Drinking Water Research Division, Cincinnati, Ohio

American Water Works Association staff who reviewed the manuscript included Ed Baruth, George Craft, Bruce Elms, Robert Lamson, Joe McDonald, Steve Posavec, and Elizabeth Ralph.

Publication of the first edition was made possible through a grant from the US Environmental Protection Agency, Office of Drinking Water, under Grant No. T900632-01. Principal authors of portions of the original manuscript were Jack W. Hoffbuhr, Michael D. Curry, Ralph W. Leidholdt, Nancy E. McTigue, and Thomas E. Braidech.

The following individuals are credited with participating in the review of the first edition: Donald B. Anderson, Charles R. Beer, James O. Bryant Jr., Hugh T. Hansen, James T. Harvey, William R. Hill, Kenneth D. Kerri, Russell W. Lane, Jack E. Layne, O. Thomas Love, Andrew J. Piatek Jr., Donald C. Renner, John F. Rieman, Frank W. Sollo, Alan A. Stevens, James M. Symons, and Robert L. Wubbena.

INTRODUCTION

Today, many of the water sources used to supply drinking water to the public are contaminated. The contamination may be from natural pollutants or manufactured chemicals that have been disposed of carelessly. Drinking water can also be contaminated if a water treatment system is improperly operated.

For these reasons, most public water systems treat their source water to make it safe and palatable for human use. The water that treatment systems deliver to the consumer must not contain contaminants in amounts that could cause disease or be toxic to the consumer. In addition, the taste, odor, and color of the water must be acceptable to the public.

The Operator's Role in Treatment

The delivery of safe, palatable drinking water is dependent on well-trained, knowledgeable treatment plant operators whose job is to ensure that safe water is delivered to the customer at all times. The duties of a water treatment plant operator vary widely. In general, an operator may be expected to

- operate mechanical equipment, including pumps, filters, meters, and chemical feeders
- operate electrical and electronic equipment, including motors, controllers, automatic monitors, recorders, and standby power systems
- calibrate, maintain, service, repair, and replace various mechanical, electrical, and electronic equipment
- determine proper chemical dosages and control chemical applications for the treatment processes
- inventory, order, and store chemicals
- inventory and maintain an appropriate stock of spare parts for equipment

- keep accurate and complete records of treatment operations and submit required reports to government agencies
- collect water samples for testing by state or commercial laboratories
- perform certain laboratory analyses
- maintain a safe working environment
- perform regular preventive maintenance on various types of equipment
- perform general plant maintenance and housekeeping
- keep informed on local, state, and federal regulations affecting the water system
- keep informed on new technology and investigate improved equipment or methods of operation that could improve the efficiency or safety of treatment plant operations
- recommend to superiors any repairs, replacements, or improvements that should be made to the treatment system

To perform these duties successfully, an operator must have good judgment and sufficient education in the fundamentals of mathematics, hydraulics, electricity, bacteriology, and chemistry. The operator also needs training and experience in the operation, maintenance, repair, and replacement of water treatment plant equipment.

The chapters in this book discuss specific treatment processes in detail. However, not every treatment plant will use every process. The types and arrangements of processes in a water treatment plant depend primarily on the characteristics of the raw water to be treated. The type of treatment used may also be dictated by other factors, such as the size of the water system, water availability, customer desires, and local economic conditions.

A well-informed operator should know more than just how to operate his or her local plant. The operator must understand how and why the treatment process works, and learn about other processes that may work better or may at some time have to be added to the current treatment operation.

CHAPTER 1

Water Treatment Processes

Most groundwater and surface water sources that are used to supply a public water system contain some impurities. The types and concentrations of these impurities generally dictate the types of treatment (if any) that must be performed on the water before it is provided to the public.

Why Water Requires Treatment

Removing certain impurities, adding additional substances, or otherwise changing the character of water is generally done either to protect public health or to improve the water's aesthetic qualities. Water utilities also treat water in order to comply with drinking water regulations.

Health-Related Treatment

The contaminants in drinking water that can cause sickness or death fall into the general categories of organic and inorganic chemicals, radionuclides, and various types of microorganisms. In some cases, the danger may not be noticed immediately; possible adverse health effects may occur only after years of exposure. In other cases, sickness can be caused from chemicals and harmful microorganisms as a result of only one drink of contaminated water. These contaminants must be removed from the water.

Chemicals may also be added to drinking water to benefit public health. The principal example is fluoride. The public benefit of maintaining fluoride in drinking water to reduce the incidence of cavities in children's teeth is well documented, and many states require fluoridation of municipal public water supplies.

Aesthetic-Related Treatment

Water may have aesthetic qualities that make it unpalatable to the consumer. Water with an unpleasant taste, odor, or color is simply not acceptable. Water may also have qualities that make it undesirable for use in the home, such as extreme hardness, high levels of dissolved solids, or the tendency to cause stains in laundry or on fixtures.

While these undesirable qualities do not generally pose a health threat to consumers, water systems try to furnish the best possible water within economic constraints. If customers are dissatisfied with the quality of their water, they can generally bring pressure on the water utility to make treatment changes that will improve the quality.

Treatment Required by Regulations

All contaminants considered by the US Environmental Protection Agency (USEPA) to pose a danger to public health in drinking water are governed by primary drinking water regulations. The regulations specify a maximum contaminant level (MCL) for most regulated contaminants. If a public water system in the United States is found to exceed any of the limits, the system must take steps to provide treatment to lower the level of contamination or change its source of water. In a few instances, rather than establishing an MCL, the USEPA provides a treatment regulation. The regulation dictates a type of treatment that will minimize the level of contamination without the need to specifically measure the contaminant level.

The secondary drinking water standards provided by the USEPA set limits for impurities that do not pose a known health threat. Although these secondary MCLs are not mandatory, they are strongly recommended; if these limits are exceeded, the public generally finds the aesthetic qualities of the water disagreeable. Many states have requirements for aesthetic quality that are more stringent than the federal standards.

Additional details of drinking water regulations are provided in *Water Quality*, also part of this series.

Types of Treatment

Because groundwater and surface water come from different sources, they require different types of treatment. These types of treatment are discussed in the following sections.

Groundwater Treatment

Groundwater usually contains little or no turbidity (cloudiness) and few harmful microorganisms. In addition, it usually has a relatively uniform mineral content and constant temperature.

However, groundwater often has relatively high hardness and objectionable amounts of contaminants, including iron, manganese, hydrogen sulfide, and radionuclides. In addition, many wells have been found in recent years to be contaminated by synthetic chemicals, such as pesticides, herbicides, and industrial solvents. A unique problem experienced by some groundwater systems is discolored water caused by iron bacteria.

In the past, many groundwater systems have supplied the public with water of acceptable quality without providing any treatment at all. However, increasing numbers of systems are now adding treatment for one or more of the following reasons:

- to meet federal and state requirements for disinfection and maintenance of a chlorine residual in the distribution system
- to remove contaminants identified as posing a threat to public health
- to satisfy a demand by the public for improved aesthetic quality of the water

Surface Water Treatment

By USEPA directive, all surface water is assumed to be contaminated by harmful microorganisms. The water must therefore receive adequate treatment to ensure that harmful organisms are effectively killed or inactivated. In addition, surface water usually has turbidity at levels in excess of federal regulations. This turbidity, which may be caused by microbiological contamination, sand, silt, algae, or other suspended matter, must be reduced. A reduction in turbidity is necessary because it is aesthetically unacceptable to customers and because it interferes with adequate disinfection.

Surface water usually has varying temperature, mineral content, and levels of contamination due to natural causes as well as human activities. This variability requires flexibility in the treatment process in order to provide the highest-quality water at the least possible treatment cost.

Selection of Treatment Methods

The types of water treatment that are regularly used to improve water quality, remove microorganisms, and reduce the level of toxic substances fall into the following general categories:

- aeration
- coagulation, sedimentation, and filtration processes
- lime softening
- ion exchange processes
- membrane processes
- chemical oxidation (disinfection)
- adsorption

In most instances, water quality can be improved using one or more of these methods. The general effectiveness of these treatment methods is summarized in Table 1-1. These methods will be discussed in detail in later chapters. The process of choosing the treatment method or methods to be used can, at times, be quite involved. Questions to ask when making the decision include

- Will a groundwater or surface water source be used? As noted earlier, water from different sources requires different types of treatment.
- What is the system size? Some treatment methods are best suited for small systems for reasons such as initial cost and simplicity of operation. Methods with the lowest operating cost may be applicable only to a large, complex operation.
- What water quality improvements are required? If only one type of treatment is needed to meet regulations and make the water acceptable by the public, the lowest-cost method of achieving it is usually chosen. If two or more types of improvement are necessary, they must be considered together in deciding on the best treatment methods to use.
- Are there other water quality improvements that are desirable? There are often some improvements to quality that are desirable but hard to justify as an absolute necessity. These can often be accomplished along with the required quality improvements at little or no extra cost if the proper treatment method is selected.

TABLE 1-1 General effectiveness of water treatment processes for contaminant removal

Contaminant Category	Aeration and Stripping	Coagulation, Sedimentation, Filtration	Lime Softening	Ion Exchange		Membrane Processes			Chemical Oxidation, Disinfection	Adsorption		
				Anion	Cation	Reverse Osmosis	Ultrafiltration	Electrodialysis		Granular Activated Carbon	Powdered Activated Carbon	Activated Alumina
A. Primary Contaminants												
1. Microbial and turbidity												
Total coliforms	P	G–E	G–E	P	P	E	E	—	E	F	P	P–F
Giardia lamblia	P	G–E	G–E	P	P	E	E	—	E	F	P	P–F
Viruses	P	G–E	G–E	P	P	E	E	—	E	F	P	P–F
Legionella	P	G–E	G–E	P	P	E	E	—	E	P	P	P–F
Turbidity	P	E	G	F	F	E	E	—	P	F	P	P–F
2. Inorganics												
Arsenic (+3)	P	F–G	F–G	G–E	P	F–G	—	F–G	P	F–G	P–F	G–E
Arsenic (+5)	P	G–E	G–E	G–E	P	G–E	—	G–E	P	F–G	P–F	E
Asbestos	P	G–E	—	—	—	—	—	—	P	—	—	—
Barium	P	P–F	G–E	P	E	E	—	G–E	P	P	P	P
Cadmium	P	G–E	E	P	E	E	—	E	P	P–F	P	P
Chromium (+3)	P	G–E	G–E	P	E	E	—	E	F	F–G	F	P
Chromium (+6)	P	P	P	E	P	G–E	—	G–E	P	F–G	F	P
Cyanide	P	—	—	—	—	G	—	G	E	—	—	—
Fluoride	P	F–G	P–F	P–F	P	E	—	E	P	G–E	P	E
Lead	P	E	E	P	F–G	E	—	E	P	F–G	P–F	P
Mercury (inorganic)	P	F–G	F–G	P	F–G	F–G	—	F–G	P	F–G	F	P
Nickel	P	F–G	E	P	E	E	—	E	P	F–G	P–F	P
Nitrate	P	P	P	G–E	P	G	—	G	P	P	P	P
Nitrite	F	P	P	G–E	P	G	—	G	G–E	P	P	P

NOTES: P: poor (0 to 20% removal); F: fair (20 to 60% removal); G: good (60 to 90% removal); E: excellent (90 to 100% removal); "—": not applicable/ insufficient data.

Costs and local conditions may alter a process's applicability.

Table continued next page

TABLE 1-1 General effectiveness of water treatment processes for contaminant removal (continued)

Contaminant Categories	Aeration and Stripping	Coagulation, Sedimentation, Filtration	Lime Softening	Ion Exchange		Membrane Processes			Chemical Oxidation, Disinfection	Adsorption		
				Anion	Cation	Reverse Osmosis	Ultrafiltration	Electrodialysis		Granular Activated Carbon	Powdered Activated Carbon	Activated Alumina
2. Inorganics (continued)												
Radium (226 and 228)	P	P–F	G–E	P	E	E	—	G–E	P	P–F	P	P–F
Selenium (+6)	P	P	P	G–E	P	E	—	E	P	P	P	G–E
Selenium (+4)	P	F–G	F	G–E	P	E	—	E	P	P	P	G–E
3. Organics												
Volatile organic chemicals	G–E	P	P–F	P	P	F–E	F–E	F–E	P–G	F–E	P–G	P
Synthetic organic chemicals	P–F	P–G	P–F	P	P	F–E	F–E	F–E	P–G	F–E	P–E	P–G
Pesticides	P–F	P–G	P–F	P	P	F–E	F–E	F–E	P–G	G–E	G–E	P–G
Trihalomethanes	G–E	P	P	P	P	F–G	F–G	F–G	P–G	F–E	P–F	P
Trihalomethane precursors	P	F–G	P–F	F–G	—	G–E	F–E	G–E	F–G	F–E	P–F	P–F
B. Secondary contaminants												
Hardness	P	P	E	P	E	E	G–E	E	P	P	P	P
Iron		F–G	F–E	E	P	G–E	G–E	G	G–E	G–E	P	PP
Manganese	P–F	F–E	E	P	G–E	G–E	G	G–E	F–E	P	P	P
Color	P	F–G	F–G	P–G	—	—	—	—	F–E	E	G–E	G
Taste and odor	F–E	P–F	P–F	P–G	—	—	—	—	F–E	G–E	G–E	P–F
Total dissolved solids	P	P	P–F	P	P	G–E	P–F	G–E	P	P	P	P
Chloride	P	P	P	F–G	P	G–E	P	G–E	P	P	P	—

NOTES: P: poor (0 to 20% removal); F: fair (20 to 60% removal); G: good (60 to 90% removal); E: excellent (90 to 100% removal); "—": not applicable/ insufficient data.

Costs and local conditions may alter a process's applicability.

Table continued next page

TABLE 1-1 General effectiveness of water treatment processes for contaminant removal (continued)

Contaminant Categories	Aeration and Stripping	Coagulation, Sedimentation, Filtration	Lime Softening	Ion Exchange		Membrane Processes			Chemical Oxidation, Disinfection	Adsorption		
				Anion	Cation	Reverse Osmosis	Ultrafiltration	Electrodialysis		Granular Activated Carbon	Powdered Activated Carbon	Activated Alumina
B. Secondary Contaminants (continued)												
Copper	P	G	G–E	P	F–G	E	—	E	P–F	F–G	P	—
Sulfate	P	P	P	G–E	P	E	P	E	P	P	P	G–E
Zinc		P	F–G	G–E	P	G–E	E	—	E	P	—	—
Total organic carbon	F	P–F	G	—	G–E	G	G–E	P–G	G–E	F	F–G	—
Carbon dioxide	G–E	P–F	E	P	P	P	P	P	P	P	P	P
Hydrogen sulfide	F–E	P	F–G	P	P	P	P	P	F–E	F–G	P	P
Methane	G–E	P–E	P	P	P	P	P	P	P	P	P	P
C. Proposed Contaminants												
Volatile organic chemicals	G–E	P	P–F	P	P	F–E	F–E	F–E	P–G	F–E	P–G	P
Synthetic organic chemicals	P–F	P–G	P–F	P	P	F–E	F–E	F–E	P–G	F–E	P–E	P–G
Disinfection by-products	—	P–E	P–F	P–F	—	P	F–G	F–G	F–G	F–E	P–G	—
Radon	G–E	P	P	P	P	P	P	P	P	E	P–F	P
Uranium	P	G–E	G–E	E	G–E	E	—	E	P	F	P–F	G–E
Aluminum	P	F	F–G	P	G–E	E	—	E	P	—	—	—
Silver	F–G	G–E	P	G	—	—	—	P	F–G	P–F	—	—

Source: Water Quality and Treatment *1990.*

NOTES: P: poor (0 to 20% removal); F: fair (20 to 60% removal); G: good (60 to 90% removal); E: excellent (90 to 100% removal); "—": not applicable/ insufficient data.

Costs and local conditions may alter a process's applicability.

- Are there any adverse side effects of any treatment methods? In some instances, there are adverse side effects. For example, disinfection may cause the creation of disinfection by-products (DBPs) in excess of the MCL. Treatment to control DBPs can then increase the corrosivity of the water.

Disposal of Treatment Wastes

Prior to enactment of the Clean Water Act in 1987 and the environmental protection laws that followed, wastes created in water treatment processes were generally disposed of in the easiest way possible, at a relatively low cost. Under increasingly stringent regulations and public scrutiny, disposal of water treatment wastes is now a major consideration in the selection of which processes to use, and it is a significant cost of plant operation.

The disposal of water treatment plant wastes by almost any means other than discharge to the sanitary sewer system will usually require a discharge permit. State authorities should be consulted on allowable methods of disposal, permits required, and monitoring requirements before a new disposal method is used. Environmental laws are strict, and violators may be subject to heavy fines and possible liability for any environmental problems they create.

Central Versus Point-of-Use Treatment

Property owners and businesses can use a variety of treatment equipment to further treat water delivered by a public water system. The most common example is the home water softener. Treatment installed at the point where water enters the building is commonly called point-of-entry (POE) treatment. Treatment provided at only one plumbing fixture or faucet is usually referred to as point-of-use (POU) treatment.

POE and POU equipment is designed primarily for voluntary use by property owners. However, in some situations, a water system may be able to avoid central treatment and meet state and federal drinking water regulations by ensuring that the water is properly treated by every customer. Further discussion of POE and POU devices is provided in appendix B.

Selected Supplementary Readings

Back to Basics Guide to Groundwater Treatment. 1992. Denver, Colo.: American Water Works Association.

Back to Basics Guide to Safe Drinking Water. 1990. Denver, Colo.: American Water Works Association.

Back to Basics Guide to Surface Water Treatment. 1992. Denver, Colo.: American Water Works Association.

Design and Construction of Small Water Systems — A Guide for Managers. 1984. Denver, Colo.: American Water Works Association.

Geldreich, E.E., J.A. Goodrich, and R.M. Clark. 1990. Characterization of Unfiltered Water. ***Jour. AWWA***, 82(12):40.

Manual of Water Utility Operations. 8th ed. 1988. Austin, Texas: Texas Water Utilities Association.

Symons, J.M. 1994. ***Plain Talk About Drinking Water.*** Denver, Colo.: American Water Works Association.

Water Quality and Treatment. 4th ed. 1990. New York: McGraw-Hill and American Water Works Association (available from AWWA).

CHAPTER 2

Treatment of Water at the Source

The treatment applied to raw water to make it safe and palatable for public use is generally performed in a treatment plant, where the process can easily be applied, controlled, and closely monitored. There are a few circumstances in which it is more economical or practical to provide treatment at the source (in situ treatment).

In Situ Treatment

The addition of chemicals into a well should be considered carefully. If similar results can be obtained by treating the water after it leaves the well, then that approach may be better because it can usually be controlled more easily. Consideration must also be given to any adverse effects that added chemicals may have in terms of disintegrating or clogging the well pump and piping.

The most frequently used in-well treatment is the application of an oxidant to wells that have an infestation of iron bacteria. A weak chlorine solution or other oxidant fed at a point near the pump can help reduce the red color in the water caused by iron bacteria. In situ control of iron can also be accomplished by feeding polyphosphate into the well to sequester the iron before it has a chance to oxidize. Additional details of iron and manganese control are covered in chapter 10.

In-well chemical application may also be justified when the pH of the groundwater is such that damage may be done to the pump and piping unless immediate pH correction is made.

If in situ treatment is considered, recommendations concerning its advantages and disadvantages and usage procedures should be obtained from well installation firms and chemical suppliers. Final approval should then be obtained from the state before the treatment is begun.

Aquatic Plant Control

Excessive growth of aquatic plants can cause problems for the operation of a surface water treatment plant. However, it is important to remember that these aquatic plants are normal inhabitants of the aquatic environment and have a definite role in maintaining the ecological balance in lakes, ponds, and streams. For example, algae help purify the water by adding oxygen during the process of photosynthesis. Algae and rooted aquatic plants (water weeds), in moderate amounts, are essential in the food chains of fish and waterfowl.

In addition, water weeds remove nutrients from the water that would otherwise be available for the growth of algae. Therefore, overcontrol of aquatic plants may result in algae problems more serious than the weed problem.

The problems encountered at water treatment facilities from aquatic plants usually arise from the overproduction of a few types of plants at certain times of the year. The goal of a well-planned control program is to control aquatic plants only to the extent necessary to prevent water quality and treatment problems.

Aquatic plants may be classified into the following categories based on their growth form and location in the water:

- algae
- floating weeds
- submerged weeds
- emergent weeds

The taste, odor, color, and mechanical problems caused by aquatic plants can usually be treated or dealt with in a treatment plant if necessary. However, it is usually much easier, less expensive, and more effective to control the growth in the source water.

Algae

Algae are primitive plants that have no true leaves, stems, or root systems and reproduce by means of spores, cell division, or fragmentation. About 17,500 species of algae have been identified, and there are probably many more. The four major groups of algae of interest in water treatment are blue-green algae, green algae, diatoms, and pigmented flagellates.

Heavy concentrations of algae may cause the following operational problems in a water system:

- taste, odor, and color
- toxicity
- filter clogging
- slime accumulation on structures
- corrosion of structures
- interference with other treatment processes
- formation of trihalomethanes (THMs) when the water is chlorinated

Some of the algae that may cause problems in water treatment are illustrated in *Basic Science Concepts and Applications* (also part of this series), appendix D, Algae Color Plates. Table 2-1 lists the problems caused by various types of algae that are experienced by water suppliers.

Taste, Odor, and Color

Although the exact mechanism of taste-and-odor production by algae is not completely understood, the problems are probably caused by certain complex organic compounds that are by-products of their life cycle. The tastes caused from the presence of algae have been categorized as sweet, bitter, and sour. Algae-caused tongue sensations are categorized as oily or slick, metallic or dry, and harsh or astringent. Odors are frequently described as musty, earthy, fishy, grassy, hay-like, or septic. Color can be caused by algal by-products and is usually an indicator that taste-and-odor problems will also occur. Colors range from yellow-green to green, blue-green, red, and brown.

Toxicity

Several types of freshwater algae are somewhat toxic (poisonous). Effects on humans from various toxic algae include skin irritation, promotion of hayfever allergies, and outbreaks of gastrointestinal illness. Blooms of blue-green algae on ponds have also been known to cause fish kills and livestock poisoning.

Filter Clogging

Algae can shorten filter runs by forming a mat on the filter's surface, a process known as "blinding the filter." Filters must then be backwashed more frequently to restore their filtering capacity, which adds to the plant operating costs. Diatoms are usually the primary group associated with this problem.

TABLE 2-1 Problems caused by algae in water supplies

Problem and Algae	Color of water	Algal Group
Slime-Producing Algae		
Anacystis (Aphanocapsa, Gloeocapsa)		Blue-green
Batrachospermum		Red
Chaetophora		Green
Cymbella		Diatom
Euglena sanguinea var. furcata		Flagellate
Euglena velata		Flagellate
Gloeotrichia		Blue-green
Gomphonema		Diatom
Oscillatoria		Blue-green
Palmella		Green
Phormidium		Blue-green
Spirogyra		Green
Tetraspora		Green
Algae Causing Coloration of Water		
Anacystis	Blue-green	Blue-green
Ceratium	Rusty brown	Flagellate
Chlamydomonas	Green	Flagellate
Chlorella	Green	Green
Cosmarium	Green	Green
Euglena orientalis	Red	Flagellate
Euglena rubra	Red	Flagellate
Euglena sanguinea	Red	Flagellate
Oscillatoria prolifica	Purple	Blue-green
Oscillatoria rubescens	Red	Blue-green
Algae Causing Corrosion of Concrete		
Anacystis (Chroococcus)		Blue-green
Chaetophora		Green
Diatoma		Diatom
Euglena		Flagellate
Phormidium		Blue-green
Phytoconis (Protococcus)		Green
Algae Causing Corrosion of Steel		
Oscillatoria		Blue-green

NOTE: Many of these algae are pictured in appendix D of *Basic Science Concepts and Applications*, also part of this series.

Table continued next page

TABLE 2-1 Problems caused by algae in water supplies (continued)

Problem and Algae	Algal Group
Algae Persistent in Distribution Systems	
Anacystis	Blue-green
Asterionella	Diatom
Chlorella	Green
Chlorococcum	Green
Closterium	Green
Coelastrum	Green
Cosmarium	Green
Cyclotella	Diatom
Dinobryon	Flagellate
Elaktothrix gelatinosa	Green
Epithemia	Diatom
Euglena	Flagellate
Gomphosphaeria aponina	Blue-green
Scenedesmus	Green
Synedra	Diatom
Algae Interfering With Coagulation	
Anabaena	Blue-green
Asterionella	Diatom
Euglena	Flagellate
Gomphosphaeria	Blue-green
Synedra	Diatom
Algae Causing Natural Softening of Water	
Anabaena	Blue-green
Aphanizomenon	Blue-green
Cosmarium	Green
Scenedesmus	Green
Synedra	Diatom
Toxic Marine Algae	
Caulerpa serrulata	Green
Egregia laevigata	Brown
Gelidium cartilagineum var. robustum	Red
Gonyaulax catenella	Dinoflagellate
Gonyaulax polyedra	Dinoflagellate
Gonyaulax tamarensis	Dinoflagellate

NOTE: Many of these algae are pictured in appendix D of *Basic Science Concepts and Applications,* also part of this series.

Table continued next page

TABLE 2-1 Problems caused by algae in water supplies (continued)

Problem and Algae	Algal Group
Toxic Marine Algae (continued)	
Gymnodinium brevis	Dinoflagellate
Gymnodinium veneficum	Dinoflagellate
Hesperophycus harveyanus	Brown
Hornellia marina	Flagellate
Lyngbya aestuarii	Blue-green
Lyngbya majuscula	Blue-green
Macrocystis pyrifera	Brown
Pelvetia fastigiata	Brown
Prymnesium parvum	Flagellate
Pyrodinium phoneus	Dinoflagellate
Trichodesmium erythraeum	Blue-green
Toxic Freshwater Algae	
Anabaena	Blue-green
Anabaena circinalls	Blue-green
Anabaena flos-aquae	Blue-green
Anabaena lemmermanni	Blue-green
Anacystis (Microcystis)	Blue-green
Anacystis cyanea (Microcystis aeruginosa)	Blue-green
Anacystis cyanea (Microcystis flos-aquae)	Blue-green
Anacystis cyanea (Microcystis toxica)	Blue-green
Aphanizomenon flos-aquae	Blue-green
Gloeotrichia echinulata	Blue-green
Gomphosphaeria laeustris (Coelosphaerium kuetzingianum)	Blue-green
Lyngbya contorta	Blue-green
Nodularia spumigena	Blue-green
Parasitic Aquatic Algae	
Oodinium limneticum	Dinoflagellate
Oodinium ocellatum	Dinoflagellate

Source: Algae in Water Supplies. *1962. US Public Health Service.*

NOTE: Many of these algae are pictured in appendix D of *Basic Science Concepts and Applications*.

Slimes

Slimes come from the layer that surrounds the algal cell. The slime from algae can form a slimy, slippery layer that is unsightly, has a bad odor, and can be dangerous if it accumulates on walking surfaces. Because most algae require sunlight to grow, slime that accumulates in dark portions of treatment plants and distribution systems is not caused by algae but is usually due to bacteria.

Corrosion

Algae may contribute to the corrosion of concrete and metal structures, either directly on surfaces where they grow or indirectly by changing the water physically or chemically. Algae are not usually the direct cause of corrosion of iron or steel pipes because most algae are not capable of growth in the absence of light.

Interference With Other Treatment Processes

As algae grow and die, they may cause changes in the water's pH, alkalinity, hardness, level of dissolved oxygen, and concentration of organic matter. These changes can interfere with normal treatment processes, for example, by increasing the chlorine demand and chemical dosages necessary for adequate coagulation. Inadequate control of algae may also require the use of carbon adsorption treatment to reduce taste, odor, or color to acceptable levels.

Trihalomethane Formation

Free chlorine will react with certain organic substances in the water to produce THMs (see chapter 7 for more details). The chlorination of water that has a high concentration of algae may produce particularly high levels of THMs.

Algae Control

Before any algae control procedures are attempted, the operator should establish a routine program of collecting raw-water samples at least once a week and then counting the numbers of the different types of algae in each sample. Based on this information, a decision can be made concerning the best time to initiate algae control procedures.

Several biological and chemical methods have been evaluated for algae control in large water bodies. However, two effective methods commonly used for drinking water supplies are copper sulfate or powdered activated carbon (PAC).

Copper Sulfate

Algae have been controlled using copper sulfate since 1904. Not all algae are effectively killed by this chemical, so it is important that the problem-causing algae be accurately identified. Table 2-2 summarizes the effectiveness of copper sulfate against various types of algae.

The effectiveness of copper sulfate treatment also depends on its ability to dissolve in water, which in turn depends on pH and alkalinity. The required dosage therefore depends on the chemical characteristics of the water to be treated. The best and most lasting control will result if the water has a total alkalinity less than or equal to approximately 50 mg/L as calcium carbonate ($CaCO_3$) and a pH between 8 and 9.

The following suggested dosages for copper sulfate are general recommendations only and may not be the best dosages for every situation:

- Bodies of water with a total methyl-orange alkalinity equal to or greater than 50 mg/L as $CaCO_3$ are usually treated at a dosage of 1 mg/L, calculated for the volume of water in the upper 2 ft (0.6 m) of the lake, regardless of the lake's depth. This converts to about 5.4 lb (2.4 kg) of commercial copper sulfate per acre of surface area. The 2-ft (0.6-m) depth has been determined as the effective range of surface application of copper sulfate in those waters. The chemical tends to precipitate below this application depth.
- For water bodies having a total methyl-orange alkalinity less than 50 mg/L as $CaCO_3$, a dosage of 0.3 mg/L is recommended. This dosage is based on the total lake volume and converts to about 0.9 lb (0.4 kg) of commercial copper sulfate per acre-foot of volume.

The minimum copper sulfate dosage depends on the alkalinity of the water; the maximum safe dosage depends on the toxic effect on fish life. A safe dosage for most fish is 0.5 mg/L; however, trout are very sensitive and can be killed by dosages greater than 0.14 mg/L. The state department of game and fish should be consulted to determine if any special precautions need to be taken.

The simplest way to apply copper sulfate to control algae in small lakes, ponds, or reservoirs is to drag burlap bags of the chemical behind a motorboat. The boat is guided in a zigzag course for overlapping coverage over the water, as shown in Figures 2-1 and 2-2. Lakes are also sometimes treated by power spray application from shore (Figure 2-3) or from a motorboat.

TABLE 2-2 Relative toxicity of copper sulfate to algae

Group	Very Susceptible	Susceptible	Resistant	Very Resistant
Blue-green	*Anabaena, Anacystis, Gomphosphaeria, Rivularia*	*Cylindrospermum, Oscillatoria, Plectonema*	*Lyngbya, Nostoc, Phormidium*	*Calothrix, Symploca*
Green	*Hydrodictyon, Oedogonium, Rhizoclonium, Ulothrix*	*Botryococcus, Cladophora, Oscillatoria Enteromorpha, Gloeocystis, Microspora, Phytoconis, Tribonema, Zygnema*	*Characium, Chlorococcum, Clorella, Coccomyxa, Crucigenia, Desmidium, Draparnaldia, Golenkinia, Mesotaenium, Oocystis, Palmella, Pediastrum, Staurastrum, Stigeoclonium, Tetraedron*	*Ankistrodesmus, Chara, Coelastrum, Dictyosphaerium, Elakatothrix, Kirchneriella, Nitella, Pithophora, Scenesdesmus Testrastrum*
Diatoms	*Asterionella, Cyclotella, Fragilaria, Melosira*	*Gomphonema, Navicula, Nitzschia, Stephanodiscus, Synedra, Tabellaria*	*Achnanthes, Cymbella, Neidium*	
Flagellates	*Dinobryon, Synura, Uroglenopsis, Volvox*	*Ceratium, Cryptomonas, Euglena, Glenodinium, Mallomonas*	*Chlamydomonas, Peridinium, Haematococcus*	*Eudorina, Pandorina*

Source: Algae and Water Pollution. *1977. US Environmental Protection Agency.*

The effect of the copper sulfate treatment on algal populations can be noticed soon after the chemical has been added. Within a few minutes, the color of the water will change from dark green to grayish-white. At no time are all the algae in the lake entirely eliminated, but the water should be visibly free of cells for two or three days following a complete application.

FIGURE 2-1 Path taken for copper sulfate application to small water bodies

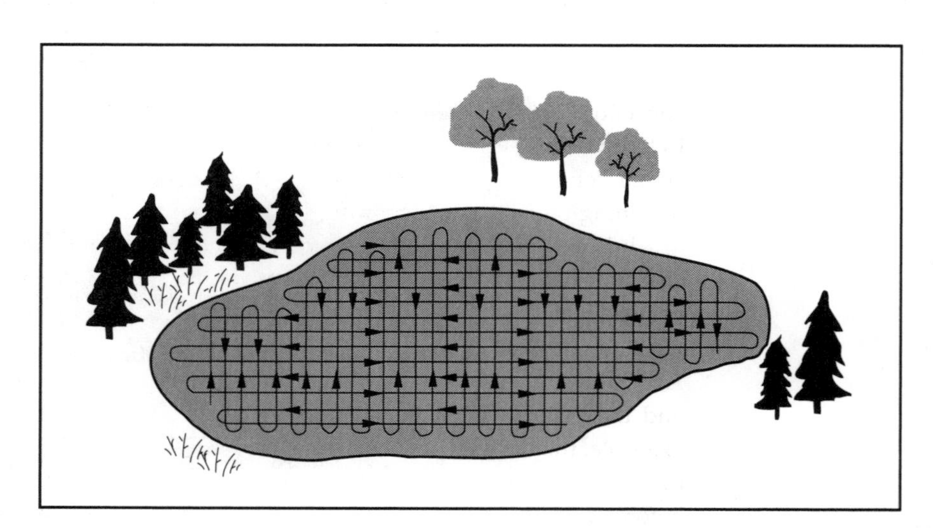

FIGURE 2-2 Path taken for copper sulfate application to large water bodies

Proper treatment ensures that most of the algae will be eliminated, and a long time will pass before the algae can again create problems. The frequency of treatment depends on local climate and the amount of nutrients in the water. Warm temperatures, plentiful sunlight, and a high nutrient

FIGURE 2-3 Power spray application of copper sulfate

Courtesy of Applied Biochemists, Inc., Milwaukee, Wis.

concentration all tend to encourage a rapid regrowth of algae. In general, one to three complete treatment applications per season should be sufficient. The actual length of time between applications can best be determined by obtaining periodic algal counts.

It is **extremely important** to note that copper sulfate, or any other active chemical used to control aquatic organisms in bodies of water, is classed as a pesticide and is therefore regulated by the US Environmental Protection Agency (USEPA) under the Federal Insecticide, Fungicide, and Rodenticide Act (FIFRA). Under this law, each package sold must be specifically labeled, and the product must be used in accordance with the label directions. For example, the current USEPA regulations require that only a portion of a body of water be treated at one time, and a time delay of 10 to 14 days is required before treating the remainder of the water. The waiting period allows for decomposition and oxygen levels to return to normal, thereby reducing stress or death to fish.

Many states also require a special permit for algae control. Operators should carefully check that all local, state, and federal requirements are met before undertaking an algae control program. Failure to comply with

FIGURE 2-4
Crew preparing to feed permanganate for algae control

Courtesy of Public Works Magazine

FIGURE 2-5
Boat equipped with a wooden hopper–feeder for applying permanganate

Courtesy of Public Works Magazine

regulations could result in a fine or liability for damages if the chemical application has an adverse effect on wildlife or the environment as a result of improper application.

Potassium Permanganate

Control of algae in reservoirs has also been accomplished by applying potassium permanganate. Although experience using it is limited, potassium permanganate has worked successfully to control algae conditions when copper sulfate has not worked well. One theory for its success is that iron is essential for the formation of chlorophyll in plants; if the iron in the water is oxidized and precipitated by the permanganate, the iron deficiency will retard the growth of the algae.

Permanganate can be applied to the water using the same technique used with copper sulfate: towing a burlap sack containing the chemical through the water. This method is only appropriate for small reservoirs because the sack will quickly be disintegrated by the permanganate. Larger reservoirs have been treated by feeding the crystals into a wooden hopper, with a screened bottom, hung over the side of a boat (Figures 2-4 and 2-5).

Permanganate is a strong oxidizer, so persons working with the chemical should wear rubber gloves and clothing and use caution in handling it.

Powdered Activated Carbon

PAC can also be used to control algae. This is not a form of chemical treatment because it operates by a physical rather than a chemical process. The activated carbon forms a black blanket over the water, cutting off sunlight, which is vital for algae growth. However, a large amount is needed to block the sun effectively, and powdered carbon is messy and difficult to handle from a boat. PAC can also be added manually or by a chemical feeder as the water enters the treatment plant, to adsorb algal by-products responsible for taste-and-odor problems.

Pond Covers

A pond cover can be used to control algae growth in smaller bodies, such as presedimentation impoundments. This greatly reduces the amount of sunlight available to the algae for photosynthesis. The cover shown in Figure 2-6 is made of a synthetic rubber fabric that floats on the surface of the water. The fabric must be specifically approved for contact with potable water. Although floating covers effectively control algae, they can interfere with normal dewatering and cleaning operations.

FIGURE 2-6
Pond cover

Photo courtesy of JPS Elastomerics Corp.

Rooted Aquatic Plants

Rooted aquatic plants (water weeds) are different from algae in that they are plants with defined leaves, stems, and root systems. They can be classified as

- emergent weeds
- floating or surface weeds
- submerged weeds

The three types of rooted aquatic plants are illustrated in Figure 2-7.

Emergent Weeds

Emergent weeds grow in shallow water on or near the shoreline. They root in the bottom mud and can extend well above the water surface. Cattails, water willows, and rushes are familiar examples. Figure 2-8 shows emergent weeds surrounding a lake.

Floating or Surface Weeds

Plants that have leaves floating on the surface of the water can be either free floating or rooted in the bottom mud. Sometimes they are mixed with

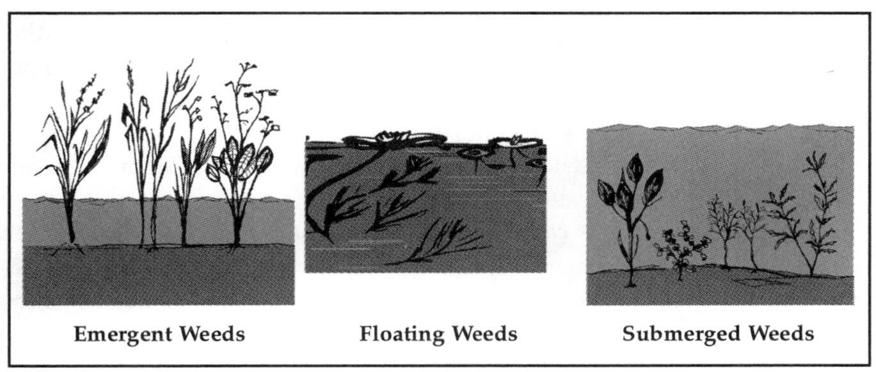

FIGURE 2-7
Types of rooted aquatic plants

Courtesy of Applied Biochemists, Inc., Milwaukee, Wis.

FIGURE 2-8
Emergent weeds around a lake

Courtesy of Applied Biochemists, Inc., Milwaukee, Wis.

emergent weeds. The most common example of the rooted types of these plants is water lilies. Duckweed and similar plants are free floating.

Submerged Weeds

Submerged weeds grow entirely underwater and are rooted in the bottom mud. The depth to which they will grow is limited primarily by the depth of sunlight penetration. The clearer the water, the taller the plants are likely to grow. Coontail and blatterwort are examples of submerged weeds.

Effects of Water Weeds

Water weeds can cause the same problems — color, tastes, and odors — for water system operations and treatment as algae. In addition, floating plants can clog intakes and pumps. Rooted aquatic vegetation can also serve as a habitat (breeding area) for disease-causing and nuisance insects, which can create problems for both operators and nearby residents.

However, water weeds do serve many beneficial functions. They provide shelter and attachment surfaces for small beneficial organisms, they provide spawning and schooling areas for fish, and they produce dissolved oxygen (DO). They also consume and temporarily store nutrients such as phosphorus that could otherwise support the growth of algae.

Control of Rooted Aquatic Plants

Procedures to control rooted aquatic plants should begin whenever the plants start to cause operational control problems at the treatment plant or add color, taste, or odor to the raw water.

The three methods for controlling rooted aquatic plants in lakes, reservoirs, and other surface water are

- physical treatment
- biological treatment
- chemical treatment

Physical Methods

Methods for physically controlling the growth of aquatic plants include

- harvesting
- dewatering
- dredging
- shading
- lining

Harvesting. Methods used for harvesting depend on the extent and amount of control desired. The technique employed can vary from hand pulling or hand cutting and raking, to using power-driven harvesting machines.

Dewatering. Completely or partially draining a body of water is known as dewatering. It can be effective in killing water weeds. It is most practical if the weed problem is in a small impoundment, such as a presedimentation basin. It may not be practical if the weeds are in a major lake or reservoir. During the dewatering process, the water level is lowered and the impoundment is allowed to dry. Then, if the lake bottom is stable enough to support heavy equipment, a scraper or front-end loader is used to clear away the dried plant material; otherwise, the dewatered area is left exposed for a period of time (several months) until the plants' root systems dehydrate.

Dredging. A clamshell crane, drag line, or hydraulic dredge can be used to control aquatic weeds because dredging the bottom mud removes any plants that are growing there. In addition, dredging the near-shore areas of a water body increases the water depth, thereby decreasing the area of suitable habitat for the plants.

Shading. Two methods of shading (i.e., limiting the amount of sunlight reaching the bottom of a water body) have been used for controlling water weed growth. One method raises the turbidity of the water by adding clay to form a colloidal suspension. This allows less sunlight penetration into the water and thereby reduces the amount of growth. Shading can also be accomplished by placing sheets of black plastic on the lake bottom. This effectively shades areas of the bottom from the sun and reduces plant growth.

Lining. Ponds can also be lined with a synthetic rubber material to prevent the growth of rooted aquatic plants. The lining also prevents water loss due to seepage through the bottom of the pond (Figure 2-9).

Biological Methods

Biological controls using specific species of crayfish, snails, and fish have proven extremely effective for control of rooted aquatic plants. However, the state department of game and fish must be contacted before such a program is instituted, to determine if they allow the species to be transplanted.

Chemical Methods

Chemical control of aquatic plants by the use of herbicides should be undertaken only when the problem becomes unmanageable by other means. The chemicals commonly used are diquat and endothall. Both have been registered for use in drinking water supplies and are relatively safe if the application instructions are followed. The main drawback to these herbicides is that a waiting period of several days after application is required before the water can be safely used for human consumption. Diquat has a recommended 10-day waiting requirement and endothall a 7-day requirement.

FIGURE 2-9 Pond lining to control algal growth

Courtesy of JPS Elastomeric Corp.

State regulations should be checked to determine if they differ from the manufacturer's recommendations. The length of the waiting period would make it difficult for a utility with only one raw-water storage reservoir to use the chemical for aquatic plant control.

Small areas near shore can be treated by distributing dry chemical herbicides by hand or spraying the chemicals through a shoulder-strap sprayer. For large areas, herbicides can most effectively be applied using a power sprayer from shore or from a motorboat.

Suggested Records for Algae and Weed Control Programs

An important part of an algae and water weed control program is keeping good records. Data for threshold odor test results and for any complaints of tastes and odors registered by consumers should be recorded daily. When chemical treatment is performed, records should also include

- the reason for pretreating, such as taste-and-odor problems or filter clogging
- type of algae or weed treated
- algal count or estimated weed coverage

- chemical used, concentration, and dosage
- date of pretreatment
- length of time since last treatment
- weather conditions
- other water conditions, such as temperature, pH, and alkalinity
- method of application
- names of personnel involved
- results of pretreatment, such as taste and odor following final treatment or filter conditions

These types of records can be used to solve similar problems in the future. With good records, the operator will be able to reproduce previous results and, as a result, save considerable time. Residual algae counts or estimates of weed coverage can serve as reliable guides in deciding when the next treatment will be needed.

Destratification of Reservoirs

Stratification of the water in lakes and reservoirs occurs when a warm layer of water overlies a colder layer. In temperate zones, stratification occurs during the spring and summer when the air temperature is higher than the water temperature. Water has its greatest density at 39°F (4°C), so the colder water sinks to the bottom and the lighter, warm water stays at the surface.

The Effects of Stratification

Stratification occurs when there are three water layers with different temperatures. The upper, warmer layer is called the epilimnion; the lower layer is the hypolimnion; and the temperature transition zone in between is the thermocline. In the hypolimnion, the water is stagnant and frequently becomes anaerobic (completely void of oxygen). The decomposition of matter under anaerobic conditions can result in a reduction in pH and the production of hydrogen sulfide gas, which can cause tastes and odors. In addition, the hydrogen sulfide may reduce iron and manganese in the soil of the lake's bottom to a soluble form, which may then have to be removed in the treatment plant to avoid complaints of rusty water by water customers.

While the water is stratified, acceptable quality water can often be obtained from the epilimnion because it is mixed by the wind. But when the upper layer cools later in the year and its density becomes the same as the bottom water, destratification occurs — or, as it is often termed, "the lake turns over." When this happens, water quality can be extremely bad for a number of days. An intense storm over a reservoir can occasionally generate enough energy to mix the water in a stratified reservoir, which can also result in poor water quality.

In colder climates, lakes may become stratified because the ice-covered upper layer is lighter than deeper, warmer water. In this case the lower layer, or even the entire lake, may become anaerobic. Figure 2-10 shows conditions in a stratified lake at the spring and fall turnovers and destratification using compressed air.

Methods of Destratification

Destratification of a lake is often accomplished by pumping compressed air through a hose to the low point on the lake bottom and releasing it through a diffuser. The rising air bubbles add air to the water and also serve to bring the cold water up from the bottom. Destratification is also successfully done in lakes by using an electrically driven floating aerator that is moored over the deep part of the lake.

The destratification equipment must start operating in early spring, whenever the average air temperature is higher than the water temperature, and continue until late fall. In addition to eliminating the problems caused by anaerobic bottom water, destratification of lakes often reduces the growth of objectionable algae.

Asiatic Clams and Zebra Mussels

The Asiatic clam (*Corbicula fluminea*) was introduced to the United States from Southeast Asia in 1938 and has become a significant pest in almost every river system south of 40° latitude. If not controlled, the clams can infest raw-water intake pipelines and treatment facilities, resulting in a reduction in flow capacities and a clogging of mechanical equipment.

The zebra mussel (*Dreissena polymorpha*) is a freshwater shellfish that grows to about 1.5 in. (3.8 cm) long (Figures 2-11 and 2-12). These mussels have recently invaded the Great Lakes, having first been found in Lake St. Clair in 1988. The mussel is native to the Black and Caspian Seas and is thought to have been brought over from Europe in the ballast water of a freighter around 1985. The zebra mussel population in the Great Lakes has increased very rapidly, and it is predicted that it will eventually spread to most waters of North America.

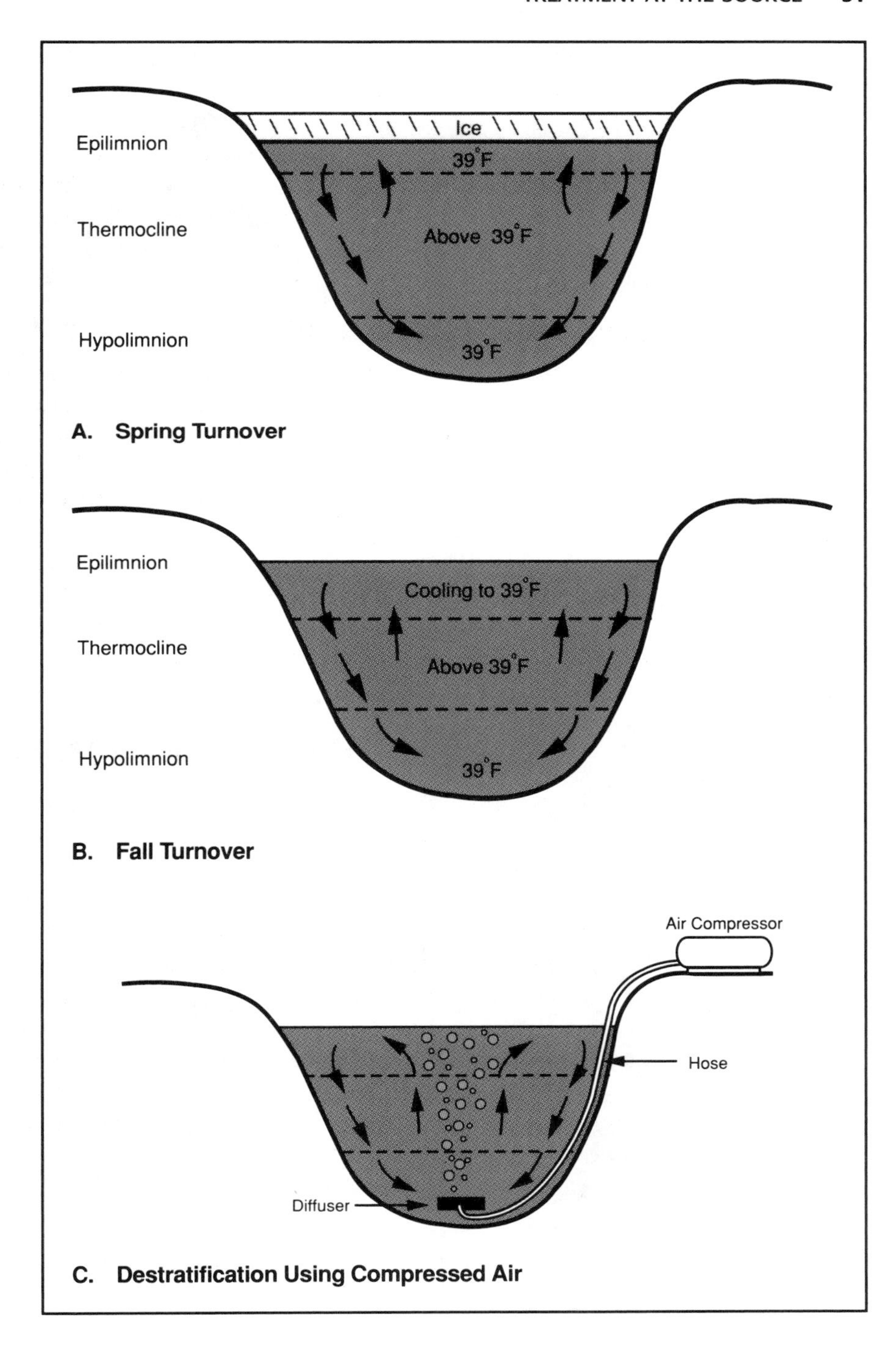

FIGURE 2-10
Lake turnover and destratification

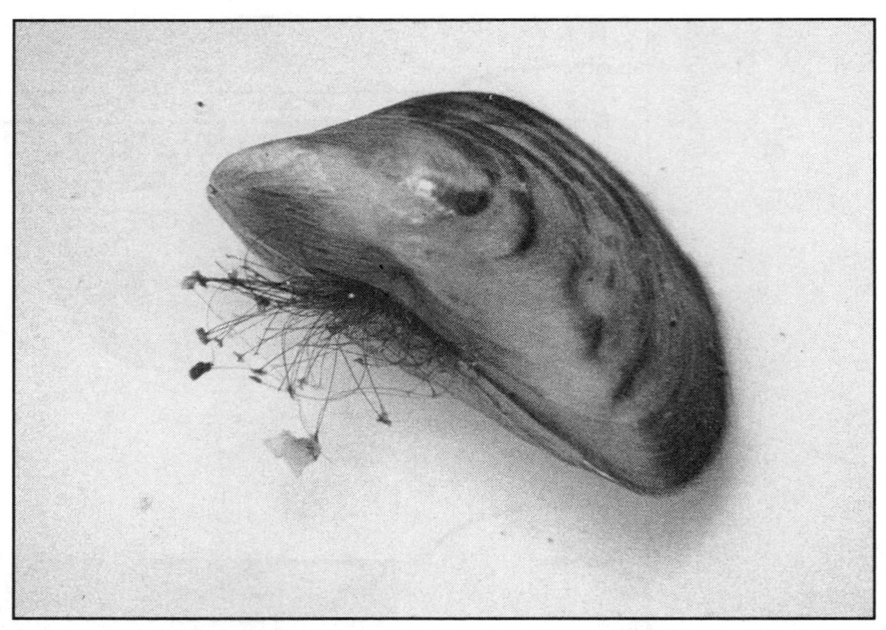

FIGURE 2-11 An adult zebra mussel

Courtesy of Fred Snyder, Ohio Sea Grant

FIGURE 2-12 An accumulation of zebra mussels

Here are some reasons why the zebra mussel multiplies and spreads so rapidly:

- A female zebra mussel may release 30,000 to 40,000 eggs in a single season, and a mussel's lifespan is about four or five years.
- The larvae (veligers) that develop after the external fertilization of the eggs are free-swimming for several weeks before they attach themselves; thus, they can be transported a considerable distance by water currents.
- They commonly adhere to the bottom of boats, so the adults are readily moved around to disperse eggs throughout a body of water.
- Adult mussels can survive out of the water for several days, so they are liable to be transferred from one body of water to another on the bottom of a transported boat.

The proliferation of zebra mussels in a lake can have great ecological consequences. The mussels obtain their food by filtering algae and other matter that passes by. The volume of water filtered by one mussel may be small, but the number of mussels can be so great that it is estimated that the entire population may be able to completely filter the volume of a large lake once every 11 days. The consequence is that the food source is taken away from other filter-feeding organisms such as native mussels and fish. In addition, there is an increased deposition of organic material on the lake bottom and a destruction of fish-spawning areas.

The mussels attach themselves to virtually any hard surface, such as rocks and pilings. They particularly favor water intake structures and the interior of water intake pipes because they can filter the flowing water. The buildup of mussel shells can completely block an intake screen, and it can greatly decrease flow through a pipe as a result of the increased roughness of the pipe surface. In extreme cases, the interior diameter of an intake pipe can be materially reduced.

Another problem is that a sudden die-off of mussels near an intake can cause serious taste-and-odor problems. It has also been reported that clumps of older shells can break off and block intake screens.

Control of Asiatic Clams and Zebra Mussels

Asiatic clams and zebra mussels have a few predators, but they are relatively insignificant for controlling the growth of these pests in US waters. Once Asiatic clams and zebra mussels have attached themselves to a structure and grown to adult size, they are relatively difficult to kill. They can sense adverse conditions and close up in their shells for a period of time. The only way to kill the adults is to feed chemicals that will destroy them continuously for several days at a time. When adults are killed, it is likely that offensive tastes and odors will be created. In addition, the shells of the dead mussels continue to adhere to the intake or pipe and can only be removed mechanically. It reportedly requires a water pressure of at least 80 psi (550 kPa) to dislodge shells that have accumulated on intake screens.

Because ridding structures of the adults is so difficult, initiating treatment prior to invasion is advisable. Zebra mussels breed only when the water temperature is above about 53.6°F (12°C) (which is about June through October in the central United States), so control measures need to be practiced only during the reproductive season if desired.

Application of low-voltage electric current to a metallic pipeline has been tried and is reported to discourage attachment, but it also accelerates corrosion of the pipe. Another control method is to backflush the intake periodically with a high concentration of chlorine or some other oxidant. Only larvae will be killed by a short dose, so the treatment must be done often enough to kill them before they develop protective shells.

A control method used by many systems in the United States is to feed a control chemical at the intake. The equipment for this is difficult and expensive to install in intakes that are a long distance from shore, but it seems to be one of the best alternatives. Chemicals that have been used include chlorine, potassium permanganate, copper sulfate, or one of several biocides that are approved for addition to potable water. The most common chemical used in Europe is chlorine; however, the concern in the United States over generating excessive levels of trihalomethanes makes it necessary to consider other alternatives. Any North American water system installing new intakes should consider installing control methods for Asiatic clams and zebra mussels even if they are not currently a threat in the area. It appears that the mussels will eventually invade most surface waters, and the cost of installing control equipment during original construction is much lower than it would be later.

Selected Supplementary Readings

Belanger, S.E., et al. 1991. Sensitivity of the Asiatic Clam to Various Biocidal Control Agents. ***Jour. AWWA,*** 83(10):79.

Cameron, G.N., et al. 1989. Minimizing THM Formation During Control of the Asiatic Clam: A Comparison of Biocides. ***Jour. AWWA,*** 81(10):53.

Casitas Municipal Water District. 1987. ***Current Methodology for the Control of Algae in Surface Reservoirs.*** Denver, Colo.: American Water Works Association Research Foundation and American Water Works Association.

Cobban, B. 1991. Z-E-B-R-A Spells Trouble for Treatment Operators. ***Opflow,*** 17(10):1.

Cooke, G.D., and R.E. Carlson. 1989. ***Reservoir Management for Water Quality and THM Precursor Control.*** Denver, Colo.: American Water Works Association Research Foundation and American Water Works Association.

Gohlke, A.F. ***Algae Control in Farm Ponds With Copper Sulfate.*** Booklet No. 11. El Paso, Texas: Phelps Dodge Refining Corporation.

How to Identify and Control Water Weeds and Algae. Milwaukee, Wis.: Applied Biochemists, Inc.

Klerks, P.L., and P.C. Fraleigh. 1991. Controlling Adult Zebra Mussels With Oxidants. ***Jour. AWWA,*** 83(12):92.

Manual M7, Problem Organisms in Water: Identification and Treatment. 1995. Denver, Colo.: American Water Works Association.

Manual of Water Utility Operations. 8th ed. 1988. Austin, Texas: Texas Water Utilities Association.

Nalepa, T.F., and D.W. Schloesser. 1992. ***Zebra Mussels: Biology, Impacts and Control.*** Boca Raton, Fla.: Lewis Publishers.

Raman, R.K., and B.R. Arbuckle. 1989. Long-Term Destratification in an Illinois Lake. ***Jour. AWWA,*** 81(6):66.

Robbins, R.W., J.L. Glicker, D.M. Bloem, and B.M. Niss. 1991. ***Effective Watershed Management for Surface Water Supplies.*** Denver, Colo.: American Water Works Association Research Foundation and American Water Works Association.

Smith, S.A. 1992. ***Methods for Monitoring Iron and Manganese Biofouling in Water Wells.*** Denver, Colo.: American Water Works Association Research Foundation and American Water Works Association.

Water Quality and Treatment. 4th ed. 1990. New York: McGraw-Hill and American Water Works Association (available from AWWA).

CHAPTER 3

Preliminary Treatment

Preliminary treatment, also known as pretreatment, is generally used when a water source contains large quantities of sticks, weeds, leaves, or other floating debris as well as gravel, sand, or other gritty substances. If not removed, this material can jam equipment, damage pumps and piping, and greatly add to the loading on the normal treatment processes. Preliminary treatment is also used when the raw water contains unusually heavy concentrations of sediment. The preliminary treatment processes discussed in this chapter are

- screening
- presedimentation
- microstraining

Screening

The first pretreatment provided in most surface water treatment systems is screening. Coarse screens located on an intake structure are usually called trash racks or debris racks. Their function is to prevent clogging of the intake by removing sticks, logs, and other large debris in a river, lake, or reservoir. Finer screens may then be used at the point where the water enters the treatment system to remove smaller debris that has passed the trash racks.

The two basic types of screens used by water systems are bar screens and wire-mesh screens. Both types are available in models that are manually cleaned or automatically cleaned by mechanical equipment.

Bar Screens

Bar screens are made of straight steel bars, welded at both ends to two horizontal steel members. The screens are generally ranked by the open distance between bars as follows:

- fine: spacing of $\frac{1}{16}$ to $\frac{1}{2}$ in. (1.5 to 13 mm)
- medium: spacing of $\frac{1}{2}$ to 1 in. (13 to 25 mm)
- coarse: spacing of $1\frac{1}{4}$ to 4 in. (32 to 100 mm)

A bar screen assembly is installed in a waterway at an angle of about 60 to 80° from the horizontal (Figure 3-1). This angle is important, particularly in manually cleaned bar screens. The slope makes it convenient to rake debris up the screen and onto the concrete operating platform for drainage and eventual disposal. The slope also helps keep the screen from clogging between cleanings. As debris stacks up against the screen, the passing water

FIGURE 3-1
Bar screen assembly installed in a waterway

Courtesy of Envirex Inc., Waukesha, Wis.

lifts and pushes it up the slope, leaving the submerged part of the screen open and clear.

Small treatment plants that receive only small amounts of debris usually have manually cleaned bar screens at the raw-water intake. Automatically cleaned screens are used at plants receiving large amounts of debris and at any plant or intake structure where it is not practical or convenient to reach the screen for cleaning.

Automatically cleaned bar screens are available in a variety of styles. The bar screen shown in Figure 3-1 is equipped with an automatic rake, which is a horizontal piece of metal that moves up the face of the screen. It is pulled by a continuous chain-and-sprocket drive attached at both ends of the rake. Figure 3-2 is a side view of an automatically cleaned bar screen. Note how the

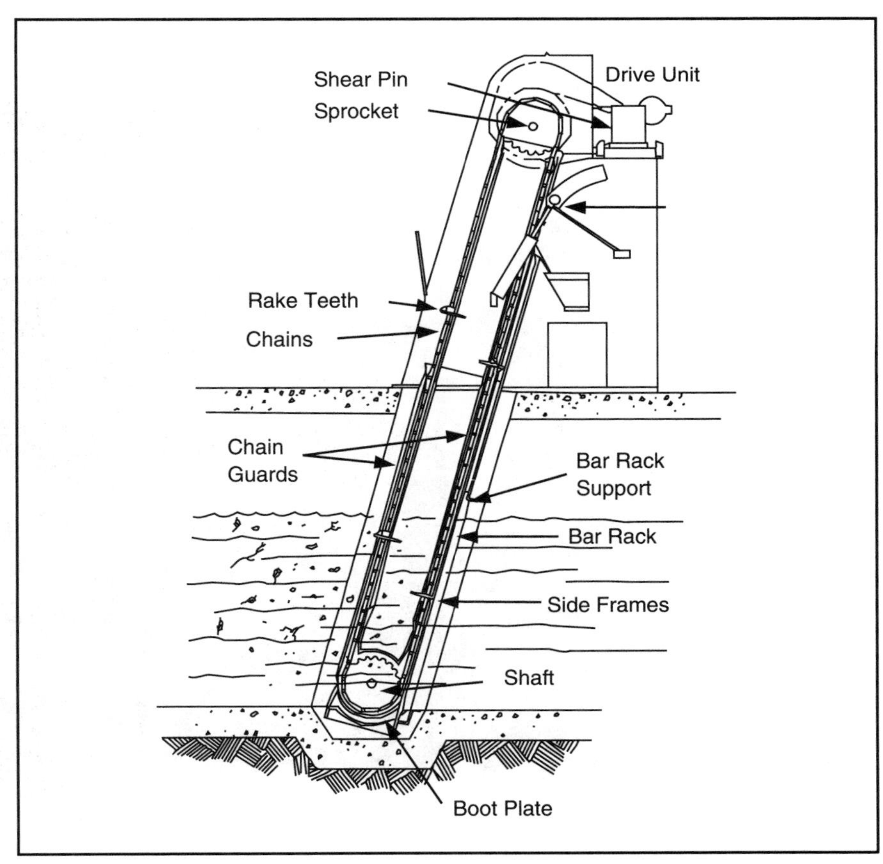

FIGURE 3-2 Side view of an automatically cleaned bar screen

Courtesy of Envirex Inc., Waukesha, Wis.

rakes move up past the screen, dump the debris (known as screenings) into the collecting hopper, and then return into the water to repeat the cycle.

Wire-Mesh Screens

Wire-mesh screens, commonly referred to as traveling water screens, are made of fabric woven from stainless steel or other corrosion-resistant, wire-like materials (Figure 3-3). The fabric may have openings as wide as ⅜ in. (10 mm) or as narrow as $1/60$ in. (0.4 mm).

Screens can be cleaned manually if they do not require frequent cleaning. Screen segments can be lifted out of the water and cleaned with a brush or hose. However, because debris can accumulate quickly on wire mesh, automatically and continuously cleaned wire-mesh screens are usually favored over manually cleaned units. Figure 3-4 shows an automatically cleaned wire-mesh screen. It is mounted vertically in the water and moves continuously, while spray nozzles located in the head terminal are used to wash away the screenings. The screenings and wash water then fall away from the screen and are conveyed to the disposal area.

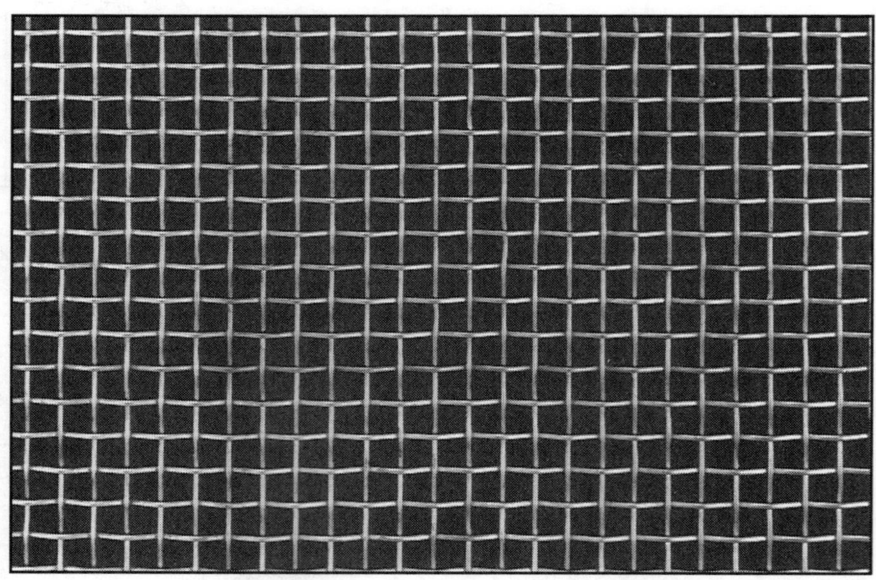

FIGURE 3-3 **Wire-mesh screen material**

Courtesy of FMC Corporation, MHS Division

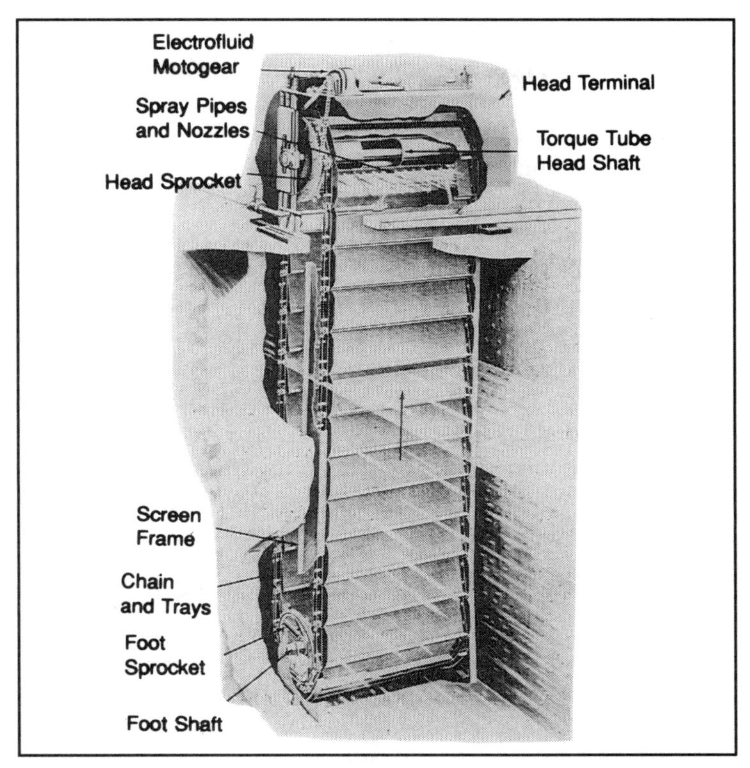

FIGURE 3-4 Automatically cleaned wire-mesh screen

Courtesy of FMC Corporation, MHS Division

Maintenance of Screening Equipment

Clogging and corrosion are the principal problems associated with screening. To prevent these problems, routine cleaning and inspection are required.

Manually cleaned screens must be checked and cleaned frequently. The frequency of inspections depends on weather conditions and the type of watershed. The largest amount of debris is usually encountered during autumn, when trees lose their foliage. Large amounts of debris may also be deposited on screens during the spring and rainy seasons, when high water transports debris, leaves, and branches along riverbanks. Heavily wooded watersheds will usually produce a very heavy loading of debris at certain times of the year.

Automatic screening devices should not be neglected. Routine inspection of the units is essential. The mechanical and electrical systems are usually equipped with protection devices designed to shut down operation in case of emergency or jamming. A mechanical protection device is incorporated into the unit (shown in Figure 3-2); if a piece of debris becomes jammed between the screen and the rake, a shear pin is designed to break and disconnect the motor.

Electrical protection devices include circuit breakers that automatically shut down the system when there is an electrical malfunction or the motor is overloaded. If possible, an alarm should be provided to alert the plant operator of any malfunction of the screening equipment. If there is no alarm system, visual inspection must be made more frequently to check for possible clogging or equipment failure.

The constant wetting and drying of the screen equipment create ideal conditions for accelerated corrosion. Screening equipment should therefore be inspected at least monthly for signs of corrosion. A stock of replacement parts should be available so that repairs can be made promptly in the event of equipment failure.

Suggested Records for Screening Equipment

Records should be kept of the type and quantity of screened material that is removed. Reference to these records will help identify an appropriate schedule of inspection and cleaning frequency. The records may also help identify activities in the watershed that are causing excessive amounts of debris. An investigation can then be made to see if there is any way to reduce the amount of debris reaching the water intake.

Complete, up-to-date operating records should include

- date of inspection
- amount of material removed from screens (in cubic feet or cubic meters)
- notations regarding unusual or unexpected types of debris or water conditions

Maintenance records, which identify the type and location of equipment, should include a list of required spare parts, a checklist of spare parts on hand, and the date and description of maintenance performed.

Presedimentation

Silt and other gritty material are present to some extent in most surface water supplies. In many sources, this material is particularly extensive after storms or heavy rains that stir up the water. Presedimentation is a pretreatment process used to remove gravel, sand, and silt from the raw water before it enters the main treatment facility. Sand and gravel must be removed because they could jam equipment and wear down pump impellers and other moving parts. In addition, reducing the heavy silt loading of river water can greatly reduce the load on the coagulation and sedimentation facilities. A well-designed presedimentation system can remove up to 60 percent of the settleable material.

Sand is also a problem in groundwater supplies. Sand can typically be prevented from entering wells through the use of proper grouting, screening, and well development. If sand is getting into a well through a broken screen, for instance, it will quickly damage the pump impellers. And if it is pumped into the distribution system, it can eventually block the water mains and could be quite expensive to remove. Sand pumping by some wells cannot be avoided, so sand traps or other removal equipment must be installed to remove the sand before the water enters the distribution system.

The three types of presedimentation systems are

- presedimentation impoundments
- sand traps
- mechanical sand-and-grit removal devices

Presedimentation Impoundments

Impoundments are commonly used for river supplies that have a heavy loading of silt. An impoundment can be a simple earthen reservoir or a concrete structure. The storage capacity can range from one day to several months of actual water plant use. Although presedimentation impoundments may be constructed for the primary purpose of allowing sediment to settle out of the water, they also serve other functions.

An important extra benefit of an impoundment is that it stores raw water that can be used if, for some reason, it is undesirable to draw water from the source. For example, there might be a toxic chemical spill upstream on a river or a particularly bad taste or odor problem in the source water.

A major problem with impoundments is that water held for an extended period of time may develop an excessive growth of algae or aquatic plants. Chlorine cannot normally be used for controlling the aquatic plants because it will generate excessive trihalomethanes (THMs), but the use of other chemicals may be possible. Research is continuing on by-product formation from the use of alternative oxidants. Growth of rooted plants can be controlled by lining the impoundment, and algal growth can be controlled by installing a cover over the impoundment.

It is generally best to have two or more impoundments. Individual basins can then be taken out of service one at a time to control aquatic life by using chemicals or dewatering the impoundment for a short period of time.

Sand Traps

A sand trap is a depression in the bottom of a structure, for example, the bottom of a wet well (Figure 3-5). Because the wet well is much larger than the inlet pipe, the water slows down as it enters, and the suspended sand and gritty material settle to the bottom. A baffle is usually installed in sand traps to prevent short-circuiting and to direct the incoming flow downward.

A drain valve installed at the bottom of the wet well is used to periodically flush out the accumulated sand. Sand traps must be cleaned manually and have relatively small holding capacity, so they are best suited for raw water that contains relatively little sand and grit (less than 100 mg/L).

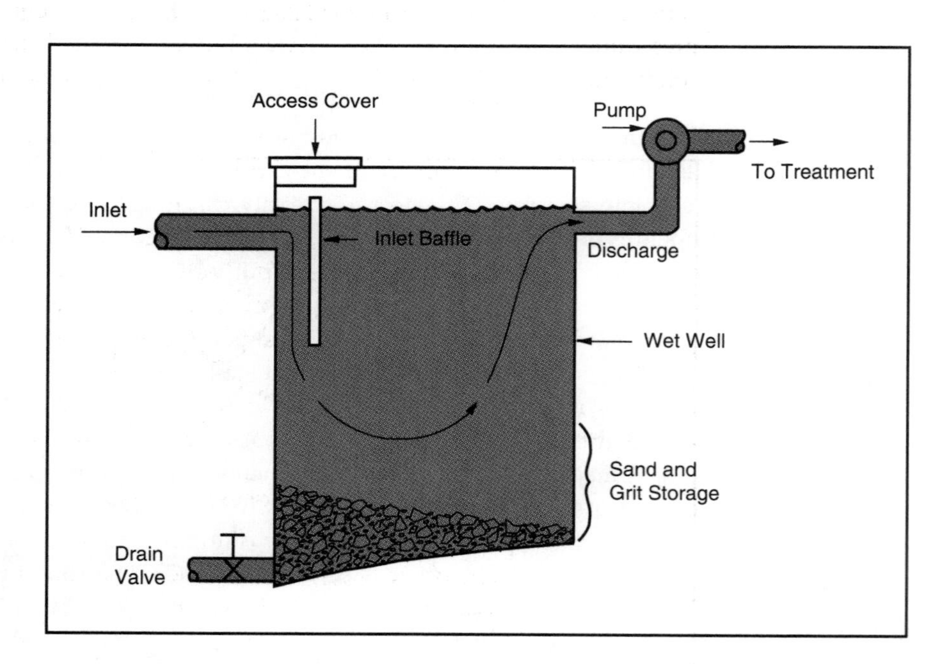

FIGURE 3-5 Sand trap at the bottom of a wet well

Mechanical Sand-and-Grit Removal Devices

Mechanical devices are typically used when the raw water contains large amounts of suspended solids. A centrifugal sand-and-grit removal device, often called a cyclone degritter, is illustrated in Figure 3-6. As sand-laden water enters the unit, it begins to travel in a spiral path inside the cylindrical section. Centrifugal force several times the force of gravity develops and throws the sand particles toward the cylinder wall. The sand particles then move in a spiral toward the small end of the cone, where they are discharged, along with some water, into the sand accumulator tank. The clean water leaves the unit through the vortex finder with almost all sand removed.

Figure 3-7 illustrates the simplicity of a typical cyclone installation. These devices have no moving parts, so they require relatively little maintenance even though they handle relatively abrasive material.

Plate and Tube Settlers

Several types of shallow-depth settling units are designed to achieve settling of suspended solids much more rapidly than in open basins (see chapter 5 for more detail). Several water systems that use very turbid water sources have, in recent years, installed plate or tube settlers to improve the presedimentation process. These devices provide elevated surfaces on which solids can settle, rather than having to fall all the way to the basin's bottom.

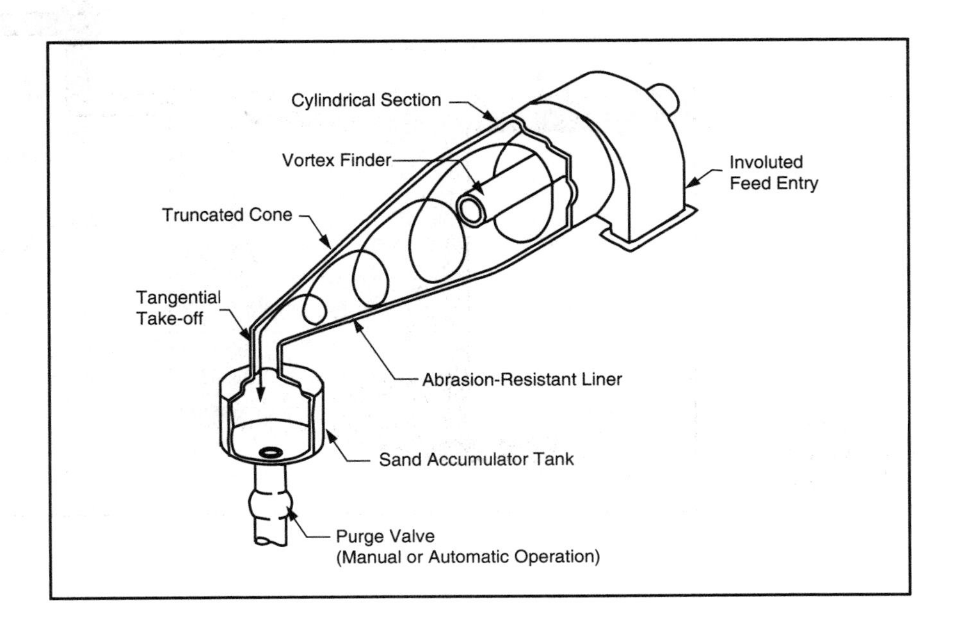

FIGURE 3-6 Centrifugal sand-and-grit removal device

FIGURE 3-7 Typical cyclone separator installation

Courtesy of LAKOS Filtration Systems, Claude Laval Corp.

In the past, these systems used large open basins to decrease the solids loading of the water before it entered the plant, but these basins generally have problems. If the water is held for long periods of time, algae will grow. Chlorine cannot be used to control the algae because it will increase the level of THMs. Also, if coagulants are added to the raw water to achieve reasonable settling, the sludge accumulation from these basins cannot be returned to the water source because of the coagulant contamination.

Systems that have installed plate or tube settlers have found that the solids loading of the water is quickly reduced without the need for any chemical addition. The installation also occupies much less space than a typical open basin, which is an additional advantage for plants located in congested areas.

Operation of Presedimentation Systems

To ensure successful removal of sand and grit, presedimentation systems must be tested regularly and cleaned routinely. Influent and effluent samples must be collected and tested for settleable solids.

The frequency of sampling and testing varies from plant to plant, depending on the amount of sand and gritty material in the raw water. During peak flow periods, sampling and testing may be required daily because of the rapidly increasing sand and grit loads; at other times, weekly or even monthly testing may be adequate.

All presedimentation systems must be cleaned routinely to prevent the water flowing through the removal system from mixing the grit and sand back into suspension and carrying it into the treatment plant or water distribution system. Deposits can also become anaerobic (lacking free oxygen), resulting in tastes and odors.

A wet-well type of sand trap is cleaned by allowing accumulations to discharge through a drain line. An access cover is provided so that the wet well can be hosed down during the draining.

Cyclone separators are cleaned automatically and continuously. The operator must periodically check that the sand-and-grit discharge storage bin or hopper is emptied and that the material is properly disposed of.

To clean a presedimentation impoundment, it must be completely drained and dried. The accumulated material can then be removed by scrapers, dozers, or front-end loaders. If the impoundment has a floating cover, the cover must be carefully rolled out of the way of the cleaning equipment.

Suggested Records for Presedimentation Systems

Detailed record keeping is important. Information about the type of grit and the amount removed by presedimentation helps to determine the frequency of sampling and testing. It also provides a record of what time of year sand and grit can be expected to be a problem.

Records are also necessary for monitoring the continued efficiency of sand-and-grit removal. A gradual decrease in removal efficiency may signal the need to clean accumulated deposits or perform other maintenance work. Detailed records dealing with presedimentation should cover

- date of sampling and testing
- the amount of suspended solids in the raw water, in milligrams per liter, or milliliters per liter if a settleability test is used
- the amount of suspended solids in presedimentation effluent
- cleaning date, time required, and estimated quantity of removed material

Microstraining

A microstrainer (Figure 3-8) is a very fine screen used primarily to remove algae, other aquatic organisms, and small debris that can clog treatment plant filters.

Process Equipment

The most common type of microstraining unit consists of a rotating drum lined with finely woven material, such as stainless-steel wire fabric.

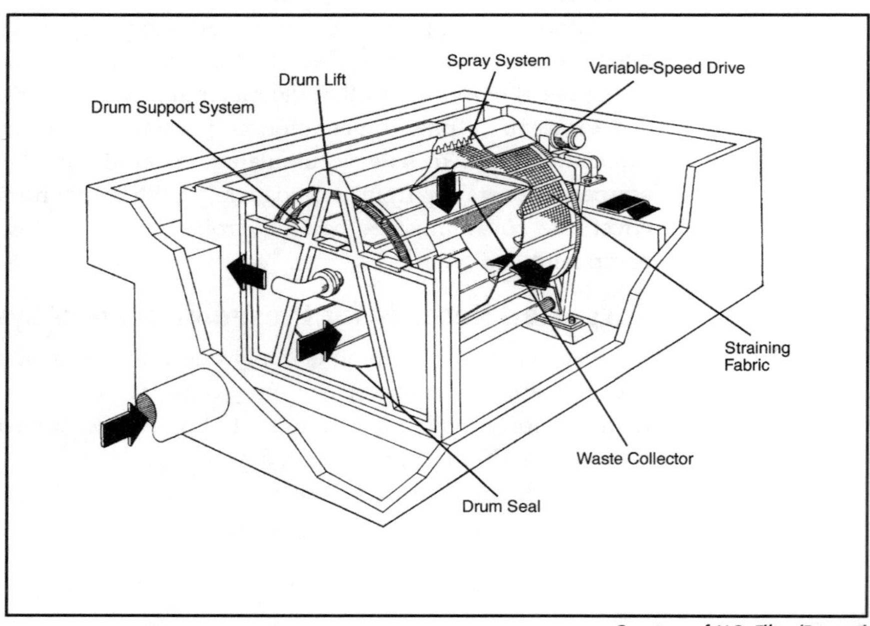

FIGURE 3-8 Fine-screened microstrainer

Courtesy of U.S. Filter/Permutit

One commonly used fabric has about 160,000 openings per square inch (250 openings per square millimeter), which is about the same as tightly woven clothing fabric. Other fabrics may have larger or smaller openings.

The microstrainer drum rotates slowly, usually 4 to 7 revolutions per minute (rpm), as water enters the inside of the drum and flows outward through the fabric. Algae and other aquatic organisms deposited on the inside of the fabric form a mat of debris, which adheres to the fabric and rotates up to the backwash hood area. At the top of the rotating drum, high-pressure (25–50-psi or 172–345-kPa) jets spray the back side of the fabric, causing the matted debris to break away. The debris and backwash water fall by gravity into a debris trough inside the drum and flow either directly to a disposal point or to a pond or tank that separates the debris from the water.

Advantages and Disadvantages

A major advantage of microstrainers is the improvement they make in the operation of sand filters. They generally remove from 50 to 90 percent of the filter-clogging material from the water, depending on the type of algae present. Because the load on sand filters is reduced, the filters can be operated for a longer time without backwashing. This saves backwash water and increases the amount of filtered water available to the water system.

Although there are definite advantages to using microstrainers in certain situations, there are also limitations. Straining removes only relatively coarse particles from the water. Microstrainers cannot remove all algae, and they do not remove bacteria, viruses, *Cryptosporidium, Giardia lamblia,* or most suspended matter that contributes to turbidity. Even eggs of tiny aquatic animals can pass through the fabric. In addition, microstrainers have no effect on the removal of dissolved substances such as inorganic and organic chemicals. Therefore, microstrainers should not be considered or used as a substitute for coagulation, flocculation, and filtration.

Although microstrainers are made to resist corrosion, the constant wet–dry conditions under which they operate cause them to require quite a bit of maintenance. This includes painting, lubrication, replacement of worn parts, and fabric repair and replacement.

Finally, it is recommended that chlorine not be added before the water enters a microstrainer for the following reasons:

- Live algae are easier to clean off the mesh than dead algae.
- If there is soluble iron in the raw water, it may oxidize to ferric hydroxide on the mesh, which will form a sticky, jellylike coating.
- Chlorine reaction with algae may accentuate tastes and odors.
- The amount of chlorine required (chlorine demand) will be lower if applied after the microstrainer.
- Free chlorine may corrode the mesh.

Selected Supplementary Readings

Mallevialle, J., and I.H. Suffet, eds. 1987. ***Identification and Treatment of Tastes and Odors in Drinking Water.*** Denver, Colo.: American Water Works Association Research Foundation and American Water Works Association.

Manual of Water Utility Operations. 8th ed. 1988. Austin, Texas: Texas Water Utilities Association.

Suffet, I.H., J. Mallevialle, and E. Kawczynski, eds. 1995. ***Advances in Taste-and-Odor Treatment and Control.*** Denver, Colo.: American Water Works Association Research Foundation and American Water Works Association.

CHAPTER 4

Coagulation and Flocculation

The large particles of suspended matter in raw water can be removed by allowing them to settle out in a presedimentation basin. But there are smaller particles in almost all surface water and some groundwater that will not settle out within a reasonable time without some help to accelerate the process. The common term for this suspended matter is *nonsettleable solids,* which usually consist of a combination of biological organisms, bacteria, viruses, protozoans, color, organic matter, and inorganic solids. The term applied to all suspended matter in water is *turbidity*.

Not only is visible turbidity in drinking water objectionable to customers, but harmful bacteria, viruses, and protozoans (such as *Giardia lamblia*, *Cryptosporidium*, and amoebas) are likely to be present. These pathogens may be protected from contact with the disinfectant by the suspended matter, so disinfection of surface water without removing its turbidity cannot produce consistently safe water.

The treatment most frequently used for treating water to remove turbidity is known as *conventional treatment*. This is a combination of the following steps:

1. Coagulation: adding and rapid mixing of chemical coagulants into the raw water.
2. Flocculation: slow mixing of the chemicals with the water to assist in building up particles of floc.
3. Sedimentation: allowing the floc to settle out of the water.
4. Filtration: removing almost all of the suspended matter that remains by passing the water through filters.

Process Description

In the coagulation and flocculation processes, nonsettleable solids are converted into large and heavier settleable solids by physical–chemical changes brought about by adding and mixing coagulant chemicals into the raw water. The settleable solids can then be removed by the sedimentation and filtration processes. Nonsettleable solids resist settling for the following two reasons:

- particle size
- natural forces between particles

Particle Size

Untreated, natural water contains the following three types of nonsettleable solids:

- suspended solids
- colloidal solids
- dissolved solids

Suspended Solids

The particles held in suspension by the natural action of flowing water are called suspended solids. The smallest suspended solids do not settle quickly, and for purposes of water treatment are called nonsettleable. Denser and heavier suspended solids are referred to as settleable solids because they will settle unaided to the bottom of a sedimentation basin within 4 hours.

One reason that nonsettleable particles resist settling is their small size. Consider a particle of coarse sand in the shape of a cube, 1 mm on a side; it would have a surface area of 6 mm^2, which is about the area of the head of a large pin. This particle could be expected to settle quickly, dropping about 1 ft (0.3 m) every 3 seconds. If this same particle were to be ground down by erosion to particles that are only 0.000001 mm on a side, the exposed surface area of all particles would be increased to 6 m^2, which is about the area of two pool tables. The increase in surface area greatly increases the drag forces that resist settling, and it would now take about 60 years for these tiny particles to settle the same 1-ft (0.3-m) distance from natural forces alone. The natural settling rates for small particles are listed in Table 4-1.

TABLE 4-1 Natural settling rates for small particles

Particle Diameter, *mm*	Representative Particle	Time Required to Settle in 1-ft (0.3-m) Depth
		Settleable
10	Gravel	0.3 seconds
1	Coarse sand	3 seconds
0.1	Fine sand	38 seconds
0.01	Silt	33 minutes
		Considered Nonsettleable
0.001	Bacteria	55 hours
0.0001	Color	230 days
0.00001	Colloidal particles	6.3 years
0.000001	Colloidal particles	63-year minimum

Source: Water Quality and Treatment. *3rd ed. 1971.*

Colloidal Solids

Very fine silts, bacteria, color-causing particles, and viruses that do not settle in a reasonable time are called colloidal solids. Although individual colloidal solids cannot be seen with the naked eye, their combined effect is often seen as color or turbidity in water. These particles are small enough to pass through later treatment processes if not properly coagulated and flocculated.

Dissolved Solids

Any particles of organic or inorganic matter that are dissolved in water — such as salts, chemicals, or gases — are referred to as dissolved solids. A dissolved solid is the size of a molecule and is invisible to the naked eye. Most of the trace metals and organic chemicals found in water are dissolved. They are nonsettleable and can cause public health or aesthetic problems such as taste, odor, or color if not removed. Unless converted to a precipitate by chemical or physical means, they cannot be removed from the water. The size ranges of the various types of solids are shown in Figure 4-1.

Natural Forces

Particles in water usually carry a negative electrical charge. Just as like poles of a magnet repel each other, there is a repelling force between any two particles of like charge. In water treatment, this natural repelling electrical

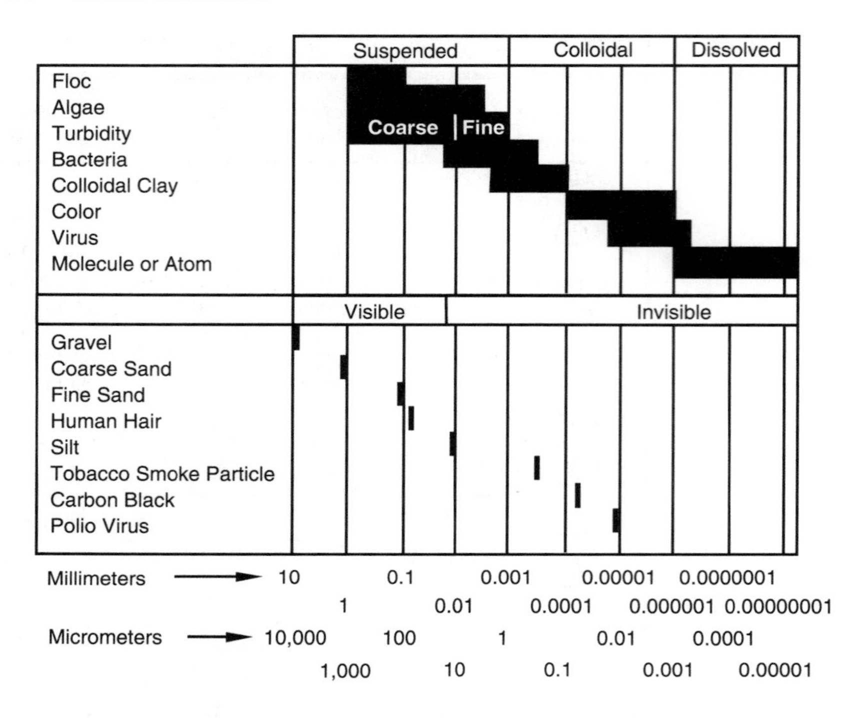

FIGURE 4-1 Size range of solids

force is called *zeta potential*. The force is strong enough to hold the very small, colloidal particles apart and keep them in suspension.

The van der Waals force is an attraction that exists between all particles in nature and tends to pull any two particles together. This attracting force acts opposite to the zeta potential. So long as the zeta potential is stronger than the van der Waals force, the particles will stay in suspension.

Effect of Coagulation and Flocculation

The coagulation–flocculation process neutralizes or reduces the zeta potential of nonsettleable solids so that the van der Waals force of attraction can start pulling particles together. These particles are then able to gather into small groups of microfloc, as shown in Figure 4-2. Although these particles are larger than the original colloids, they are held together rather weakly. Individual particles are invisible to the naked eye and are still nonsettleable. However, the gentle stirring action created by the flocculation process brings the microfloc particles together to form large and relatively heavy floc particles (macrofloc), which can be settled or filtered. The jellylike floc particles are usually visible and will look like small tufts of cotton or snowflakes in the water.

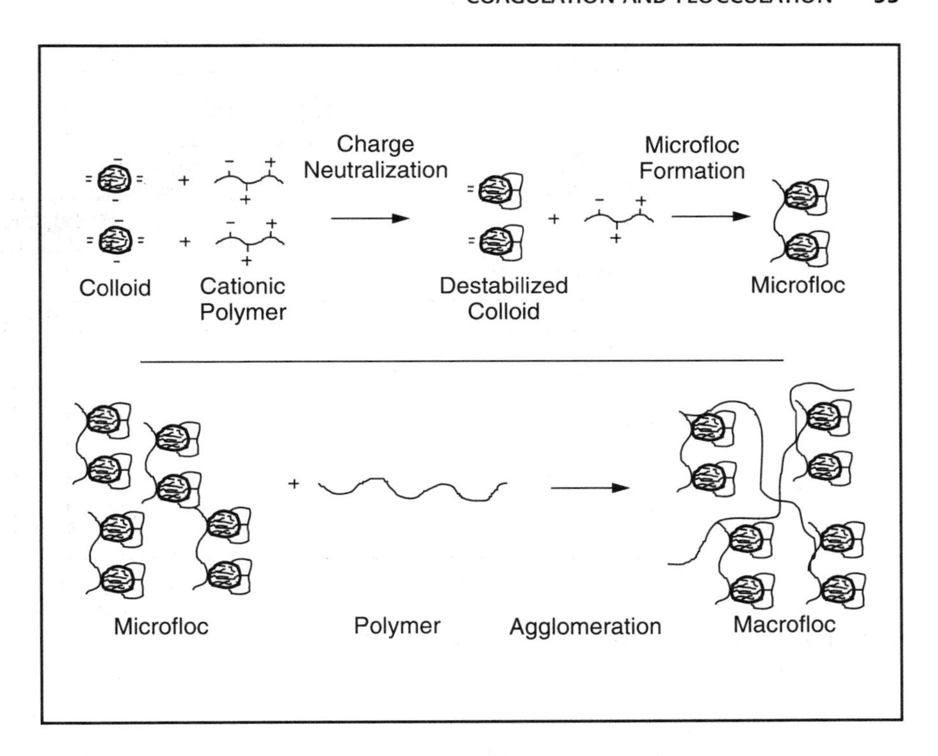

FIGURE 4-2 Microfloc and macrofloc formation

Coagulant Chemicals and Feed Equipment

There are two types of chemicals used in the coagulation process: coagulants and coagulant aids. The dosages of coagulants and coagulant aids must be closely monitored to ensure that effective coagulation is occurring. Tests for control of the coagulation and flocculation process are described later in this chapter.

Coagulants

The troublesome particles to be removed from water are usually negatively charged, and the coagulants used in water treatment normally consist of positively charged ions. The positive charges neutralize the negative charges and promote coagulation.

Some coagulants contain ions with more positive charges than others. Those consisting of trivalent ions, such as aluminum and iron, are 50 to 60 times more effective than chemicals with bivalent ions, such as calcium. They are 700 to 1,000 times more effective than coagulants with monovalent ions, such as sodium. (See *Basic Science Concepts and Applications*, part of this series, for a discussion of valences.)

Alum

Because alum is the most common coagulant used for water treatment, it is important to understand how it promotes settling. The process that takes place is as follows:

1. Alum added to raw water reacts with the alkalinity naturally present to form jellylike floc particles of aluminum hydroxide, $Al(OH)_3$. A certain level of alkalinity is necessary for the reaction to occur. If not enough is naturally present, the alkalinity of the water must be increased.
2. The positively charged trivalent aluminum ion neutralizes the negatively charged particles of color or turbidity. This occurs within 1 or 2 seconds after the chemical is added to the water, which is why rapid, thorough mixing is critical to good coagulation.
3. Within a few seconds, the particles begin to attach to each other to form larger particles.
4. The floc that is first formed is made up of microfloc that still has a positive charge from the coagulant; the floc particles continue to neutralize negatively charged particles until they become neutral particles themselves.
5. Finally, the microfloc particles begin to collide and stick together (agglomerate) to form larger, settleable floc particles.

Many physical and chemical factors can affect the success of a coagulant, including mixing conditions; pH, alkalinity, and turbidity levels; and water temperature. Alum works best in a pH range of about 5.8 to 8.5. If it is used outside this range, the floc either does not form completely or it may form and then dissolve back into the water.

Iron Salts

Iron salts such as ferric chloride and ferric sulfate can operate effectively over a wider range of pH values than alum. However, they are quite corrosive and require special facilities for storage and handling.

Both alum and ferric sulfate are affected by the alkalinity of the raw water. If the alkalinity is not high enough, an effective floc will not form. If floc is not completely formed because of insufficient alkalinity or a pH value outside the optimal range, but the alkalinity or pH is later changed during treatment or in the system, the floc can re-form in the distribution system. This will, of course, cause customer complaints and problems due to a buildup of sediment in the system piping.

Other Coagulant Chemicals

In general, if pH and alkalinity are at proper levels, coagulation can be improved by increases in turbidity, temperature, and mixing energy. Table 4-2 provides a listing of commonly used coagulants. Other coagulant chemicals that are occasionally used for special conditions are aluminum ammonium sulfate and aluminum potassium sulfate. Table 4-3 gives typical dosage ratios for combinations of coagulants. In most cases, the secondary chemical is added to adjust the pH or otherwise enhance the effectiveness of the primary chemical.

It has also been found that the use of ozone as an initial disinfectant often improves coagulation. This has the effect of lowering the cost of coagulant chemicals, reducing sludge disposal costs, and lengthening filter runs.

Additional details on the chemistry of floc formation can be found in *Basic Science Concepts and Applications* (part of this series), Chemistry Section.

TABLE 4-2 Common coagulation chemicals

Common Name	Chemical Formula	Comments
Aluminum sulfate	$Al_2(SO_4)_3 \cdot 14(H_2O)$	Most common coagulant in the United States; often used with cationic polymers
Ferric chloride	$FeCl_3$	May be more effective than alum in some applications
Ferric sulfate	$Fe_2(SO_4)_3$	Often used with lime softening
Ferrous sulfate	$Fe_2(SO_4)_3 \cdot 7H_2O$	Less pH dependent than alum
Aluminum polymers	—	Include polyaluminum chloride and polyaluminum sulfates
Cationic polymers	—	Synthetic polyelectrolytes; large molecules
Sodium aluminate	$Na_2Al_2O_4$	Used with alum to improve coagulation
Sodium silicate	$Na_2O \cdot (SiO_2)_x$	x can range from 0.5 to 4.0; ingredient of activated silica coagulant aids

Source: Adapted from *Water Treatment Plant Design.* 1990.

TABLE 4-3 Coagulant combinations and ratios

Coagulants	Typical Dosage Ratio (First to Second Coagulant)
Aluminum sulfate + caustic soda	3:1
Aluminum sulfate + hydrated lime	3:1
Aluminum sulfate + sodium aluminate	4:3
Aluminum sulfate + sodium carbonate	1:1 to 2:1
Ferric sulfate + hydrated lime	5:2
Ferrous sulfate + hydrated lime	4:1
Ferrous sulfate + chlorine	8:1
Sodium aluminate + ferric chloride	1:1

Coagulant Aids

A coagulant aid is a chemical added during coagulation to achieve one or more of the following results:

- to improve coagulation
- to build stronger, more settleable floc
- to overcome the effect of temperature drops that slow coagulation
- to reduce the amount of coagulant needed
- to reduce the amount of sludge produced

The use of coagulant aids can significantly reduce the amount of alum used and, accordingly, the amount of sludge produced. Because alum sludge is difficult to dewater and dispose of, a reduction in sludge is often the prime consideration in the decision to use a coagulant aid. The three general types of coagulant aids are

- activated silica
- weighting agents
- polyelectrolytes

Activated Silica

Activated silica has been used as a coagulant aid with alum since the late 1930s and remains in wide use today. Used properly in dosages from 7 to 11 percent of the coagulant dosage, activated silica will increase the rate of coagulation, reduce the coagulant dosage needed, and widen the pH range for effective coagulation. Activated silica must be prepared by the operator at the plant. The chemical actually delivered to the plant is sodium silicate, Na_2SiO_3. The operator "activates" the sodium silicate by adding an acid, typically hypochlorous acid, to reduce the alkalinity.

The chief advantage of using activated silica is that it strengthens the floc, making it less likely to break apart during sedimentation or filtration. In addition, the resulting floc is larger and denser and settles more quickly. Improved color removal and better floc formation at low temperatures can also result. Activated silica is usually added after the coagulant, but adding it before can also be successful, especially with low-turbidity water. It should never be added directly with the alum because they react with each other.

A major disadvantage of using activated silica is the precise control required during the activation step to produce a solution that will not gel. Too much silica will actually slow the formation of floc and cause filter clogging.

Weighting Agents

When some natural materials are added to water, they form additional particles that enhance floc formation. Weighting agents are principally used to treat water high in color, low in turbidity, and low in mineral content. This type of water would otherwise produce small, slowly settling floc.

Bentonite clay is a common weighting agent. Dosages in the range of 10 to 50 mg/L usually produce rapidly settling floc. In water with low turbidity, the clay increases turbidity, which speeds formation of floc by increasing the number of chance collisions between particles. Powdered limestone and powdered silica can also be used as weighting agents.

Polyelectrolytes

Polyelectrolytes (also called polymers) have extremely large molecules that, when dissolved in water, produce highly charged ions. The following are three basic polyelectrolyte classifications that may be either natural or synthetic materials:

- cationic polyelectrolytes
- anionic polyelectrolytes
- nonionic polyelectrolytes

Cationic polyelectrolytes. Cationic polyelectrolytes are polymers that produce positively charged ions when dissolved in water. They are widely used because the suspended and colloidal solids commonly found in water are generally negatively charged. They can be used as the primary coagulant or as an aid to such coagulants as alum or ferric sulfate. For the most effective turbidity removal, the polymer is generally used in combination with a coagulant.

Advantages to using cationic polyelectrolytes include the following: the amount of coagulant can be reduced, the floc settles better, there is less sensitivity to pH, and the flocculation of living organisms such as bacteria and algae is improved.

Anionic polyelectrolytes. Anionic polyelectrolytes are polymers that dissolve to form negatively charged ions; they are used to remove positively charged solids. Anionic polyelectrolytes are used primarily with aluminum and iron coagulants. Advantages include increased floc size, improved settling, and generally stronger floc. They are not materially affected by pH, alkalinity, hardness, or turbidity.

Nonionic polyelectrolytes. Nonionic polyelectrolytes are polymers having a balanced, or neutral, charge. Upon dissolving, they release both positively and negatively charged ions. Although nonionic polyelectrolytes must be added in larger doses than other types, they are less expensive.

Compared with other coagulant aids, the required dosages of polyelectrolytes are extremely small. The normal dosage range of cationic and anionic polymers is 0.1 to 1.0 mg/L. For nonionic polymers the dosage range is 1 to 10 mg/L.

Chemicals Used to Raise Alkalinity

Increasing the alkalinity of water often enhances the effect of coagulants. The chemicals principally used to increase alkalinity are lime, soda ash, caustic soda, and sodium bicarbonate.

Lime. Lime ($CaCO_3$) is often used in conjunction with aluminum or ferrous sulfate to provide artificial alkalinity in water; in some waters it may be used alone. It is used as either quicklime or hydrated lime.

Soda ash, caustic soda, and sodium bicarbonate. Soda ash (Na_2CO_3), caustic soda (NaOH), and sodium bicarbonate ($NaHCO_3$) are also used to raise alkalinity in order to enhance the effectiveness of other coagulants. Although they are more expensive per unit weight than lime, they are often preferred because of easier feeding and handling. More information on feeding and handling lime, soda ash, and caustic soda is included in chapter 11.

Approval of Coagulant Chemicals

All chemicals used in water treatment must be approved by a US or state regulatory agency as safe for addition to potable water. The label on any container must note that the product is acceptable and that application instructions provided by the supplier are followed. Further details on the acceptability of drinking water additives are provided in appendix A.

Chemical Storage and Handling

Coagulant chemicals are available both in dry granules or powder and in liquid form. Dry chemicals are available in various packaging sizes from 50-lb (23-kg) bags to drums. In bulk, they can be delivered by transport truck or railroad car.

Chemicals should always be stored in a dry area at a moderate and fairly uniform temperature. Most chemicals will harden and cake if exposed to moisture, so bags should be stored on pallets to allow air circulation beneath them. If possible, the storage area for dry chemicals should be located over the feed machines so that they can be dumped directly into feed hoppers. The types of equipment used to handle dry chemicals include hand trucks, overhead monorails, hoists, elevators, mechanical conveyors, and pneumatic conveyors.

Chemicals in liquid form are becoming increasingly popular, particularly at larger plants where it is practical to purchase chemicals in tank truck loads. Advantages of purchasing chemicals in liquid form include greatly simplified chemical storage, handling, and feeding; less dirt and mess in the plant; less required storage space; and reduced safety hazards.

Chemical Feed Equipment

Coagulant chemicals can be fed in dry or liquid form.

Dry Chemical Feeders

There are two general types of dry chemical feeders: volumetric and gravimetric. Volumetric feeders measure out the chemical from the hopper to the mixing tank by volume per unit time. An example of this type of feeder is shown in Figure 4-3. Gravimetric feeders feed the chemical by weight, as illustrated in Figure 4-4. Volumetric feeders are generally less costly but also less accurate than gravimetric ones. The types of dry feeding mechanisms include a rotating disk, an oscillating disk, a rotary gate, a belt, and a screw. The volume or weight of chemical to be fed may be set manually or adjusted automatically based on water flow rate, turbidity, zeta potential, or signal from a streaming current meter.

All chemicals must be in solution prior to mixing with the water to be treated. Most coagulants and coagulant aids dissolve with some difficulty, so it is necessary to provide good agitation and sufficient time for a solution to

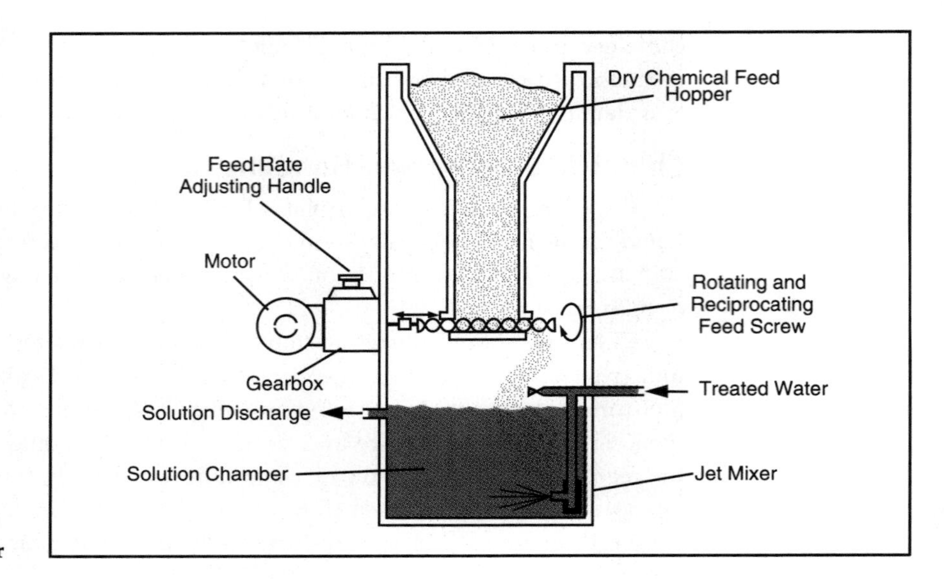

FIGURE 4-3
Volumetric feeder

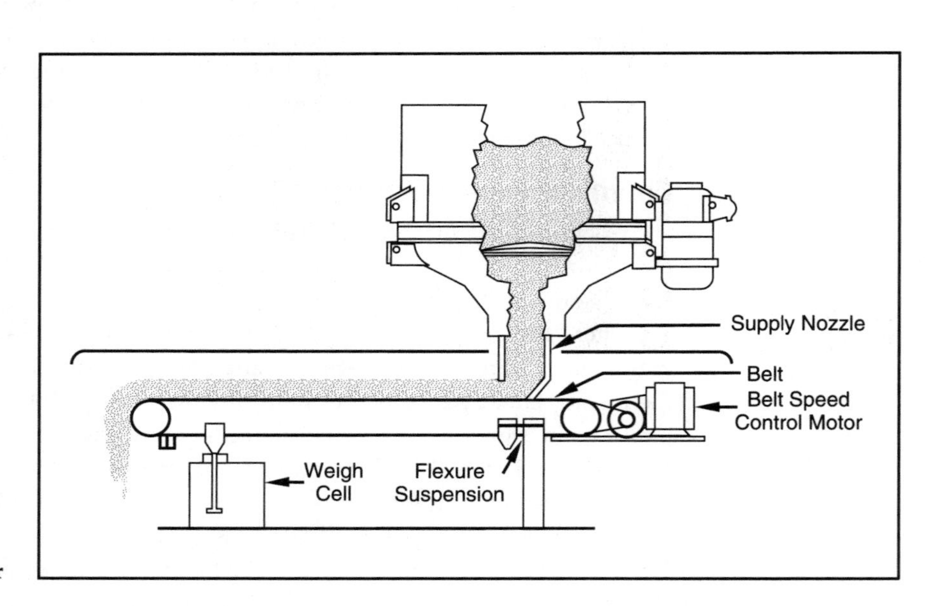

FIGURE 4-4
Gravimetric feeder

FIGURE 4-5
Solution mixer

Courtesy of Lightnin

be formed as the dry chemical is added to water. Dry chemical feeders generally feed the chemical into a mixing tank where it is put into solution. In some instances, chemical solution water should be softened to avoid the creation of unwanted precipitates. Small quantities of chemicals can also be put into solution in batches using a small tank and mixer, as shown in Figure 4-5.

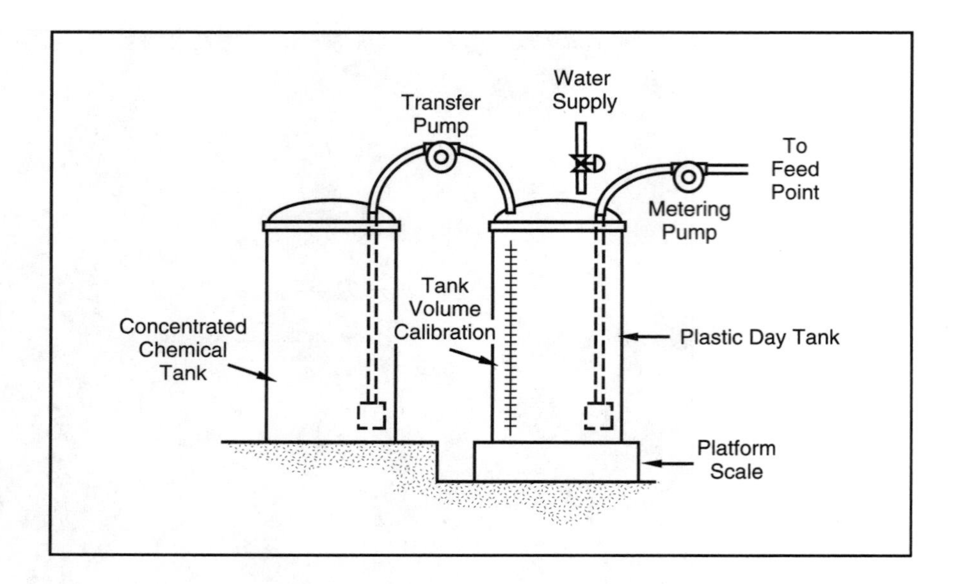

FIGURE 4-6 Chemical diluted in a day tank for feeding

Solution Feeders

Liquid coagulants and coagulant aids are fed by solution feeders. The feeders may draw directly from the liquid chemical storage tank or the chemical may be diluted in a smaller tank before it is fed, as illustrated in Figure 4-6.

The most common type of solution feeder is the metering pump. This is a positive-displacement pump that delivers a precise volume of solution with each stroke or rotation of the pumping mechanism. Figure 4-7 is a diagram of a typical metering pump. These pumps usually have variable-speed motors or drives, which can be manually adjusted or automatically controlled in response to a control signal.

Peristaltic pumps function by having a roller mechanism that squeezes a quantity of chemical through a loop of flexible hose. Their advantages are their reasonable price and their capability of handling almost any nonflammable liquid. Their principal disadvantage is that the flexible tube must be periodically replaced.

Solution feeders may also be of the decanting type, consisting of a solution tank with a drawoff pipe that is lowered or raised at a controlled rate by a variable-speed motor; a revolving dipper type, which rotates at a controllable speed; and a progressive cavity pump.

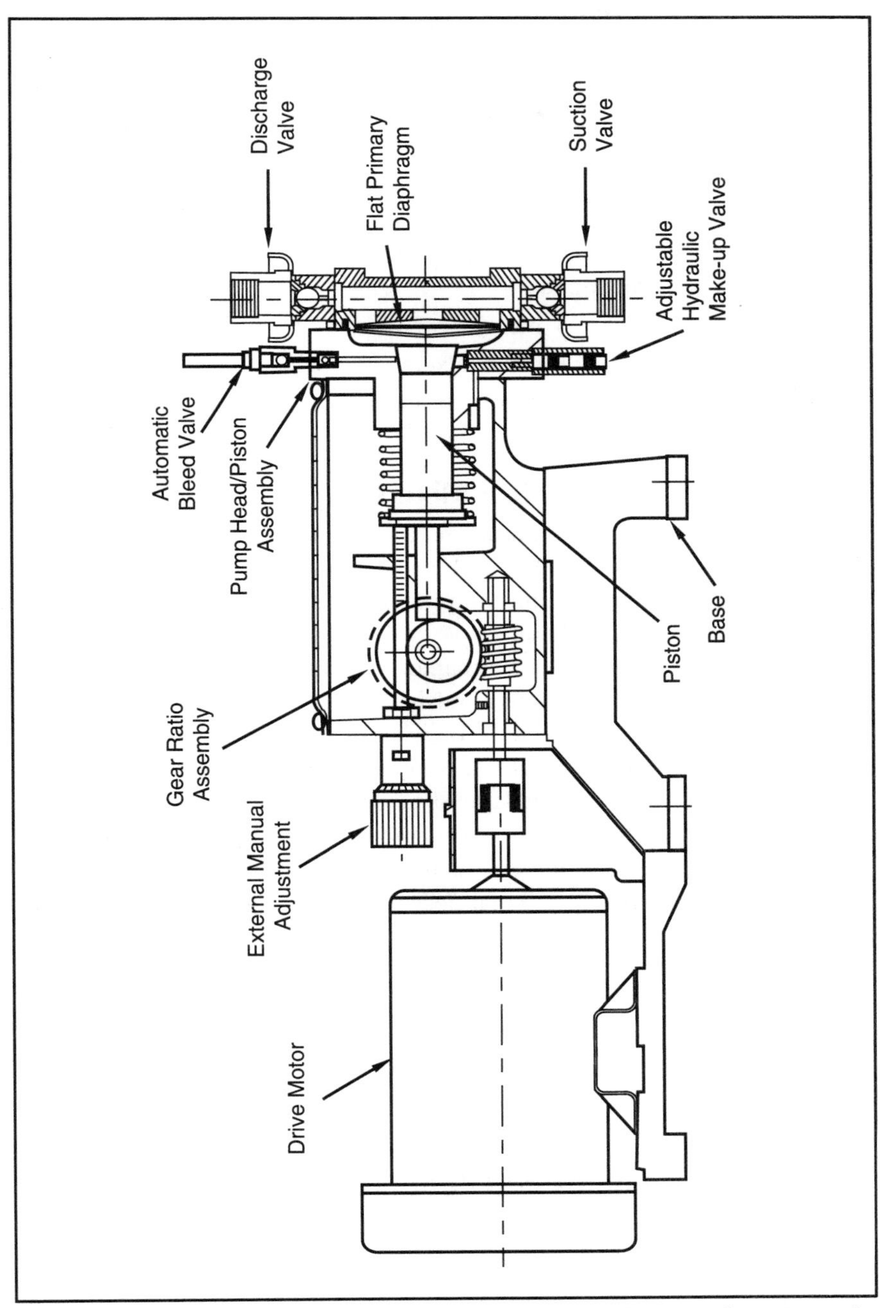

FIGURE 4-7
Diaphragm-type metering pump

Courtesy of Pulsafeeder, a Unit of IDEX Corporation

Rapid-Mix Facilities

Once coagulant chemicals have been added to raw water, it is essential to provide rapid agitation to distribute the coagulant evenly throughout the water. This is particularly true when alum or ferric salts are being used. The water must be briefly and violently agitated to encourage the greatest number of collisions between suspended particles.

The more common types of facilities used for rapid mixing (or flash mixing) are

- mechanical mixers
- static mixers
- pumps and conduits
- baffled chambers

Mechanical Mixers

Mechanical mixers are widely used for rapid mixing because of their good control features. They are usually placed in a small chamber or tank and include the propeller, impeller, or turbine type. The detention time in these chambers is designed to be very short. Figures 4-8 and 4-9 are two designs for mechanical mixers in chambers.

Mechanical mixers can also be mounted directly into a pipeline; they are then referred to as in-line mixers. This type of unit provides good instantaneous mixing with little short-circuiting, costs much less than a conventional rapid-mixing installation, and still allows for adjustment to provide the correct amount of mixing energy. Figure 4-10 illustrates a typical in-line mixer. All in-line mixers must be located close to the flocculation chamber so that flocculation and settling will not occur within the pipeline.

Static Mixers

Static (or motionless) mixers produce turbulence and mixing through use of the fixed sloping vanes within the mixer, as illustrated in Figure 4-11. They are effective and economical to install and operate; however, head loss through a static mixer is significant. Another caution is that mixing efficiency is directly related to flow rate, so there is no way to adjust the mixing energy.

Pumps and Conduits

Coagulant chemicals can also be added to the suction side of a low-lift pump to use the turbulence in the pump as a mixing mechanism (Figure 4-12). The amount of mixing is determined by the speed of the pump, so the

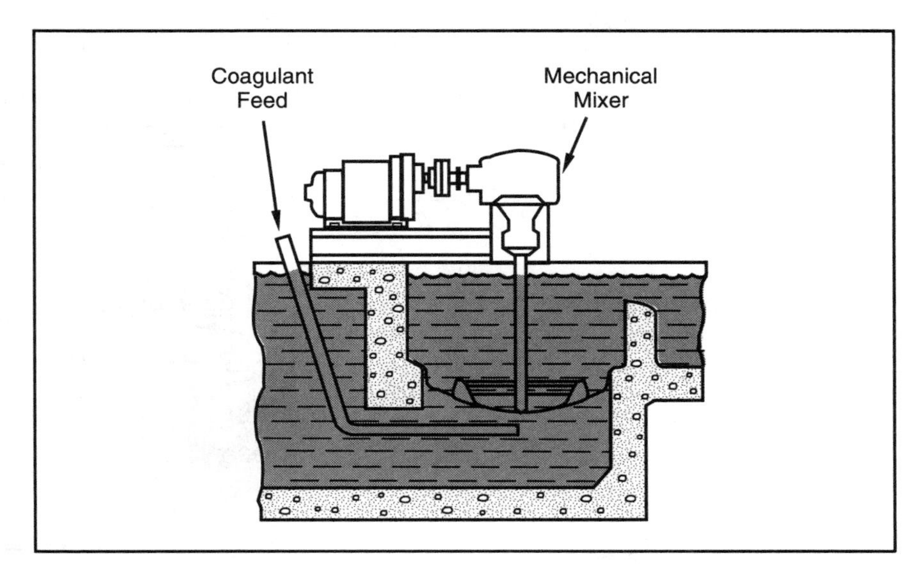

FIGURE 4-8
Mechanical mixing chamber—single-blade mixer

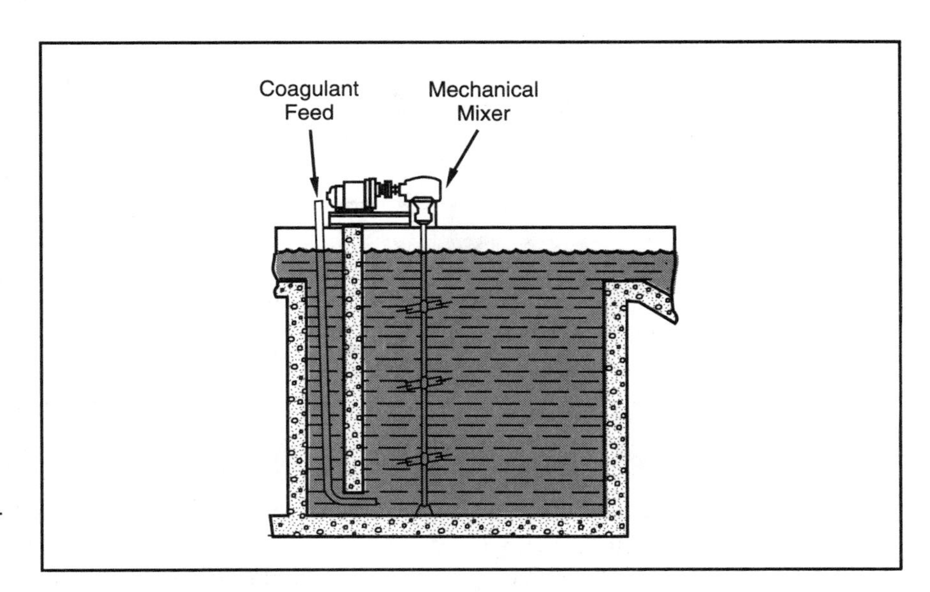

FIGURE 4-9
Mechanical mixing chamber—multiple-blade mixer

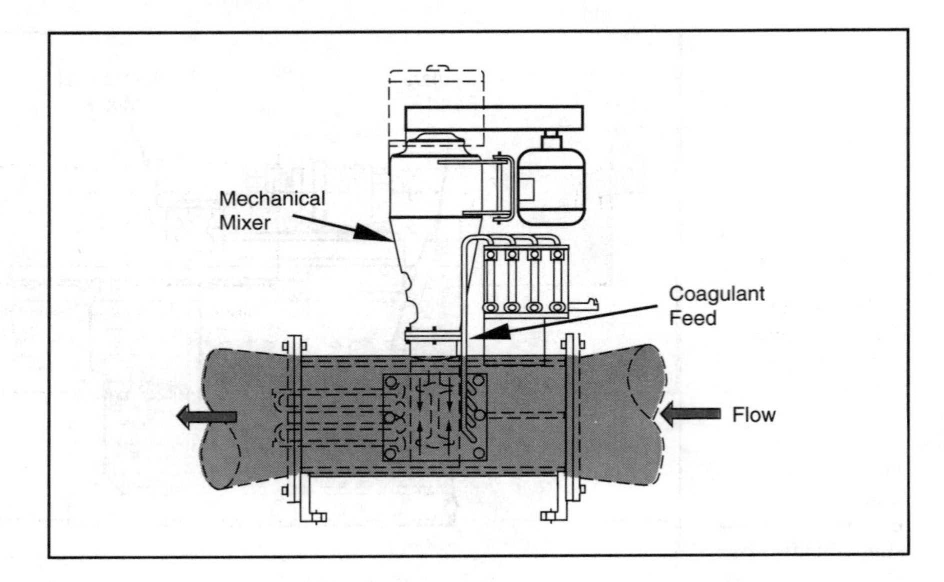

FIGURE 4-10
In-line mixer

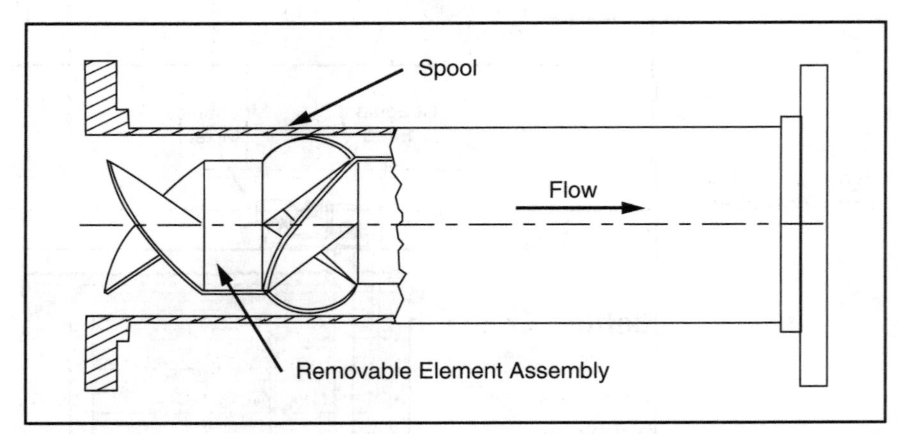

FIGURE 4-11
Section view of a static mixer

Source: Water Quality and Treatment *(1990)*

turbulence required for proper flash mixing may or may not be provided. If the turbulence is adequate, it can be used without investing in special equipment and without increasing the system head.

Disadvantages include the fact that there is little or no opportunity to adjust the operation to suit treatment needs and that coagulant may cause pump corrosion.

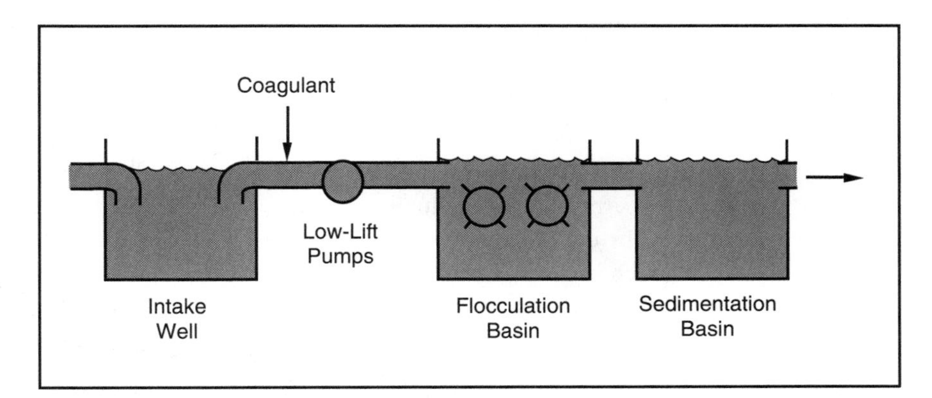

FIGURE 4-12 Low-lift pumps used for rapid mixing

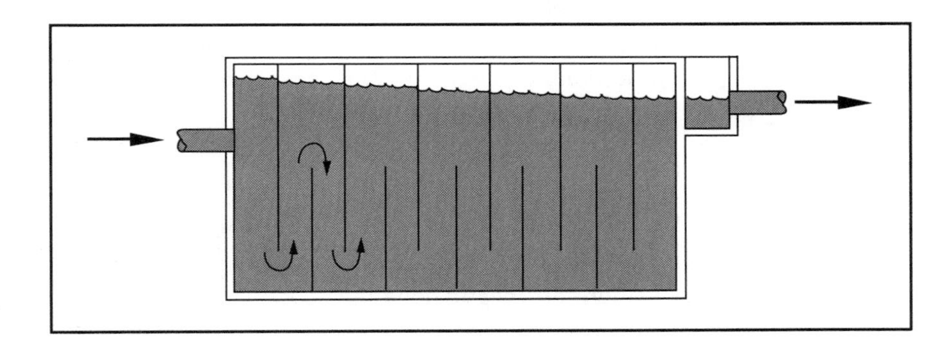

FIGURE 4-13 Section view of a baffled chamber

Baffled Chambers

The baffled chamber shown in Figure 4-13 provides turbulence to the water flowing over and under the baffles. The primary problem is that turbulence is determined by the rate of flow and generally cannot be controlled.

Flocculation Facilities

Flocculation Basins

Flocculation follows coagulation and usually takes place in a chamber that provides a slow, gentle agitation of the water. In the flocculation stage, physical processes transform the smaller particles of floc formed by the rapid mix into larger aggregates of floc. The rate of aggregation is determined by the rate at which the particles collide. But, as the aggregates grow in size, they become more fragile, so the mixing force applied must not be so great as to cause the floc particles to break up or shear. Baffles are usually provided in the basin to slow down the water flow and reduce short-circuiting. Figures 4-14A through 4-14E illustrate various designs for mechanical flocculators that are in use. Most flocculation basins are designed for tapered flocculation, which involves a reduction in velocity gradient as the water passes through the basin.

At least two basins should be provided so that one may be removed from service at a time. The units should also be designed for operation either in series or in parallel. The flow-through velocity should normally be between 0.5 and 1.5 ft/min (0.0025 and 0.0076 m/s).

Solids-Contact Basins

Solids-contact basins combine the processes of flocculation and upflow clarification in a single structure. These units are described in greater detail in chapter 5.

Regulations

Two regulations recently promulgated by the US Environmental Protection Agency (USEPA) can have a significant impact on the design and operation of coagulation and flocculation processes: the Surface Water Treatment Rule (SWTR) and requirements for reducing disinfection by-products (DBPs).

The Surface Water Treatment Rule

The SWTR was enacted by USEPA in 1989. The regulation requires most public water systems using surface water sources to provide filtration treatment. The only exceptions are those systems that can demonstrate that their raw-water quality meets very exacting quality requirements.

The regulation also imposes new requirements on many groundwater systems that are designated by their state as having the potential to become contaminated by surface water. In some cases, these designated systems must install filtration treatment. As a result of these new requirements, many water systems that were previously not providing filtration may have to add flocculation, coagulation, sedimentation, and filtration equipment to their water treatment plant in the future.

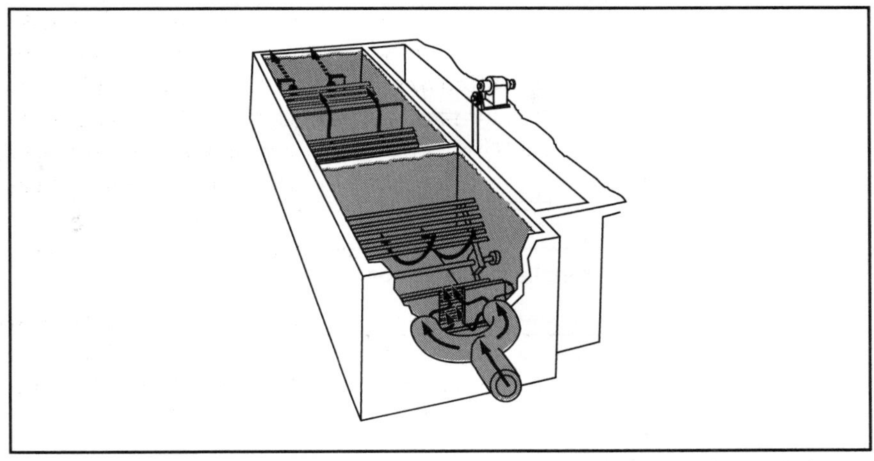

Courtesy of Envirex Inc., Waukesha, Wis.

A. Baffles and Horizontal Paddle-Wheel Flocculator

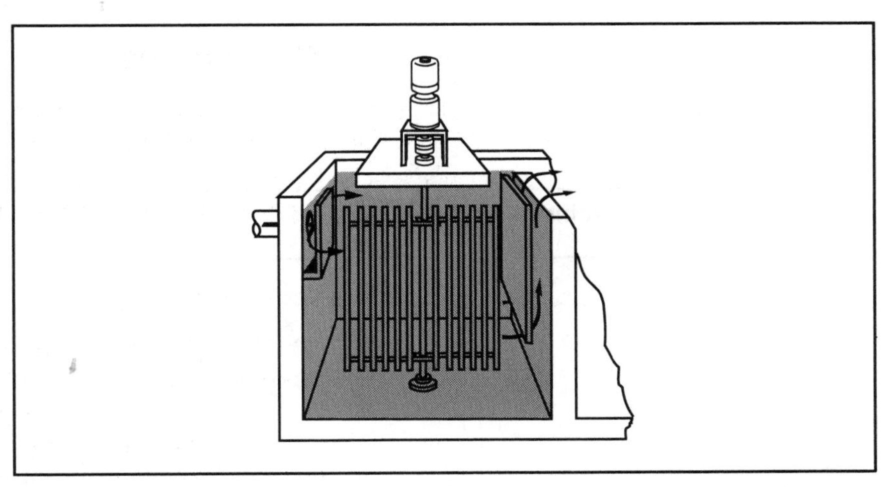

Courtesy of Envirex Inc., Waukesha, Wis.

B. Paddle-Wheel Flocculator, Vertical Type

Figure continued next page

FIGURE 4-14
Various designs
for mechanical
flocculators

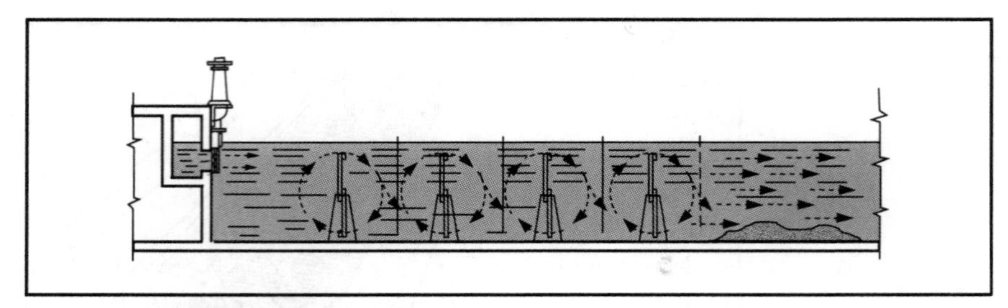

C. Propeller Flocculator

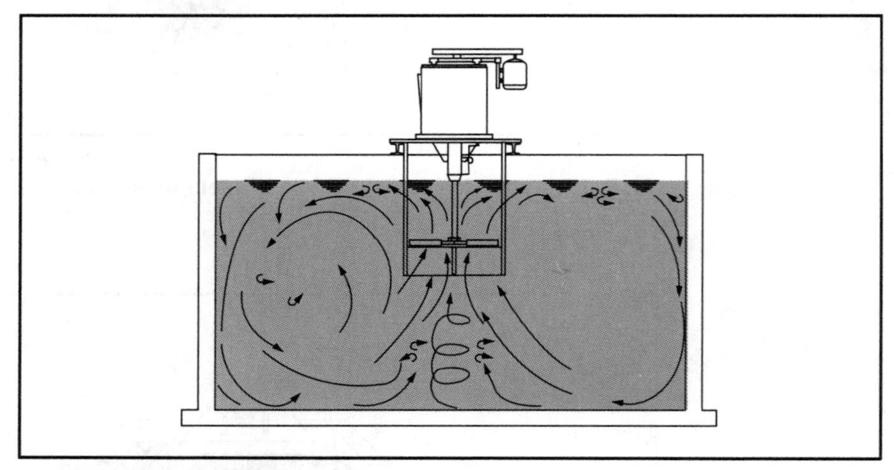

D. Turbine Flocculator

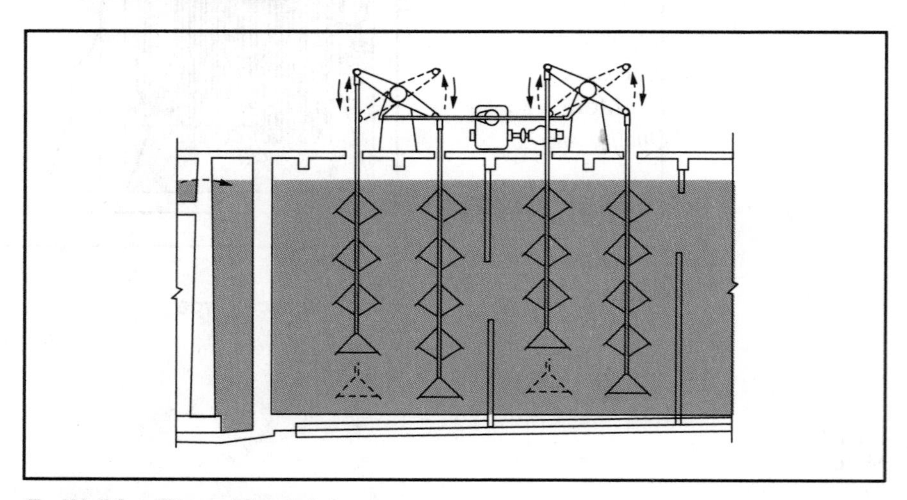

E. Walking-Beam Flocculator

FIGURE 4-14 Various designs for mechanical flocculators (continued)

The SWTR also requires surface water systems to maintain specified $C \times T$ values (disinfectant concentration × contact time) in processing the water in the treatment system. Critical to achieving the required contact time between the water and the disinfectant before the water leaves the plant are the points at which disinfectant is fed and the contact time in the various treatment basins.

Many older flocculation and sedimentation basins were constructed with insufficient baffling, which results in short-circuiting. New tests now required by the SWTR to measure the actual contact time often reveal that baffling is clearly inadequate in many plants. Many water treatment plants will have to add baffles or make other alterations to treatment basins to comply with the new requirements. Likewise, the design of new treatment basins must factor in not only the requirements for clarification but also the need to achieve the required disinfectant contact time in the system.

Requirements to Reduce Disinfection By-products

Existing and future USEPA requirements to limit the formation of total trihalomethanes (TTHMs) and other DBPs in finished water will affect the design and operation of coagulation, flocculation, and sedimentation processes in many water systems. The requirements may also affect the choice of coagulants and coagulant aids that are used. DBPs are formed by the reaction of disinfectants with precursors (natural organic substances) in the water, and they can be prevented or removed by several different methods. The best method is to remove the precursors to the extent possible before the disinfectant is applied. Some systems may find that present and future DBP requirements will dictate changes or improvements in their pretreatment processes.

More complete details of federal regulations are included in *Water Quality*, also part of this series.

Operation of the Processes

Operation of the coagulation and flocculation processes generally consists of the following steps:

- Consider the water characteristics affecting the selection of chemicals to be used.
- Apply the chemicals.
- Monitor the effectiveness of the process.

Water Characteristics Affecting Chemical Selection

The selection of chemical coagulants and coagulant aids is a continuing process of trial and evaluation. To do a thorough job of chemical selection, the following characteristics of the raw water to be treated should be considered:

- water temperature
- pH
- alkalinity
- turbidity
- color

The effect of each characteristic on coagulation and flocculation is briefly described below.

The effectiveness of a coagulant will change as raw-water characteristics change. The effectiveness of coagulation may also change for no apparent reason, suggesting that there are other factors, not yet understood, that affect the process.

Water Temperature

Lower-temperature water usually causes poorer coagulation and flocculation and can require that more of a chemical be used to maintain acceptable results.

pH

Extreme values of pH, either high or low, can interfere with coagulation and flocculation. The optimal pH varies depending on the coagulant used.

Alkalinity

Alum and ferric sulfate interact with the chemicals that cause alkalinity in the water, thus reducing the alkalinity and forming complex aluminum or iron hydroxides that begin the coagulation process. Low alkalinity limits this reaction and results in poor coagulation; in these cases, it may be necessary to increase the alkalinity of the water.

Turbidity

The lower the turbidity, the more difficult it is to form a proper floc. Fewer particles mean fewer random collisions and hence fewer chances for floc to accumulate. It may be necessary to add a weighting agent such as clay (bentonite) to low-turbidity water.

Color

Color is caused by organic compounds in the raw water. The organics can react with the chemical coagulants, making coagulation more difficult. Pretreatment with oxidants or adsorbents may be necessary to reduce the concentration of organics. Generally, the dosage of coagulant chemical needed to coagulate color must be increased as the color concentration increases.

Applying Coagulant Chemicals

An operator should begin chemical selection by using the jar test with various chemicals, both singly and in combination. The jar test is used experimentally to determine the optimal conditions for the coagulation, flocculation, and sedimentation processes. This is done by trying various combinations of chemical dosage, mixing speed, and settling interval. Jar test results are expressed in milligrams per liter, which is converted to the equivalent daily dose required for setting the chemical feeders for plant operation. Details of performing jar tests are covered in *Water Quality*, and details of chemical feed calculations are covered in *Basic Science Concepts and Applications*; both are part of this series.

Monitoring Process Effectiveness

Although jar tests provide a good indication of the results to expect, full-scale plant operation may not match these results. The adequacy of flash mixing and flocculation is not something that can be observed directly.

The following are indicators of inadequate mixing or incorrect chemical dosage:

- very small floc (called pinpoint floc)
- high turbidity in settled water
- too-frequent filter backwashing

Operating Factors That Could Affect Floc Development

Plant operating factors that could make a difference in the proper development of floc include

- inadequate flash mixing
- improper flocculation mixing
- inadequate flocculation time

Inadequate Flash Mixing

Successful coagulation is based on rapid and complete mixing. Though coagulation occurs in less than 1 second, the chamber may provide up to 30 seconds of detention time. Mixing should be turbulent enough so that the coagulant is dispersed throughout the coagulation chamber.

Some experts maintain that during the first tenth of a second the coagulant must be thoroughly mixed with every drop of water to start an efficient floc; otherwise, after that point the efficiency of the entire process declines. If polymers are used as prime coagulants, rapid mixing is less critical, but thorough mixing remains very important in encouraging as many particle collisions as possible.

Improper Flocculation Mixing

Proper flocculation requires long, gentle mixing. Mixing energy must be high enough to bring coagulated particles constantly into contact with each other, but not so high as to break up those particles already flocculated. For this reason, flocculation basins and equipment are usually designed to provide higher mixing speeds immediately following coagulation and progressively slower speeds as the water flows through the basin.

Properly coagulated and flocculated particles will look like small snowflakes or tufts of wool suspended in very clear water. The water should not look cloudy or foggy as a result of poorly formed floc. A cloudy appearance is usually caused by an inadequate alum dosage. If the water does appear cloudy, or if it displays any of the four signs of inadequate mixing listed earlier, then the speed of the flocculators may be incorrect. Under some circumstances, floc can become as large as a quarter, but at that point it may be too buoyant and will not settle well. Some systems have found that pinhead-size floc is about optimal for settling efficiency.

Inadequate Flocculation Time.

It takes time to develop heavy floc particles. Although some systems can be operated with as little as 10 minutes of detention time, others may require up to 1 hour. The average is probably about 30 minutes. Because short-circuiting can be a major problem, a minimum of three flocculation basins in series is recommended.

Operational Control Tests and Equipment

The following tests can help control the coagulation and flocculation processes:

- jar tests
- pH tests
- turbidity tests
- filterability tests
- zeta potential measurement
- streaming current tests
- particle counting

Jar Tests

Jar tests help determine the chemical or chemical combination and the dosage that will produce the best floc. Testing should not be limited to the chemicals that are currently used in the plant. A wide variety of coagulants and coagulant aids should be tried, including new chemicals as they come on the market. The tests may identify different chemicals that are more effective or less expensive to use. Chemical suppliers are usually able to provide chemical samples for testing and advice on typical dosage rates.

pH Tests

The pH of the water is important because all coagulants and coagulant aids work best within a particular pH range. If the pH of the water shifts, because of either natural causes or the effect of other treatment chemicals, the level of coagulant chemicals may need to be adjusted.

Turbidity Tests

Measurement of the turbidity in jar-test samples provides a more exact measure than simply eyeing the sample as "poor" or "good." Turbidity tests should also be run on the untreated water and on samples taken after the sedimentation basins. The results can then be compared with the jar-test turbidity results to determine how much turbidity is actually being removed, compared with the process efficiency predicted by jar tests. Comparisons of the process efficiency should be fairly consistent.

Effective coagulation and flocculation should result in a settled water turbidity of less than 10 nephelometric turbidity units (ntu). This will result in more efficient use of the filters because they will receive a minimum loading of suspended solids. Excessive carryover of floc particles can quickly seal ("blind") a filter, increasing the required frequency of backwashing. Visible floc carryover from the sedimentation process to the filters is a clear sign that blinding may occur.

Filterability Tests

The filterability of plant water that has the added coagulant can be measured to determine how efficiently the coagulated water can be filtered. The filterability test measures the amount of water filtered in a given time when flocculated plant flow (before sedimentation) is passed through a small- diameter tube known as a *pilot filter*. The pilot filter usually contains the same type of filtering media used in the plant filters and is equipped with a recording turbidimeter to continuously monitor the filtered water effluent. The amount of water passing through the filter before turbidity breakthrough can be correlated to how well the plant filters will operate under the same coagulant dosage.

The water used in a pilot filter is the actual coagulated water from the plant, not a laboratory simulation, so the operation can be directly correlated with the type of coagulant and dosage currently in use. The water takes only a few minutes to flow through a pilot filter, as opposed to a full-scale sedimentation and filter detention time of 2 or more hours. Necessary changes in coagulant dosage can therefore be predicted and implemented as raw-water quality changes, so as to prevent any deterioration of plant effluent quality.

There is no standard for the filterability test, and there are various procedures to measure filterability. Some pilot filters use filter paper, gauze, or a membrane filter. Various equipment suppliers use different terms to refer to the test, such as filterability index, filterability number, inverted gauze filter, membrane refiltration, pilot column filtration, silting index, or surface area concentration.

Zeta Potential Tests

Zeta potential (zp) is a measure of the excess number of electrons found on the surface of all particulate matter. The magnitude of the charges determines whether colloidal-size particles in suspension will repel one another and remain in suspension or agglomerate and eventually settle. The more negative the zp, the stronger the repelling force. Zeta potential is applicable to particles that range in size from ultramicroscopic to those that are visible to the naked eye. To induce the particles to settle properly, the zp should be reduced to about zero.

TABLE 4-4 Degree of coagulation within different zeta potential ranges

Average Zeta Potential	Degree of Coagulation
+3 to 0	Maximum
−1 to −4	Excellent
−5 to −10	Fair
−11 to −20	Poor
−21 to −30	Virtually none

The instrument used for making zp measurements is called a zeta meter. Although the optimal zp varies among different water sources and plant processes, a general degree of coagulation that occurs within various ranges is shown in Table 4-4.

Streaming Current Monitors

A streaming current monitor is an instrument that passes a continuous sample of coagulated water past a streaming current detector (SCD). The detector produces a continuous readout of the measurement of the net ionic and colloidal surface charge in the sample. The measurement is similar in theory to the zeta potential determination.

The optimal SCD reading varies with pH, so if a plant has source water with variable pH, the SCD reading will vary. The advantages of a streaming current monitor are that it provides continuous monitoring for coagulation control and it can even be used for automatic control of coagulant dosage if the pH remains constant.

Particle Counters

There is increasing interest in optimizing the coagulation–flocculation–filtration process to ensure removal of *Cryptosporidium* and *Giardia lamblia* cysts. The only practical method at this time is to monitor the size and concentration of minute particles in the finished water. Turbidity measurement is inadequate for this task because the concentration of cysts considered undesirable is well below the normal range measured in turbidity tests; in addition, turbidity tests give no indication of particle size.

There are several types of particle counters that have varying capabilities and advantages. In general, they operate by passing a very fine stream of water sample past a laser or light source; the light that is blocked or diffused by passing particles is measured by a photodiode, and the signal is interpreted by a computer. Today's particle counters are expensive, must be carefully calibrated and adjusted, and generally require a trained technician to interpret the results. However, they are expected to become more widely used for optimally operating surface water plants.

Operating Problems

Three common operating problems encountered during the coagulation–flocculation process are

- low water temperature
- weak floc
- slow floc formation

Low Water Temperature

Raw-water temperature that approaches the freezing point interferes with the coagulation and flocculation processes. As the water temperature decreases, the viscosity of the water increases, which slows the rate of floc settling.

The colder temperature also slows chemical reaction rates, although this effect on coagulation is relatively insignificant. Cold-water floc also has a tendency to penetrate through the filters, indicating that floc strength has decreased. The problems caused by low water temperature can best be overcome with the following techniques:

- Operate the coagulation process as near as possible to the best pH value for the water temperature.
- Increase the coagulant dosage as the water temperature decreases. This increases the number of particles available to collide and also reduces the effect of changes in pH resulting from the drop in temperature.
- Add weighting agents, such as clay, to increase floc particle density and add other coagulant aids to increase floc strength and encourage rapid settling.

Weak Floc

Weak floc is often not noticed until it has an adverse effect on filtration. A weak floc does not adhere well to the filter media; instead, it is broken up and carried deeper into the filter until it finally breaks through, causing increased turbidity in the filter effluent. Weak floc is often the result of inadequate mixing in the rapid-mix or flocculation basins. This can be checked by varying the mixing speeds and taking samples from various points to see if the floc settles any better. Jar tests should also be used to determine if other combinations of coagulants and coagulant aids produce a better floc.

Slow Floc Formation

Slow or inadequate floc formation is often a problem in water with low turbidity. If there are fewer particles, there are fewer random collisions for floc particles to form and grow. One way to correct this problem is to recycle some previously settled sludge from the sedimentation basin to add turbidity. This is the same principle used in the solids-contact clarification process. Another way to improve floc formation is to increase turbidity artificially by adding a weighting agent.

If alum or ferric sulfate is used as the coagulant, slow floc formation could also be a result of inadequate alkalinity in the raw water. If this is the case, alkalinity in the water can be increased by adding lime or soda ash.

Safety Precautions

The following are some special safety precautions applicable to the coagulation–flocculation process:

- Most dry chemicals can irritate eyes, skin, and mucous membranes. Dry-chemical feeders should be equipped with dust-control equipment, and protective clothing, goggles, and a respirator should be worn when the chemicals are being handled.
- Liquid chemicals, particularly polymers, can create dangerous slick areas if spilled, so spills should be cleaned up promptly.
- Dry alum and quicklime when mixed together, create tremendous heat; if the temperature should reach 1,100°F (593°C), highly explosive hydrogen gas will be released. These and other chemicals should be stored and used in a manner that will prevent improper mixing.

Record Keeping

Records should be maintained of past raw-water quality and the coagulants and dosages that work best for that water. Notes should also be kept of general observations relating to the operation of the coagulation–flocculation process. This is particularly important for surface water supplies because water quality often varies in each season or in relation to natural events such as heavy rains, snowmelt discharge, or droughts. Past experience during similar water quality conditions is often a good guide to how chemical addition or equipment should be adjusted to obtain optimal treatment. Figure 4-15 is a suggested record-keeping form for the coagulation–flocculation process.

Type of Coagulant ________________ **Date Started** ________________

Item	Results (by Date)								
	Date	Date	Date	Date	Date	Date	Date	Date	Date
Coagulant Dosage									
Raw Water									
Temperature (°C)									
pH									
Alkalinity (mg/L as $CaCO_3$)									
Turbidity (ntu)									
Taste and Odor									
Color (cu)									
Suspended Solids (mg/L)									
Algae Content									
Coagulated Water									
Filterability (Volume/Time)									
Zeta Potential (mV)									
Settled-Water Turbidity									
Filtered Water									
Turbidity (ntu)									
Color (cu)									
Taste and Odor									
Algae Content									
Residual Coagulant (mg/L)									

FIGURE 4-15 Record-keeping form for the coagulation–flocculation process

Selected Supplementary Readings

Amirtharajah, A., M.M. Clark, and R.R. Trussell, eds. 1991. ***Mixing in Coagulation and Flocculation.*** Denver, Colo.: American Water Works Association Research Foundation and American Water Works Association.

Back to Basics Guide to Surface Water Treatment. 1992. Denver, Colo.: American Water Works Association.

Clark, T.F. 1992. Effective Coagulation to Meet the New Regulations. ***Opflow,*** 18(1):1.

Clark, M.M., R.M. Srivastava, J.S. Lang, R.R. Trussell, L.J. McCollum, D.Bailey, J.D. Christie, and G. Stolarik. 1994. ***Selection and Design of Mixing Processes for Coagulation.*** Denver, Colo.: American Water Works Association Research Foundation and American Water Works Association.

Cornwell, D.A., C. Vandermeyden, G. Dillow, and M. Wang. 1992. ***Landfilling of Water Treatment Plant Coagulant Sludges.*** Denver, Colo.: American Water Works Association Research Foundation and American Water Works Association.

Dentel, S.K., and K.M. Kingery. 1988. ***An Evaluation of Streaming Current Detectors.*** 1988. Denver, Colo.: American Water Works Association Research Foundation and American Water Works Association.

Dentel, S.K., and K.M. Kingery. 1989. Using Streaming Current Detectors in Water Treatment. ***Jour. AWWA,*** 81(3):85.

Dentel, S.K., B.M. Gucciardi, T.A. Bober, P.V. Shetty, and J.J. Resta. 1989. ***Procedures Manual for Polymer Selection in Water Treatment Plants.*** Denver, Colo.: American Water Works Association Research Foundation and American Water Works Association.

Edzwald, J.K., and J.P. Wasah. 1992. ***Dissolved Air Flotation: Laboratory and Pilot Plant Investigations.*** Denver, Colo.: American Water Works Association Research Foundation and American Water Works Association.

Effect of Coagulation and Ozonation on the Formation of Disinfection By-Products. 1992. WITAF Report. Denver, Colo.: American Water Works Association.

Kawamura, S. 1991. Effectiveness of Natural Polyelectrolytes in Water Treatment. ***Jour. AWWA,*** 83(10):88.

Manual of Water Utility Operations. 8th ed. 1988. Austin, Texas: Texas Water Utilities Association.

Manual M3, Safety Practices for Water Utilities. 1990. Denver, Colo.: American Water Works Association.

Manual M39, Operational Control of Coagulation and Filtration Processes. 1992. Denver, Colo.: American Water Works Association.

Peterson, T.D. 1992. Streaming Current Monitor Controls Coagulant Addition. ***Opflow,*** 18(5):3.

Reckhow, D.A., U.K. Edzwald, and J.E. Tobiasons. 1993. ***Ozone as an Aid to Coagulation and Filtration.*** Denver, Colo.: American Water Works Association Research Foundation and American Water Works Association.

Recommended Standards for Water Works. 1992. Albany, N.Y.: Health Education Services.

Singer, P.C., and S. Chang. 1989. ***Impact of Ozone on the Removal of Particles, TOC, and THM Precursors.*** Denver, Colo.: American Water Works Association Research Foundation and American Water Works Association.

Von Huben, H. 1991. ***Surface Water Treatment: The New Rules.*** Denver, Colo.: American Water Works Association.

Water Quality and Treatment. 4th ed. 1990. New York: McGraw-Hill and American Water Works Association (available from AWWA).

CHAPTER 5

Sedimentation Basins and Clarifiers

Sand, grit, chemical precipitates, pollutants, floc, and other solids are kept in suspension in water so long as the water is flowing with sufficient velocity and turbulence. Sedimentation removes these solids by reducing the velocity and turbulence. Efficient solids removal by sedimentation greatly reduces the load on filtration and other treatment processes.

Process Description

Sedimentation, which is also called clarification, is the removal of settleable solids by gravity. The process takes place in a rectangular, square, or round tank called a sedimentation or settling basin or tank.

In the conventional water treatment process, sedimentation is typically used as a step between flocculation and filtration. Sedimentation is also used to remove the large amounts of chemical precipitates formed during the lime–soda ash softening process (see chapter 11).

Basins designed for efficient sedimentation allow the water to flow very slowly, with a minimum of turbulence at the entry and exit points and with as little short-circuiting of flow as possible. Sludge, the residue of solids and water, accumulates at the bottom of the basin and must then be pumped out of the basin for disposal or reuse.

Sedimentation Facilities

This section describes different types of sedimentation basins and their associated equipment and features.

Types of Basins

Although there are many variations in design, sedimentation basins can generally be classified as either rectangular or center-feed types. Figure 5-1 shows overhead views of the flow patterns in different types of sedimentation basins. Figure 5-1A shows a rectangular settling tank, and Figures 5-1B through 5-1E show circular and square tanks. The operating principles of these basins are described below.

Conventional Rectangular Basins

Rectangular basins are generally constructed of concrete or steel and designed so that the flow is parallel to the basin's length. This type of flow is called rectilinear flow. The basins must be designed to keep the flow distributed evenly across the width of the basin to minimize the formation of currents and eddies that would keep the suspended matter from settling. The

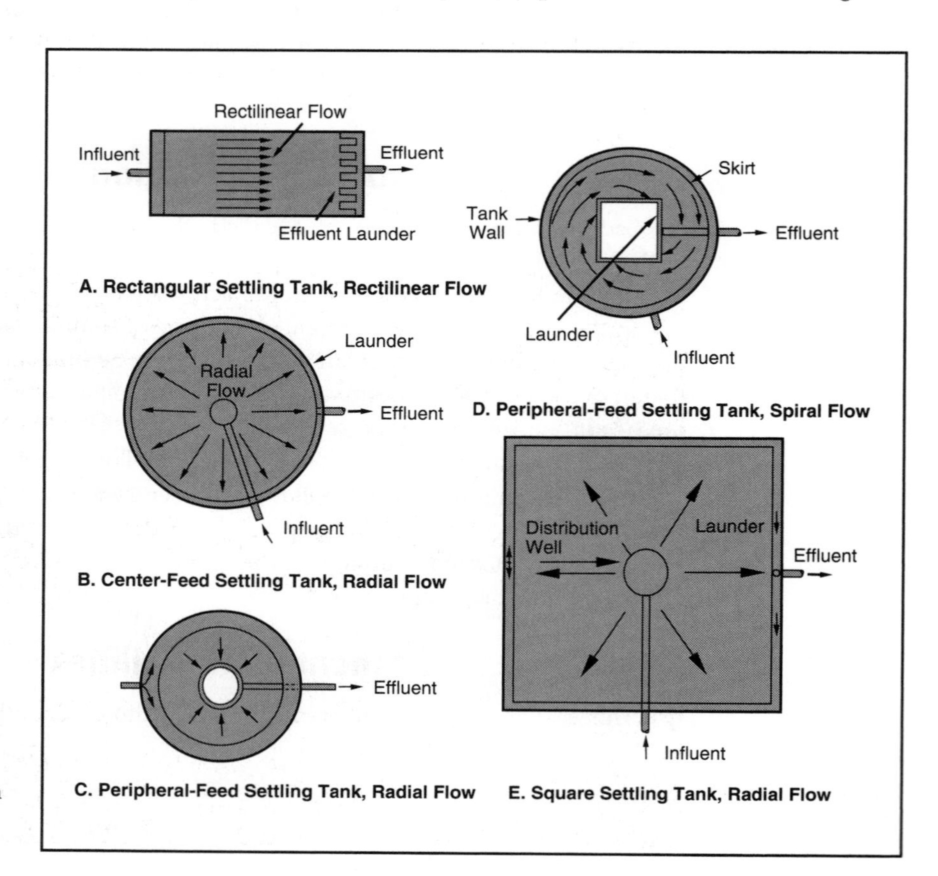

FIGURE 5-1 Overhead views of flow patterns in sedimentation basins

basins are often constructed with a bottom that slopes slightly downward toward the inlet end to make sludge removal easier. Figure 5-2 shows uncovered rectangular sedimentation basins at a large treatment plant.

Conventional Center-Feed Basins

Basins can also be constructed as either round or square, so that the water flows radially from the center to the outside. This type of flow is called radial flow. It is important that these basins also be designed to keep the velocity and flow distribution as uniform as possible. Their bottoms are generally conical and slope downward toward the center of the basin to facilitate sludge removal.

Peripheral-Feed Basins. Peripheral-feed basins are designed to feed incoming water from around the outer edge and collect it at the center. This type of basin also has radial flow. The design is otherwise similar to that of a center-feed basin.

Spiral-Flow Basins. Spiral-flow basins have one or more points around the outer edge where water is admitted at an angle. This causes the flow to circle around the basin and ultimately leave the basin at a center collector (launder).

FIGURE 5-2 Typical uncovered sedimentation basins

Basin Zones

All basin types have four zones, each with its own function. As illustrated in Figure 5-3, these zones are as follows:

- The influent zone decreases the velocity of the incoming water and distributes the flow evenly across the basin.
- The settling zone provides the calm (quiescent) area necessary for the suspended material to settle.
- The effluent zone provides a smooth transition from the settling zone to the effluent flow area. It is important that currents or eddies that could stir up any settled solids and carry them into the effluent do not develop in this zone.
- The sludge zone receives the settled solids and keeps them separated from other particles in the settling zone.

These zones are not actually as well defined as Figure 5-3 illustrates. There is normally a varying gradation of one zone into another. The settling zone is particularly affected by the other three zones, based primarily on how the basin is designed and operated. The greater the effects on this zone, the worse the effluent quality.

Parts of a Sedimentation Basin

Equipment used in conventional settling basins varies depending on the design and manufacturer. Figures 5-4 and 5-5 show the parts of typical

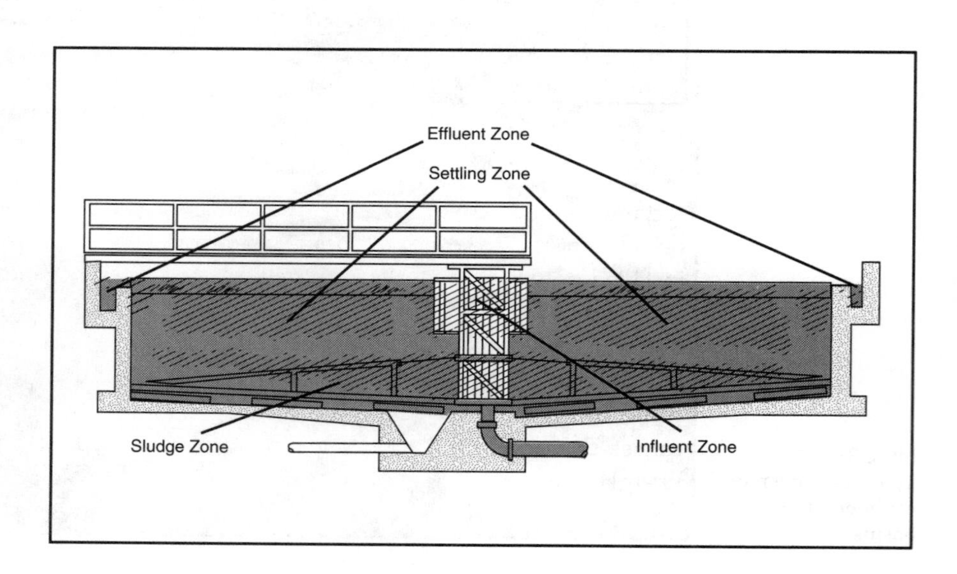

FIGURE 5-3 Zones in a sedimentation basin

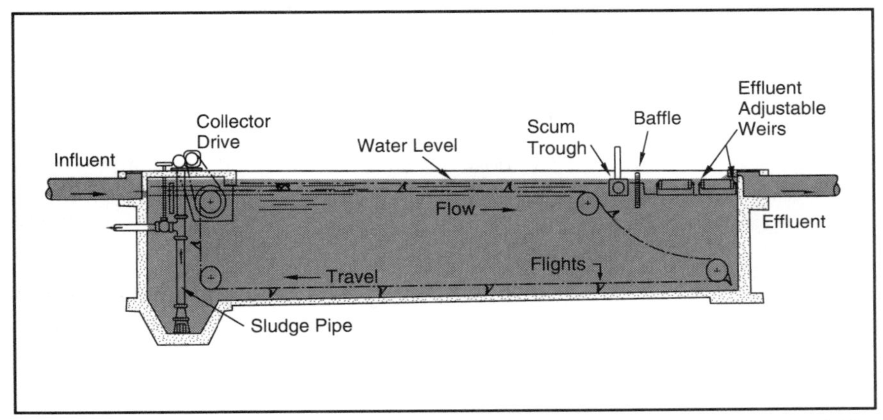

FIGURE 5-4 A typical rectangular sedimentation basin (with a continuous chain collector sludge removal system)

Courtesy of FMC Corporation, MHS Division

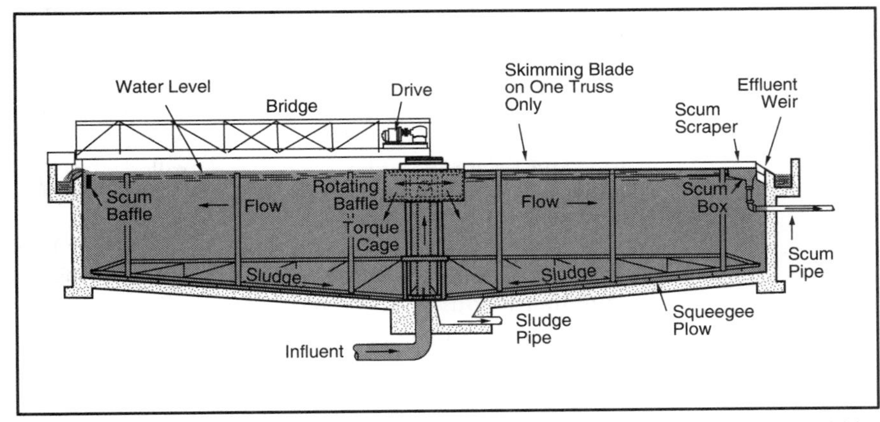

FIGURE 5-5 A typical circular sedimentation basin

Courtesy of FMC Corporation, MHS Division

rectangular and circular basins, respectively. The inlet distributes the influent evenly across (or around) the basin so the water will flow uniformly. A baffle installed downstream of the inlet reduces the velocity of the incoming water and helps produce calm, nonturbulent flow conditions for the settling zone. The water flows underneath the baffle and into the main part of the basin.

The effluent launder (also called the effluent trough) collects the settled water as it leaves the basin and channels it to the effluent pipeline, which carries the water to the next treatment process. Launders can be made of fiberglass or steel, or they may be cast concrete as a part of the tank.

The launder is equipped with an effluent (overflow) weir, which is a steel, plastic, or fiberglass plate designed to distribute the overflow evenly to all points of the launder. One of the most common types of effluent weirs is the V-notch, as illustrated in Figure 5-6. In some designs, launders can also receive the flow of water from beneath the water surface through holes in the launder wall.

Shallow-Depth Sedimentation

Shallow-depth sedimentation basins are basins designed to shorten the detention time required for sedimentation; this means smaller. Shallow basins and plate and tube settlers are two shallow-depth designs.

Shallow Basins

Basins are occasionally designed to have a fairly shallow depth in order to reduce the time necessary for the floc to settle to the bottom. Some rectangular sedimentation basins have two or three levels; the flow of water at the inlet to the basin divides into parallel flows, one over each level. These basins are designed on the principle that surface area is more important than depth. For shallow basins to work properly, it is important that coagulant doses and flash mixing be carefully controlled.

Plate and Tube Settlers

Several types of shallow-depth settling units are constructed of multiple individual modules. These modules are either plates or tubes of fiberglass, steel, or other suitable material. They are spaced only a short distance apart

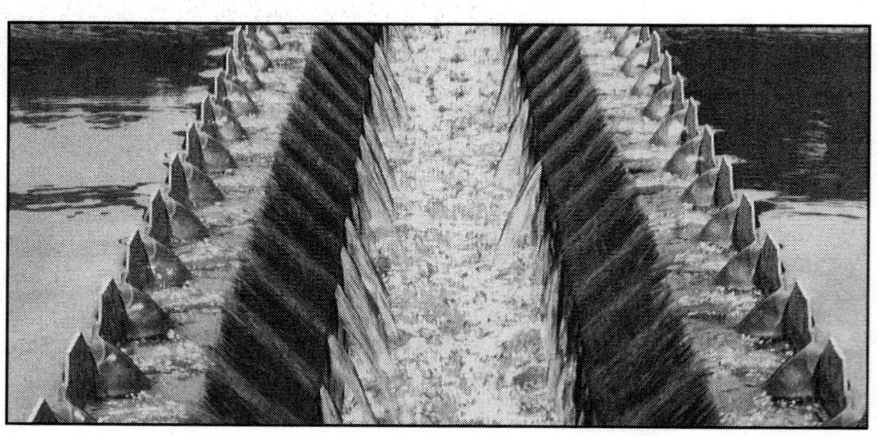

FIGURE 5-6
V-notch weir

Courtesy of Fisher Scientific

and tilted at an angle with respect to horizontal (Figures 5-7 and 5-8). If the angle is greater than 50–60°, they will be self-cleaning; in other words, the sediment will settle until it hits the plate or the tube bottom, and will then slide to the bottom of the basin. An angle as small as 7° is used when sludge is removed from the tubes or plates using periodic backflushing, possibly in conjunction with filter backwashing. A typical separation distance between the inclined surfaces of tube or plate settlers is 2 in. (50 mm); the inclined length is 3 to 6 ft (1 to 2 m).

The direction of flow through tube or plate settlers varies by the manufacturer's design. With countercurrent settling (Figure 5-9), the suspension is fed to the lower end and flows up the channels. If the angle of inclination is great enough, the solids slide down the surface, counter to the flow of the liquid. Concurrent settlers are designed so that the suspension is

FIGURE 5-7
Tube settlers

Courtesy of Wheelabrator Engineered Systems – Microfloc

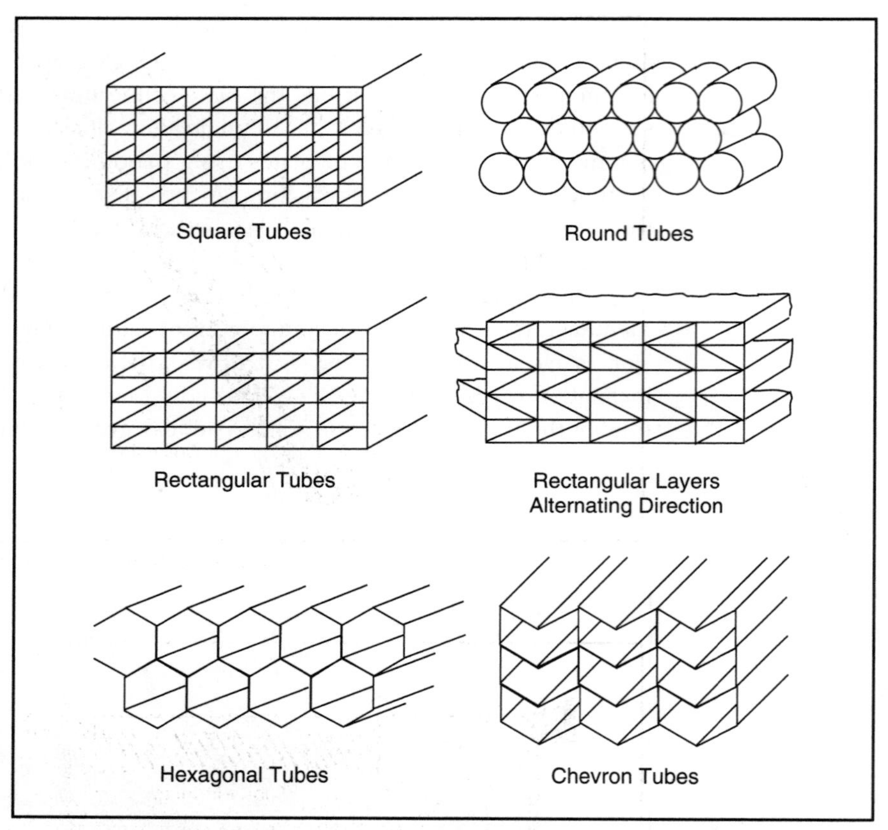

FIGURE 5-8 Various formats for tube modules

Source: Water Quality and Treatment *(1990)*

fed to the upper end and leaves at the bottom. In crossflow settlers, the flow is horizontal between the surfaces.

Tube and plate settlers are prefabricated in modules that can either be incorporated into new construction or be used to retrofit old basins so as to increase their settling efficiency. Their advantages include lightweight construction, structural rigidity, and the ability to settle a given flow rate in a much smaller basin size. As with all installations, however, the addition of tube or plate settlers to a treatment plant is not advisable unless a thorough engineering evaluation is made of the plant design to ensure they will operate properly. Figure 5-10 illustrates the installation of tube and plate settlers in rectangular and circular basins.

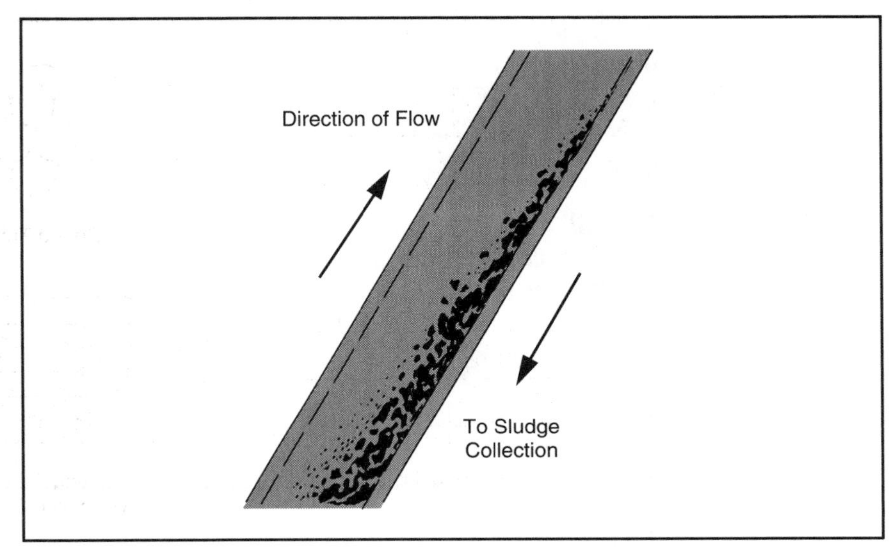

FIGURE 5-9
Countercurrent flow in tubes

Courtesy of Wheelabrator Engineered Systems – Microfloc

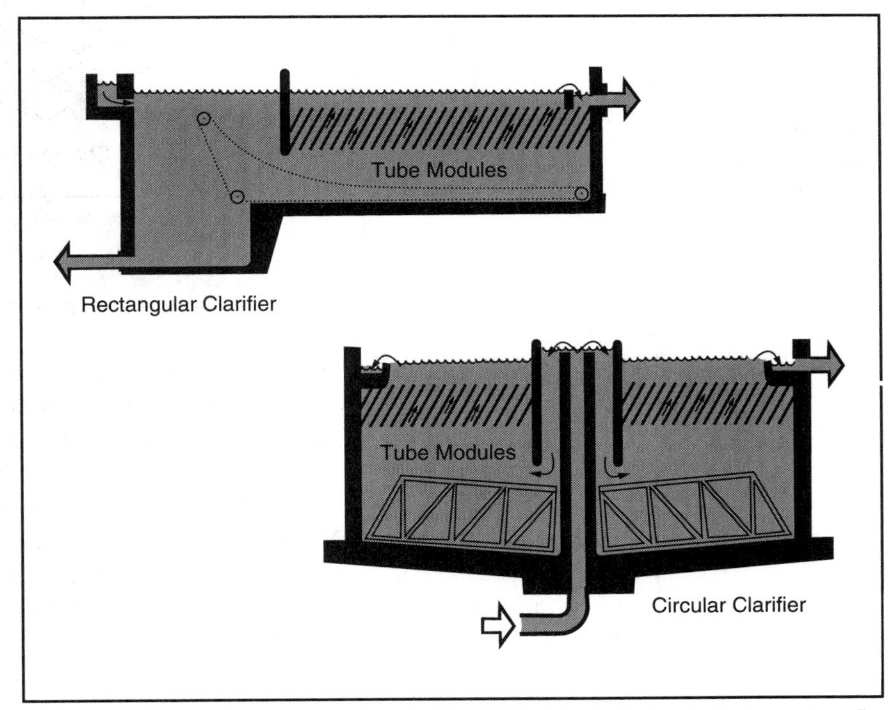

FIGURE 5-10
Tube settlers installed in sedimentation basins

Courtesy of Wheelabrator Engineered Systems – Microfloc

Another design for shallow-depth sedimentation uses inclined plates (lamellar plates), as illustrated in Figure 5-11. This design incorporates parallel plates installed at a 45° angle. In this case, the water and sludge both flow downward. The clarified water is then returned to the top of the unit by small tubes.

Sludge Removal

As solids settle to the bottom of a sedimentation basin, a sludge layer develops. If this sludge is not removed before the layer gets too thick, the solids can become resuspended or tastes and odors can develop as a result of decomposing organic matter.

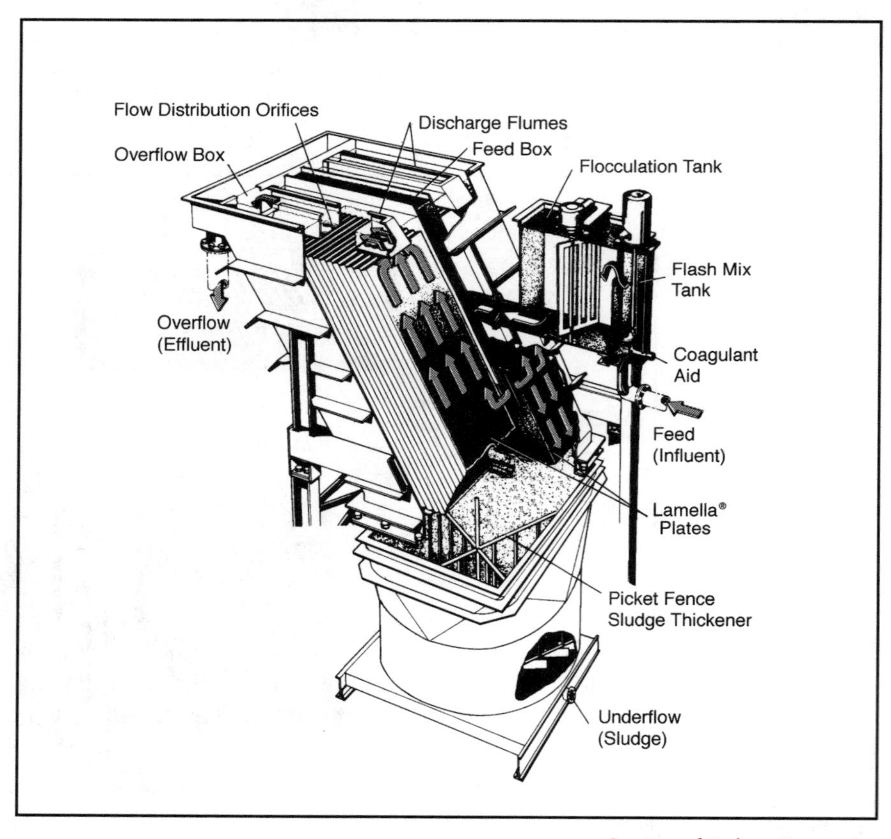

FIGURE 5-11 Lamella® plates

Courtesy of Parkson Corporation

Manual Removal

Most older sedimentation basins and newer installations where sludge accumulation is not expected to be excessive are designed to have the sludge removed manually. Many plants remove the sludge twice a year — once in spring just before heavy summer water use is expected and again in the fall after high use has ended. In most cases, fire hoses are used to wash the sludge to the drain. Basin floors are usually sloped toward the inlet end because most of the solids generally settle closer to the inlet. If sludge is not removed regularly from horizontal-flow tanks, the tank must be designed with enough depth to allow for sludge accumulation so that the sedimentation efficiency remains unaffected.

Mechanical Removal

Sludge can be removed frequently by mechanical sludge scrapers that sweep the sludge to a hopper at the end of the basin. The hopper is then periodically emptied hydraulically. For rectangular basins, the type of removal equipment used is typically one of the following:

- A chain-and-flight collector (as shown on the rectangular basin in Figure 5-4), which consists of a steel or plastic chain and redwood- or fiberglass-reinforced plastic flights (scrapers). A motor drives the chain, which pulls the flights along the basin bottom.
- A traveling-bridge collector (Figure 5-12A), which consists of a moving bridge that spans one or more basins. The mechanism has wheels that travel along rails mounted on the basin's edge. In one direction, the scraper blade moves the sludge to the hopper. In the other direction, the scraper retracts, and the mechanism skims any scum from the water's surface.
- A floating-bridge siphon collector (Figure 5-12B), which uses suction pipes or a submersible pump to withdraw the sludge from the basin. The pipes are supported by foam plastic floats, and the entire unit is drawn along the length of the basin by a motor-driven cable system.

Circular and square basins usually are equipped with scrapers or plows (labeled "squeegee plow" in Figure 5-5) that slant downward toward the center of the basin and sweep sludge to the sludge hopper or pipe.

If sludge scrapers are installed, it is not necessary to interrupt plant operations to take basins out of service for manual cleaning. Sludge scrapers also reduce personnel requirements and reduce the problems of decomposing sludge that occur under some conditions. If the sedimentation sludge is

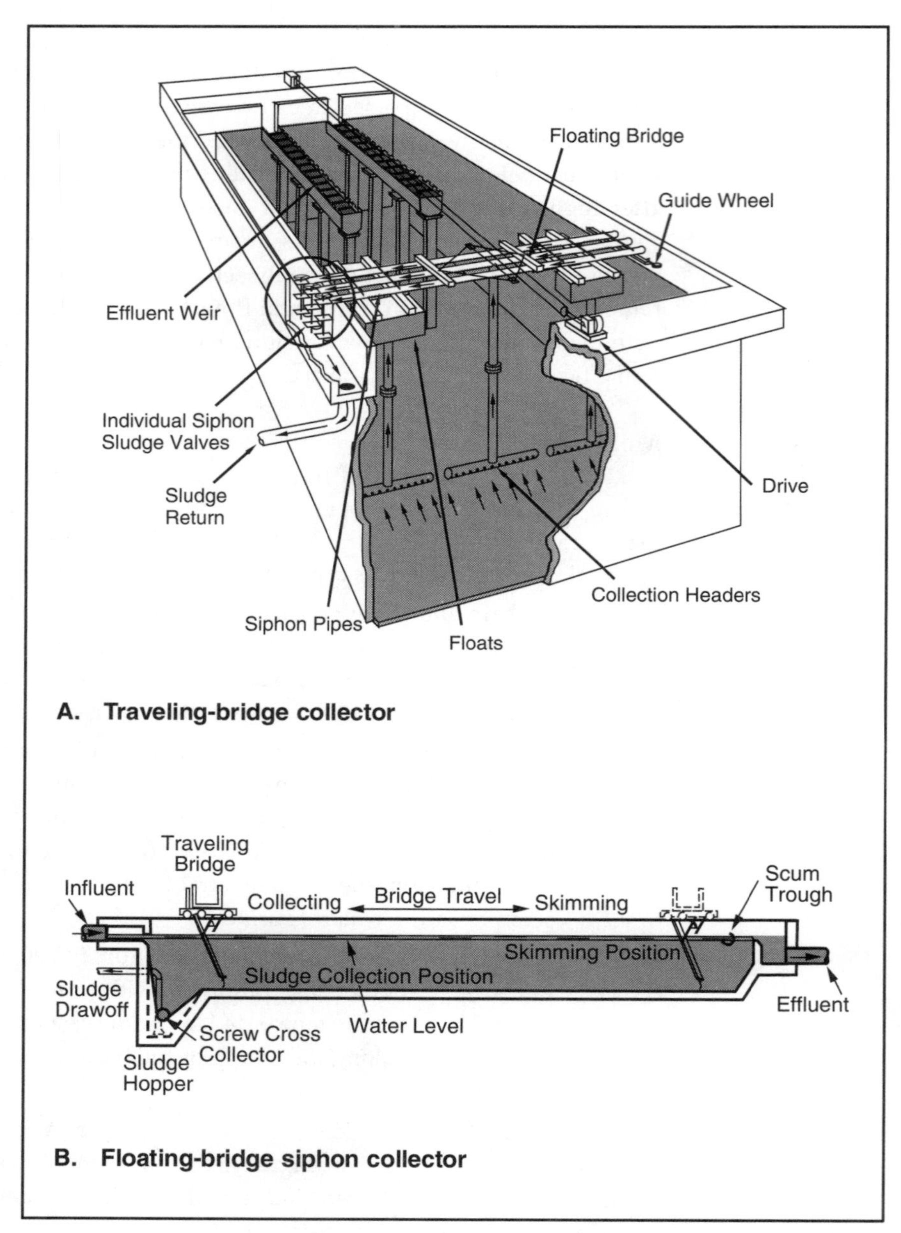

FIGURE 5-12 Two mechanical sludge removal systems used in rectangular basins

Courtesy of F.B. Leopold Company, Inc.

sent to the local wastewater treatment plant, it is best to deliver a small quantity of sludge every few days rather than large quantities less frequently.

Continuous mechanical sludge collection is also desirable when tube or plate settlers are used because more sludge is created in a smaller basin. If manual cleaning is used with tube or plate settlers, the frequency of cleaning must be increased.

Other Clarification Processes

Other clarification processes, including the use of solids-contact basins, dissolved-air flotation, and contact clarifiers, are discussed in the following section.

Solids-Contact Basins

Solids-contact basins (also called upflow clarifiers, solids-contact clarifiers, or sludge-blanket clarifiers) are generally circular and contain equipment for mixing, flow circulation, and sludge scraping. A wide variety of designs are available from different manufacturers. All are divided by baffles into two distinct zones: mixing and settling. Coagulation and flocculation take place in the mixing zone, where the raw water and coagulant chemicals are combined and slowly agitated. A typical unit is illustrated in Figure 5-13.

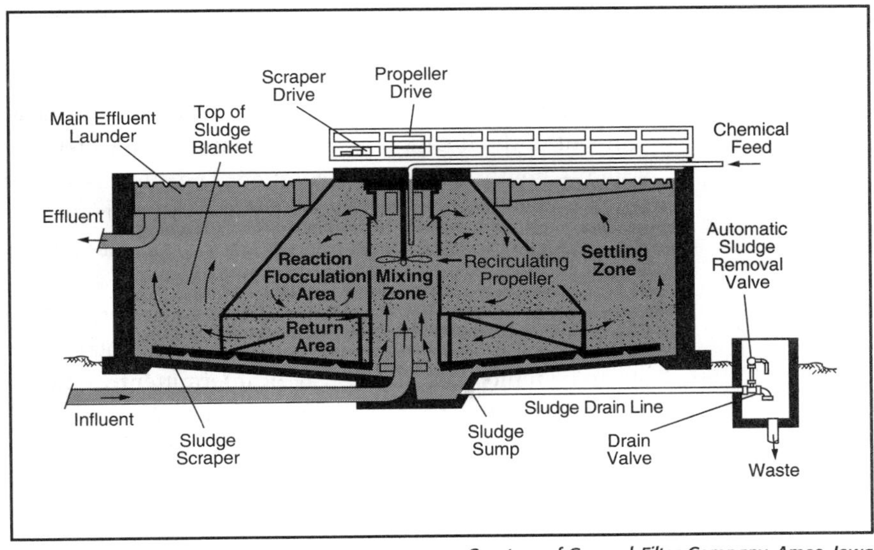

FIGURE 5-13 Sludge-blanket clarifier

Courtesy of General Filter Company, Ames, Iowa

Near the bottom of the basin, the flow is directed upward into the settling zone, which is separated from the reaction flocculation area by baffles. A point exists at which the upflow velocity can no longer support the floc particles. This point defines the top of the sludge blanket. The water has essentially been "filtered" through this sludge blanket and the clarified water flows upward into the effluent troughs. The floc particles in the sludge blanket contact other particles and grow larger until they settle to the bottom. A portion of the sludge is recycled to the mixing zone, and the remainder settles in a concentration area for periodic disposal through the blowoff system. This must be done to maintain an almost constant level of solids in the unit.

The advantage of this type of unit is that the chemical reactions in the mixing area occur more quickly and completely because of the recycled materials from the sludge blanket. This allows much shorter detention times than with conventional basins. One problem with these clarifiers is that a quick change in water temperature or a quick increase in flow can cause currents that upset the sludge blanket.

Dissolved-Air Flotation

Dissolved-air flotation (DAF) can be used in some installations in place of sedimentation. It is a process in which gas bubbles are generated so that they will attach to solid particles, causing them to rise to the surface rather than settle to the bottom.

DAF units consist of a flotation basin, saturation unit, air compressor, and recycle pumping. In operation, about 5 to 10 percent of the clarified water is recycled by the pump to the saturator, where high-pressure air is brought into contact with the water. When this water is reintroduced to the influent of the flotation basin, the gases in the water — which are supersaturated — gently come out of solution. The effect is similar to the bubbles in a carbonated beverage that rise to the top when the bottle is opened. There are several variations to this process; some are more appropriate than others for potable water treatment.

Flotation can be used in either a rectangular or a circular tank. The sludge that accumulates on the surface of the flotation tank is called *float*, and it can be removed continuously or intermittently either by flooding the basin to overflow the float or by mechanical scraping. The most widely used method is to have rubber scrapers that travel over the tank surface and push the float over a ramp, called a beach, to a trough for disposal.

Flotation is widely used for industrial water clarification and wastewater clarification. For potable water treatment, the process is particularly good for algae removal because algae naturally tend to float.

Each of the different DAF systems is named for the method of producing the gas bubbles.

Electrolytic Flotation

Electrolytic flotation uses very small bubbles of hydrogen and oxygen generated by passing a direct current between two electrodes.

Dispersed-Air Flotation

There are two different types of dispersed-air processes: foam flotation and froth flotation. These processes are not suitable for potable water treatment because high turbulence and undesirable chemicals are required to produce the froth.

Contact Clarifiers

A number of unique proprietary clarification processes have been developed in recent years and are gradually gaining acceptance. A few of them are mentioned here as examples. Additional information can be obtained from the various companies.

Pulsator Clarifier

The Pulsator, developed and marketed by Infilco Degremont, Richmond, Va., uses a unique pulsating hydraulic system to maintain a homogenous sludge layer. When flow is not introduced into the clarifier, the sludge blanket settles. Pulsation of the sludge blanket about every 40 to 50 seconds acts to maintain a uniform layer and to reduce the potential for short-circuiting of the flow through the sludge blanket. The Pulsator is often used to treat highly colored, low-turbidity water, in which the formation of settleable floc is usually difficult.

Superpulsator Clarifier

The Superpulsator (Figure 5-14), also developed and marketed by Infilco Degremont, combines the hydraulic pulsation system of the Pulsator with a series of inclined plates to help maintain high solids concentrations at increased upflow rates. The units can therefore be operated at rates two to three times greater than those of the Pulsator. The plates are inclined at an angle of 60° from the horizontal and spaced from 12 to 20 in. (300 to 510 mm) apart.

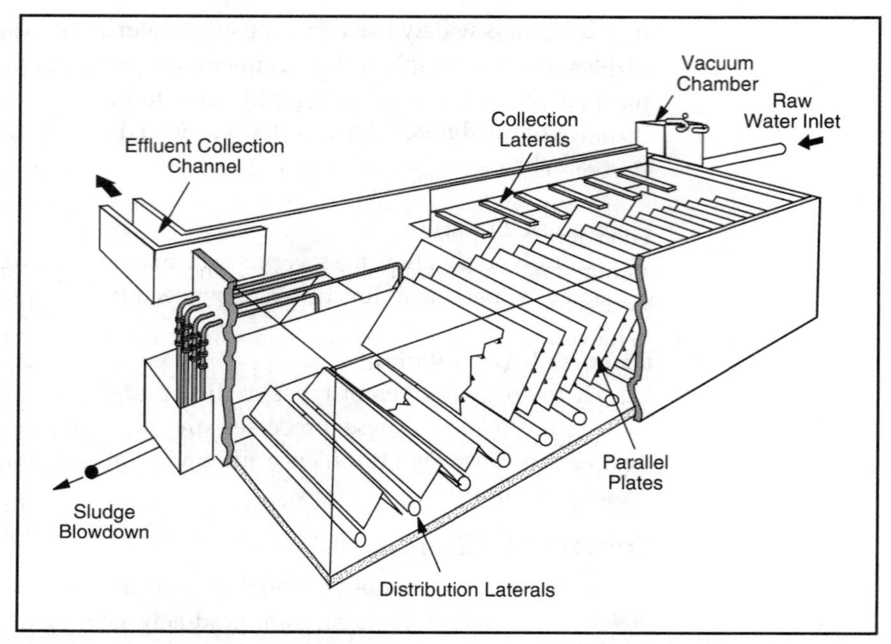

FIGURE 5-14 The Superpulsator flat-bottom clarifier with lateral flow distribution

Courtesy of Infilco Degremont Inc., Richmond, Va.

Trident Contact Adsorption Clarifier

The Trident treatment system, marketed by CPC Engineering Corporation, combines coagulation, flocculation, and clarification processes in a single upflow clarifier. The raw water, with coagulant added, enters at the bottom of the unit and passes through a bed, consisting of a plastic medium, that floats on top of the water. The solids adhere to the medium, with relatively high removal rates.

When the solids accumulation results in excessive head loss, or when effluent quality becomes unacceptable, the medium is cleaned by upflow hydraulic flushing with raw water and air. Trident operating times typically average 4 to 8 hours between backwashes.

Regulations

There are no federal regulations that specifically apply to the design of sedimentation systems. However, the requirements of the Surface Water Treatment Rule (SWTR) discussed in chapter 4 can have a significant impact on the design or operation of these facilities. The SWTR basically requires that surface water systems maintain specified $C \times T$ values (the product of disinfectant concentration and contact time) before the treated water reaches the first customer. The contact time of raw water with chlorine (or other disinfectant) while the water is in the sedimentation basin provides a major portion of the contact time required for most treatment plants.

Many older sedimentation basins were not designed with serious consideration given to avoiding dead water spaces in corners or to minimizing short-circuiting of flow. The SWTR requirements may make it necessary for some water systems to alter their sedimentation basins to achieve maximum detention. This can be done through the installation of additional baffles. An improvement in detention time can best be obtained by improving inlet and outlet baffles and improving flow patterns.

Operators who are considering adding baffles to sedimentation basins should obtain professional design assistance to ensure that the improvement will maximize detention time without causing operating problems. More complete details of federal regulations are included in *Water Quality*, also part of this series.

Operation of the Process

This section discusses certain key features associated with the operation of the sedimentation process.

Sedimentation Facility Operation

The primary function of the sedimentation process is to prepare water for effective and efficient filtration. An effective sedimentation process also removes substantial amounts of organic compounds that cause color and can be precursors for the formation of disinfection by-products. Because effective sedimentation is closely linked with proper coagulation and flocculation, the operator must ensure that the best possible floc is being formed.

Conventional Basins

The minimum detention time (that is, the theoretical time it takes a particle of water to pass through the basin) is usually designed to be 4 hours for surface water systems. The actual detention time for a basin is often shorter than the calculated period because currents and improper design features may cause the water to short-circuit through the basin. For this reason, weir overflow rates are often better indicators of the operation of sedimentation basins.

The flow rate over outlet weirs is commonly designed not to exceed 20,000 gpd per foot (250,000 L/d per meter) of weir. For light alum floc, it may have to be reduced to 14,400 gpd per foot (180,000 L/d per meter). Raw-water sources and treatment facilities are not all the same, so an operator must experiment to determine the most effective overflow rates. The most effective rate may vary at different times of the year because of changes in water quality. If water demand makes it necessary to exceed the optimal overflow rate, adjustments will have to be made elsewhere in the plant operation to compensate for the poorer-quality settled water.

A minimum of two sedimentation units should be provided, so that one can be taken out of service for cleaning or repair without disrupting plant operation. The operator must ensure that the total flow is divided evenly among the basins and that any changes in flow rate are made smoothly. This will prevent the basins from receiving surges that cause eddy currents or break up floc. The flow rate over the effluent weirs should be uniform and evenly distributed along the weir length. If it is not, the weirs may have to be adjusted.

Tube and Plate Settlers

Because the overflow rates used for tube and plate settlers are two to three times those for conventional basins, the floc must have good settling characteristics. This often requires special attention to the flocculation process, including the addition of a coagulant aid.

At times, the floc may bridge across the upper edge of the tube opening, resulting in a buildup of solids several inches (centimeters) thick. To dislodge the accumulation, the water level of the basin is usually lowered. If this is not possible, a gentle current of water must be directed across the top of the tubes with a hose or a permanent perforated header pipe mounted along the length of the settler unit. Sludge withdrawal is likely to be required more frequently for a system with tube or plate settlers than for one with conventional basins.

Solids-Contact Basins

The optimal detention time for solids-contact basins varies with equipment design and raw-water quality. The upflow rates used for turbidity removal are typically about 1 gpm/ft^2 (0.7 mm/s), which results in detention times of only 1 or 2 hours. Proper coagulation–flocculation and control of the solids concentration in the slurry or sludge blanket are essential for good results. Weir loadings normally should not exceed 10 gpm per foot (124 L/min per meter) of weir length.

To maintain a good sludge blanket, coagulant aids or weighting agents may be necessary. The solids concentration should be determined at least twice a day — more frequently if the water quality is always changing. This is done by taking samples from the taps provided on the basin and conducting settling tests as prescribed by the manufacturer. Because solids-contact basins are often used in water softening, their operation is further discussed in chapter 11.

Monitoring the Process

The primary test used to indicate proper sedimentation is the test for turbidity. The turbidity of samples taken from the raw water and the outlet of each basin should be tested at least three times a day. More frequent testing may be necessary if water quality is changing rapidly. By comparing these turbidities, an indication of the efficiency of the sedimentation process can be obtained. For example, if the raw-water turbidity is 50 ntu and effluent turbidity is 40 ntu, very little turbidity is being removed. This indicates that better coagulation–flocculation is needed or that short-circuiting is occurring. Visual examination of water samples from the effluent can also indicate if floc is being carried over onto the filters.

The turbidity of the settled water should be kept below 10 ntu. If it is above 10 ntu, the process should be checked and the operation improved. Turbidity test procedures are discussed in *Water Quality*, also part of this series.

As discussed in chapter 4, the raw-water temperature should be measured and recorded at least daily. As water becomes colder, it is more viscous, thus presenting more resistance to the settling particles. To compensate for this, the surface overflow rates may have to be reduced.

Operating Problems

The most common operating problems in sedimentation facilities are poorly formed floc, short-circuiting, density currents, wind effects, and algae growth.

Poorly Formed Floc

Poorly formed floc is characterized by small or loosely held particles that do not settle properly and are carried out of the settling basin. This is the result of inadequate rapid mixing, improper coagulants or dosages, or improper flocculation. Jar tests can generally provide the information necessary to find the specific problem. The solution may be to increase the mixing energy, use a coagulant aid, or install additional baffles in the flocculation basin.

Short-Circuiting

If a basin is not properly designed, water bypasses the normal flow path through the basin and reaches the outlet in less than the normal detention time. This occurs to some extent in every basin. It can be a serious problem in some installations, causing floc to be carried out of the basin as a result of shortened sedimentation time.

The major cause of short-circuiting is poor inlet baffling. If the influent enters the basin and hits a solid baffle, strong currents will result. A perforated baffle can successfully distribute inlet water without causing strong currents (Figure 5-15). If short-circuiting is suspected, tracer studies are the best method of determining the extent of the problem. (Publications detailing the requirements of the Safe Drinking Water Act Surface Water Treatment Rule can provide more information on tracer studies.)

Density Currents

Currents that disrupt the sedimentation process in a basin occur when the influent contains more suspended solids, and thus has a greater density, than the water in the basin. They can also occur when the influent is colder than the water in the basin. In either case, the influent sinks to the bottom of

FIGURE 5-15
Perforated baffles

the basin, where it can create upswells of sludge and short-circuits. If this is a problem, an engineering study should be conducted to determine the best solution. The problem can often be lessened by modifying the effluent weirs.

Wind Effects

Wind can create currents in open basins, thus causing short-circuiting. If wind is a problem, a barrier should be constructed around uncovered basins to lessen the wind's effect and keep debris out of the water (Figure 5-16).

Algae and Slime Growth

A problem that often occurs in open, outdoor basins is the growth of algae and slime on the basin walls, which can cause tastes and odors. If the algae and slime detach from the walls, they can clog weirs or filters.

The growths can be controlled by coating the walls with a mixture of 10 g of copper sulfate ($CuSO_4$) and 10 g of lime ($CaOH_2$) per liter of water. The basin should be drained and the mixture applied to the problem areas with a brush.

FIGURE 5-16 Barrier around an uncovered basin

Waste Disposal

Regardless of the type of basin used, sludge is probably the most troublesome operating problem. The collection and disposal methods depend in part on the nature and volume of the sludge formed. This is primarily governed by the raw-water quality and the types of coagulants used. For example, different methods of handling and disposal are typically used for alum sludge than for lime sludge. Methods of handling lime sludge are discussed in chapter 11.

In the past, sedimentation basin sludge was generally discharged into lakes and streams without treatment. This is no longer allowed under current environmental laws because the sludge can form deposits in the receiving water that are harmful to aquatic life and can produce objectionable tastes and odors.

Alum Sludge

Alum sludge is the most common form of sludge resulting from the sedimentation process, because alum is the coagulant most often used to remove turbidity. Alum sludge is a gelatinous, viscous material, typically containing only 0.1 to 2 percent solid material by weight. However, it is extremely hard to handle and dewater because much of the water is chemically bound to the aluminum hydroxide floc.

Alum sludge is commonly pumped into specially designed lagoons. When a lagoon is full, the sludge is diverted to another one, and the filled lagoon is allowed to dry until the sludge can be removed for final disposal. Water decanted from the top can be returned to the treatment plant. The process can take a year or longer, and the sludge will still contain over 90 percent of the original water, making it difficult to handle and unfit for disposal in a sanitary landfill. Therefore, it usually must be placed on land owned by the water utility. Because of the large land requirements, lagooning may not be possible for large treatment plants or where land is not available. In areas where freezing temperatures are common, the freezing and thawing of sludge can hasten the dewatering process.

Sand drying beds can accomplish more efficient dewatering of sludge than lagoons. The sludge is spread in layers over the sand, which overlies gravel and drain tiles. However, land requirements, difficulties with sludge removal, and poor performance during rainy periods are disadvantages of this method that must be considered.

If there is not much land available, dewatering can be improved using mechanical equipment. Vacuum filters, centrifuges, and filter presses can successfully dewater alum sludge to at least 20 percent solids by weight. It can then be placed in most landfills or used as a soil conditioner. Mechanical dewatering usually has a higher operation and maintenance cost than lagoons and sand drying beds.

Regardless of the mechanical equipment used, the sludge must be pretreated to make the subsequent dewatering more effective. This usually requires a sludge thickener, which is a circular tank much like a clarifier, equipped with a stirring mechanism (Figure 5-17). This mechanism breaks apart the floc particles in the sludge, allowing the water to escape and the solids to settle. A polymer is usually added to the sludge as it enters the thickener to enhance settling. A thickener can concentrate sludge up to about 5 percent solids by weight.

Because of the difficulties with dewatering and disposing of the alum sludge, many operators reduce alum dosages by effectively using coagulant aids such as polymers. This greatly reduces the volume of alum sludge that must be handled.

Other Sludges

Ferric salt coagulants also produce a sludge that is difficult to dewater. The sludge is also difficult to thicken because it has a low density, which causes it to settle slowly.

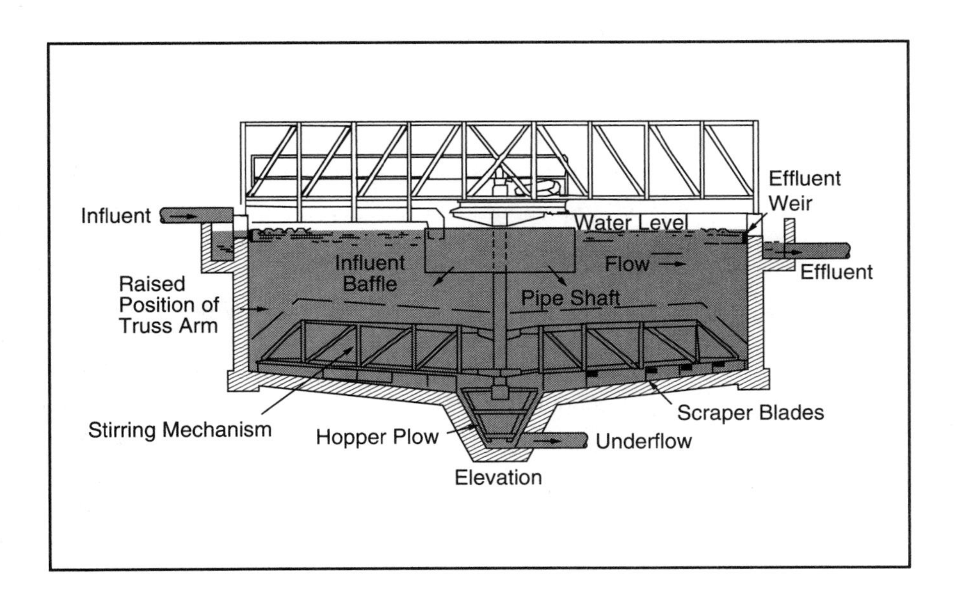

FIGURE 5-17
Sludge thickener

Disposal to a Waste Treatment Plant

A cost-effective method for disposing of treatment waste is to pump it to a sanitary sewer, for ultimate disposal by the local wastewater treatment plant. The amount of alum or ferric chloride sludge in proportion to the sanitary waste sludge is usually small enough that it does not disrupt the normal processing of sludge by the wastewater treatment plant. Some water utilities now operate their own wastewater plant for treating and disposing of water plant wastes.

Wastewater treatment authorities that will accept water treatment plant waste usually charge a fee to cover their increased operating costs. In addition, they usually have a number of restrictions, including the quantity that can be disposed of each day and a ban on any disposals during rainstorms or other times when the waste treatment facility may be operating near capacity. A study of the sewer system must be made to ensure that it can adequately handle this additional flow along with its normal domestic flow. Lime sludge cannot be discharged to a sanitary sewer system because lime deposits might build up and block the sewers.

Filter Backwash Water

The handling and disposal of filter backwash water are discussed in greater detail in chapter 6. In general, the backwash water can be combined with the sedimentation basin sludge, it can be treated separately, or it can be recycled by being added to the raw water entering the plant.

Equipment Maintenance

Manually cleaned conventional basins should be inspected at the completion of each sludge removal. Basins with mechanical sludge removal should be drained at least once a year for a general inspection.

In addition to inspection of the operation of mechanical equipment, the inlet system should be examined closely to ensure that all openings are clear of obstructions that could cause unequal flow distribution. The weirs should also be inspected to ensure they are level and not blocked by debris or chemical deposits. Uneven flow over the weirs could cause uneven flow in the basin and eddies that would disturb the settling zone.

The baffles should also be checked for deterioration that will cause short-circuiting, and any algae or slime accumulation on basin walls, weirs, and baffles should be removed.

Solids-contact basins need to be drained at least annually so that the condition of the baffles and mixing equipment can be checked. Sludge-pumping lines and equipment should be inspected routinely to ensure that they are not becoming plugged. In addition, the lines should be flushed occasionally to prevent a buildup of solids.

Safety Precautions

Open sedimentation basins should be equipped with guardrails that will prevent falls into the basins. Walkways and bridges connected to basins should also have guardrails. Life rings or poles should be kept near the basins for rescue purposes.

If basins are not covered, care must be exercised during periods of rain, ice, or snow because the walkways become very slippery. Caution is also needed when drained basins are being cleaned or inspected because growths of aquatic organisms and sludge deposits can make the surfaces slippery.

The moving parts of all machinery should be equipped with guards to prevent the machinery from catching legs, fingers, or clothing. Guards should never be removed while the machinery is operating.

Record Keeping

Good records are invaluable in helping the operator to solve problems and produce high-quality water. As raw-water quality or other conditions change, the operator can review past records to help determine what adjustments are needed. Because sedimentation is closely linked with coagulation–flocculation, both records should be maintained together. Sedimentation records should include

- surface and weir overflow rates, calculated by using the flow rate through each basin
- turbidity results for raw water and effluent from each basin
- quantity of sludge pumped or cleaned from each basin
- types of operating problems and corrective actions taken

Selected Supplementary Readings

Cornwell, D.A., and H.M.M. Koppers, eds. 1990. ***Slib, Schlamm, Sludge.*** Denver, Colo.: American Water Works Association Research Foundation and American Water Works Association.

Cornwell, D.A., M.M. Bishop, R.G. Gould, and C. Vendermeyden. 1987. ***Handbook of Practices: Water Treatment Plant Waste Management.*** Denver, Colo.: American Water Works Association Research Foundation and American Water Works Association.

Cornwell, D.A., Carel Vandermeyden, Gail Dillow, and Mark Wang. 1992. ***Landfilling of Water Treatment Plant Coagulant Sludges.*** Denver, Colo.: American Water Works Association Research Foundation and American Water Works Association.

Dulin, B.E., and W.R. Knocke. 1989. The Impact of Incorporated Organic Matter on the Dewatering Characteristics of Aluminum Hydroxide Sludges. ***Jour. AWWA,*** 81(5):74.

Edzwald, J.K., and J.P. Walsh. 1992. ***Dissolved Air Flotation: Laboratory and Pilot Plant Investigations.*** Denver, Colo.: American Water Works Association Research Foundation and American Water Works Association.

Edzwald, J.K., et al. 1992. Flocculation and Air Requirements for Dissolved Air Flotation. ***Jour. AWWA,*** 84(3):92.

Edzwald, J.K., S.C. Olson, and C.W. Tamulonis. 1994. ***Dissolved Air Flotation: Field Investigations.*** Denver, Colo.: American Water Works Association Research Foundation and American Water Works Association.

Elliott, H.A., B.A. Dempsey, D.W. Hamilton, and J.R. DeWolfe. 1990. ***Land Application of Water Treatment Sludges: Impact and Management.*** Denver, Colo.: American Water Works Association Research Foundation and American Water Works Association.

Johnson, G., et al. 1992. Optimizing Belt Filter Press Dewatering at the Skinner Filtration Plant. ***Jour. AWWA,*** 84(11):87.

Manual M3, Safety Practices for Water Utilities. 1990. Denver, Colo.: American Water Works Association.

Manual of Instruction for Water Treatment Plant Operators. 1975. Albany, N.Y.: New York State Department of Health.

Manual of Water Utility Operations. 8th ed. 1988. Austin, Texas: Texas Water Utilities Association.

Recommended Standards for Water Works. Albany, N.Y.: Health Education Services.

Riehl, M.L. ***Water Supply and Treatment.*** Arlington, Va.: National Lime Association.

Robey, Rod. 1992. Establishing a Confined-Space Entry Program. ***Opflow,*** 18(7):1.

Sludge: Handling and Disposal. 1989. Denver, Colo.: American Water Works Association.

Von Huben, H. 1991. ***Surface Water Treatment: The New Rules.*** Denver, Colo.: American Water Works Association.

Water Quality and Treatment. 4th ed. 1990. New York: McGraw-Hill and American Water Works Association (available from AWWA).

CHAPTER 6

Filtration

The removal of suspended solids by filtration plays an important role in both the natural purification of groundwater and the artificial purification of water in treatment plants.

In the natural filtration process, most suspended material is removed from groundwater as the water percolates through the soil. It is therefore not usually necessary to filter groundwater. One exception is when the water must be treated for removal of hardness, iron, or manganese. In these instances, the water usually must be filtered to remove the chemical precipitates. Another exception relates to systems designated by the state as being "under the direct influence of surface water." These systems must provide filtration treatment.

Because surface water is subject to runoff and other sources of contamination, filtration is usually necessary to remove the suspended material for both aesthetic and public health reasons.

Process Description

The suspended material that must be removed from surface water can include floc from the coagulation, flocculation, and sedimentation processes; microorganisms; and precipitates. The material is removed when water passes through a bed of granular material called filter media. The common media are sand, anthracite coal, granular activated carbon, garnet sand, or some combination of these materials.

Turbidity Removal

Visible turbidity (the presence of suspended matter) in drinking water is generally objectionable to consumers, so it should be removed for aesthetic reasons. In any event, maintaining the lowest possible turbidity in finished water is particularly important in protecting public health. This requires achieving turbidity levels that are far lower than can be seen by customers and that can be detected only with monitoring equipment.

Turbidity interferes with the disinfection process because the suspended particles shield microorganisms from the disinfectants. In addition, if not removed, the particles combine chemically with the disinfectant and leave less disinfectant to combat the microorganisms. If a sizable quantity of turbidity is allowed to enter the distribution system over a period of years, it can create tastes and odors and support bacterial growths.

The Filtration Process

The media through which water is passed in the filtration process are commonly thought of as a sieve or a microstrainer that traps suspended material between the grains of filter media. However, straining is only a minor part of the action that takes place because most suspended particles can easily pass through the spaces between the grains of the media.

As illustrated in Figure 6-1, filtration depends primarily on a combination of complex physical and chemical mechanisms, the most important being *adsorption*. As water passes through the filter bed, the suspended particles contact and adsorb (stick) onto the surface of the individual media grains or onto previously deposited material. The forces that attract and hold the particles to the grains are the same as those at work in coagulation and flocculation. In fact, some flocculation and sedimentation occur in the filter

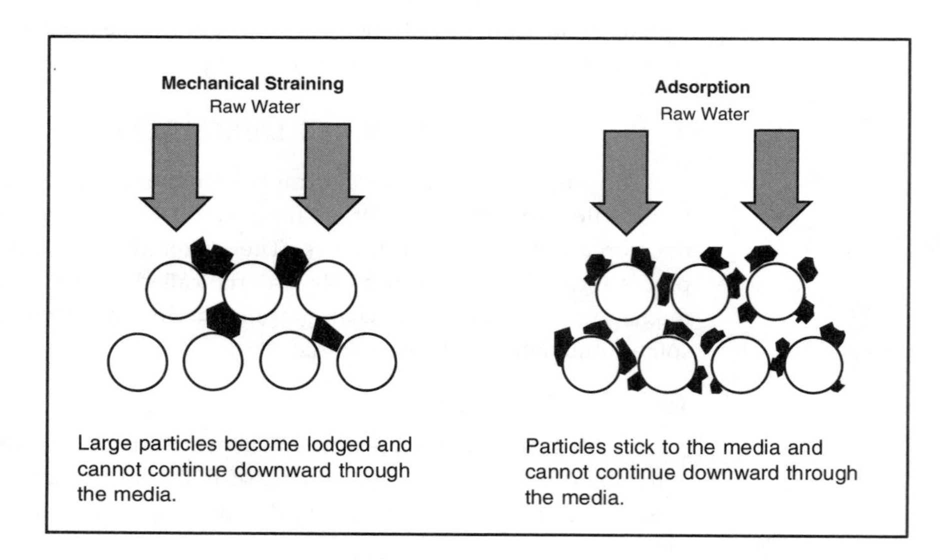

FIGURE 6-1 Filter media removal mechanisms

bed. This illustrates the importance of good chemical coagulation before filtration. Poor coagulation can cause operating problems for filters, as discussed later in this chapter.

Approaches to Filtration

Filters can be categorized as either gravity filters or pressure filters. For gravity filters, the force of gravity moves the water through the filter. For pressure filters, applied water pressure forces the water through the filter. Gravity filters are much more common, so they are discussed in more detail in this chapter.

Conventional Treatment

The combined processes of mixing, flocculation, sedimentation, and filtration shown in Figure 6-2 have formed the traditional water treatment plant design for many years. This approach is commonly called conventional treatment. This treatment has been found to provide effective removal of practically any range of raw-water turbidity.

The success of this design is due primarily to the sedimentation step, which removes most of the suspended material before the water enters the filters. After sedimentation, the water passing to the filters usually has a turbidity of 10 to 15 nephelometric turbidity units (ntu). For this reason, conventional treatment can be used regardless of raw-water turbidity and color levels. Historically, single- medium filters were used in the conventional process. However, essentially all new construction and plant conversions now use dual-media or multimedia filters to increase the process efficiency.

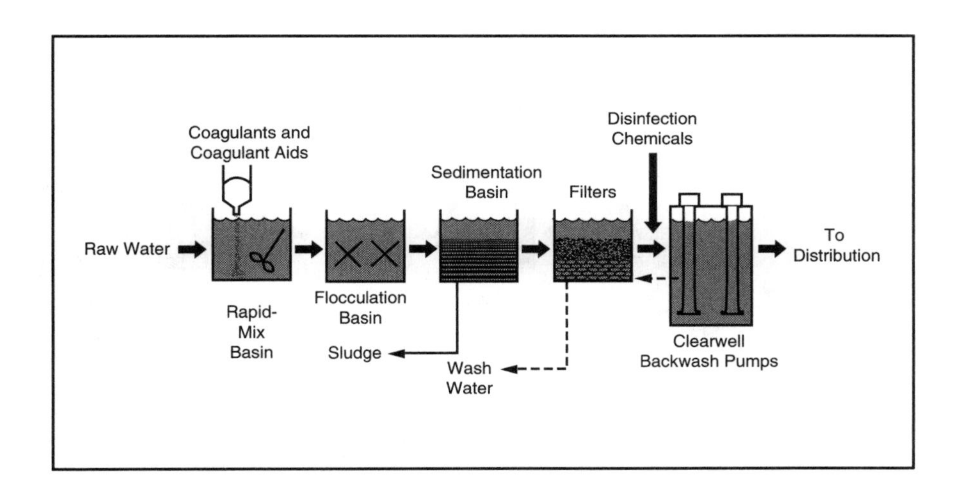

FIGURE 6-2 Conventional treatment plant

Direct Filtration

Water plants using direct filtration, as shown in Figure 6-3, do not have a sedimentation step. This type of process is generally used to treat raw water that has average turbidities below 25 ntu and color below 25 units, but it has been successfully used for water with higher turbidity and color levels. The state regulatory authority should be consulted for its requirements on the use of direct filtration. Dual-media, multimedia, or deep-bed monomedium filters should always be used for direct filtration because they can remove more suspended solids before backwashing is needed than sand-only filters.

The major advantage of direct filtration is its lower construction cost compared with that of conventional plants. However, because of the short time span between coagulant addition and filtration, and the greater load applied to the filters, this type of system must be carefully monitored to avoid turbidity breakthrough into the finished water.

Slow Sand Filtration

Slow sand filters were introduced in the United States in 1872. They were the first type of gravity filter used, but they are no longer very common. As shown in Figure 6-4, water is fed directly to the surface of slow sand filters without any chemicals being applied. The use of slow sand filters is generally limited to raw water having relatively low turbidity. The space required for slow sand filters is relatively large, so use of this process is also generally limited to small water systems in more rural areas. Slow sand filters are discussed in more detail later in this chapter.

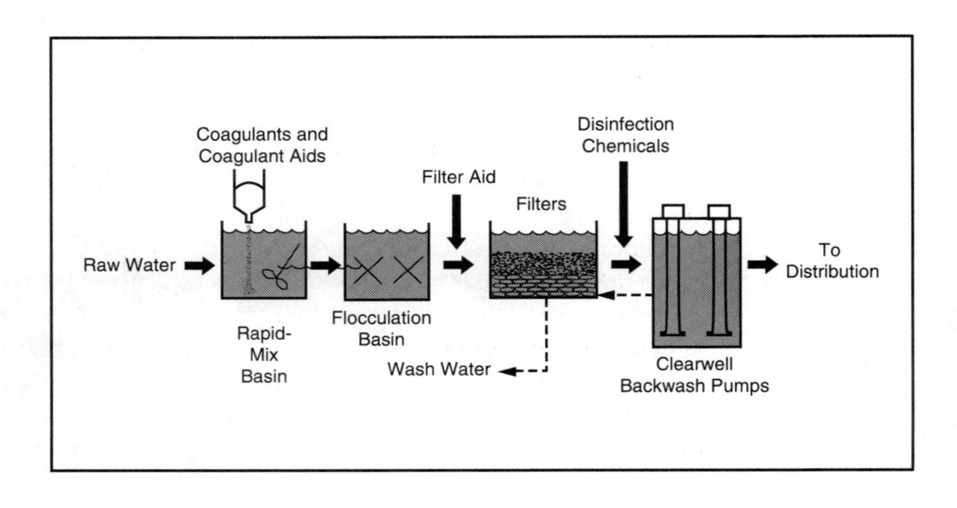

FIGURE 6-3 Direct filtration plant

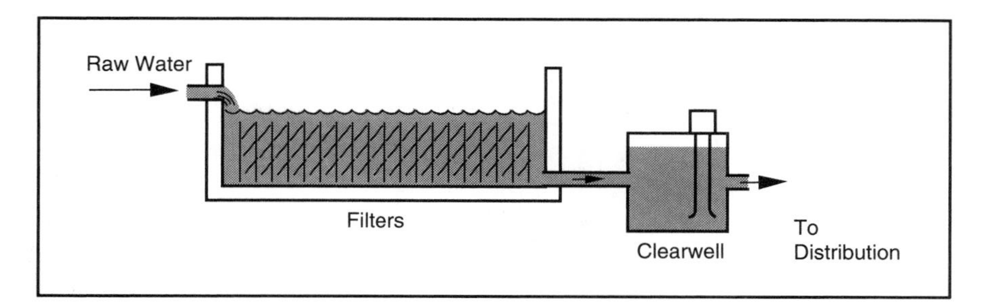

FIGURE 6-4
Slow sand filter plant

Diatomaceous Earth Filtration

In the diatomaceous earth (DE) filtration process, which utilizes pressure filtration, diatomaceous earth material must be fed to the filter unit to act as the filter medium (Figure 6-5). The process can be used only for water with low turbidity. DE filtration has a relatively low installation cost and minimal space requirements. Disadvantages include relatively high operating costs and the problem of disposing of the backwash sludge.

Package Treatment Plants

Package treatment plants are small, prefabricated units that have been designed and assembled at a factory and then shipped to the installation site. They have grown in popularity in recent years, primarily as a result of new technology that allows compact construction. Conventional treatment package systems typically include a coagulation–flocculation unit, a settling tank or a floc separation process, and a mixed-media gravity or pressure filter. A very small unit can be assembled and mounted on skids. The parts of larger installations are shipped as several units and piped together in the field (Figure 6-6). These treatment systems are often an economical solution for small utilities that must install filtration.

Potential purchasers of these systems should be aware that some units require rather careful monitoring and control to produce good-quality water consistently. Although any filtration system requires a fair amount of monitoring, operators of very small plants do not ordinarily expect to spend much time "operating" the plant. Operators contemplating the installation of a package plant should obtain assurance from the manufacturer that the system will consistently meet state and federal effluent standards. They should also obtain a realistic statement of the operation and maintenance labor that will be necessary to properly operate the facility. Pilot testing of the process is recommended before a package treatment plant is purchased.

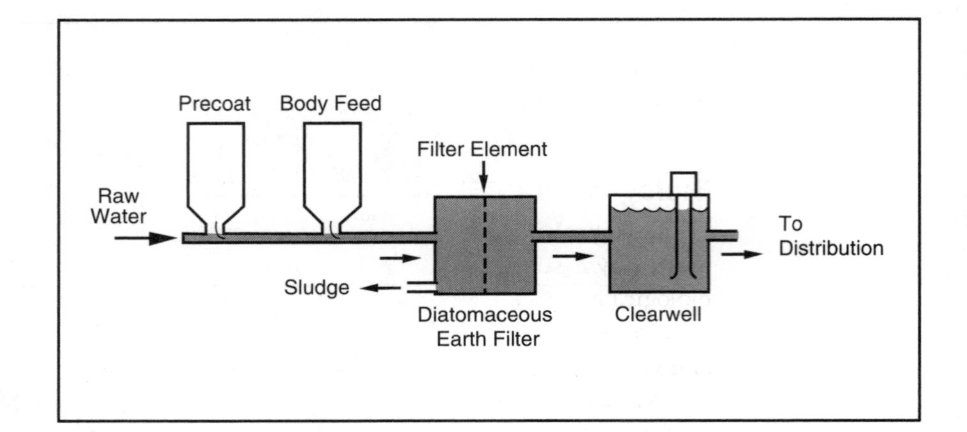

FIGURE 6-5
Diatomaceous earth filter plant

FIGURE 6-6
Typical package treatment plant

Courtesy of Infilco Degremont, Inc., Richmond, Va.

The term *package treatment plant* can no longer be assumed to mean conventional treatment. Package plants are now available to provide treatment by other methods, such as adsorption, aeration, or membrane technology.

Biological Treatment

Biological treatment of water uses microbes, not harmful to humans, to break down substances such as biodegradable organic carbon. The use of biologically active sand or carbon filters produces water that is not conducive to microbial growth in the distribution system, is free of undesirable tastes and odors, and has a reduced chlorine demand.

Numerous installations in Europe have proven that biological treatment works effectively, but there has not been much interest in the United States. This is perhaps because, if the process is not properly designed and controlled, there is a risk of introducing pathogenic microorganisms or harmful by-products into the finished water. However, future federal restrictions on the level of disinfection by-products (DBPs) in drinking water could lead to interest in and development of biological treatment as an economical and effective method of minimizing DBPs.

Types of Gravity Filters

Filters can be classified based on several different features, including the filtration rate, types of filter media, or type of operation. For the purposes of this chapter, the following classifications will be used for gravity filters:

- slow sand filters
- rapid sand filters
- high-rate filters
- deep-bed, monomedium filters

There are other variations in filter design, such as upflow and biflow filters. These have been used in only a few installations, however, so they are not discussed here. As detailed in chapter 15, membrane processes are also being developed that can be used in place of a media filter to remove turbidity and microorganisms; they may see increasing use in the future. The characteristics of the principal types of gravity filters are summarized in Table 6-1.

TABLE 6-1 Comparison of gravity filter characteristics

Characteristic	Slow Sand Filters	Conventional Rapid Sand Filters	High-Rate Filters
Filtration rate	0.05 gpm/ft^2	2 gpm/ft^2	3–8 gpm/ft^2
Media	Sand	Sand	Sand and coal or sand, coal, and garnet
Media distribution	Unstratified	Stratified: fine to coarse	Stratified: coarse to fine
Filter runs	20–60 days	12–36 hours	12–36 hours
Loss of head	0.2 ft initial to 4 ft final	1 ft initial to 8 or 9 ft final	1 ft initial to 8 or 9 ft final
Amount of backwash water used	Backwash not used	2–4% of water filtered	6% of water filtered

Slow Sand Filters

Slow sand filters were the first type of gravity filter used for water treatment. As shown in Figure 6-7, they typically consist of a layer of fine sand about 3.5 ft (1 m) thick, supported by about 1 ft (0.3 m) of graded gravel. The filtering action of the filters is dependent on fine sand and a sticky mat of suspended matter called schmutzdecke that forms on the sand surface.

The fine sand that is used has small voids that fill quickly, so slow sand filters are not normally used with waters that consistently have turbidities above 10 ntu. Chemical coagulation is not generally used to help form the schmutzdecke because it would just hasten the filling and clogging of the void spaces. Flow rates are kept quite low, with 0.05 gpm/ft^2 (0.1 m/h) being about average. Consequently, it takes a filter 0.5–1 acre (0.20–0.40 ha) in size to process 1 mil gal (4 ML) of water per day.

Slow sand filters are not backwashed. Instead they are cleaned by scraping about 1 in. (2 or 3 cm) of sand from the top. This can be done manually or with mechanical equipment. For small systems, the surface sand can be scraped off with shovels, but generally mechanical equipment is used to haul away the discarded sand. After several cleanings have reduced the sand depth to about 2 ft (0.6 m), new sand is added to bring the filter bed back to the original depth. After a filter has been cleaned, it may be necessary

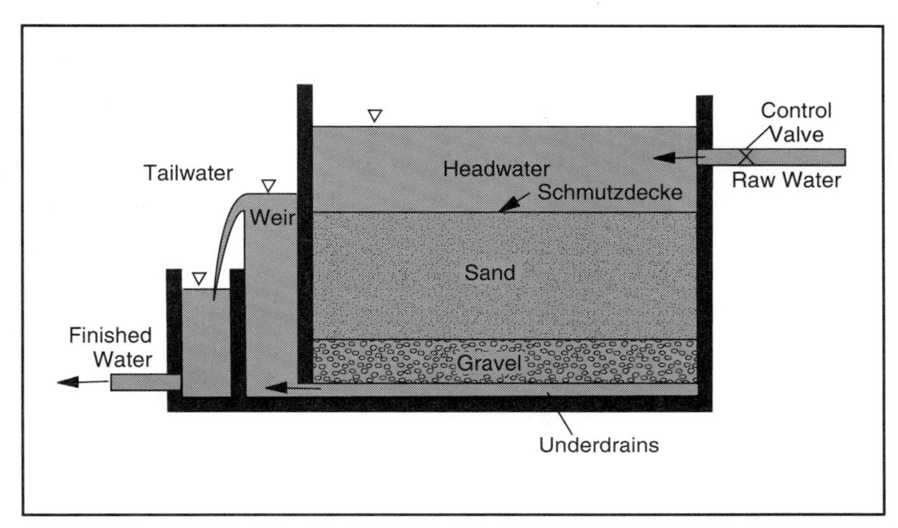

FIGURE 6-7 Schematic cross section of a slow sand filter

Source: Barrett et al. (1991)

to filter to waste (i.e., treat the filtered water as wastewater) for as long as two days before a schmutzdecke forms that will effectively remove turbidity. Because a filter may be out of service for several days as a result of scraping and filtering to waste, it is necessary for a water system to have at least two filters to provide continuous service.

The advantages of slow sand filters include low construction, maintenance, and operating costs. Disadvantages include large land area requirements for large-capacity plants and the problem of filters freezing in northern locations unless they are covered. In recent years, there has been some resurgence in the use of slow sand filters because of more modern equipment and improved design criteria. The state regulatory agency should be consulted for advice if a utility is interested in constructing a slow sand filter system.

Conventional Rapid Sand Filters

Rapid sand filters can accommodate much higher filtration rates than slow sand filters because they use coarser sand. Instead of depending on the schmutzdecke for filtering action, the filters trap suspended matter through several inches (centimeters) or more of the depth of the filter sand. Rapid sand filters are designed so that they can be backwashed in order to be cleaned and restored for use. At least two filters must be provided so that one

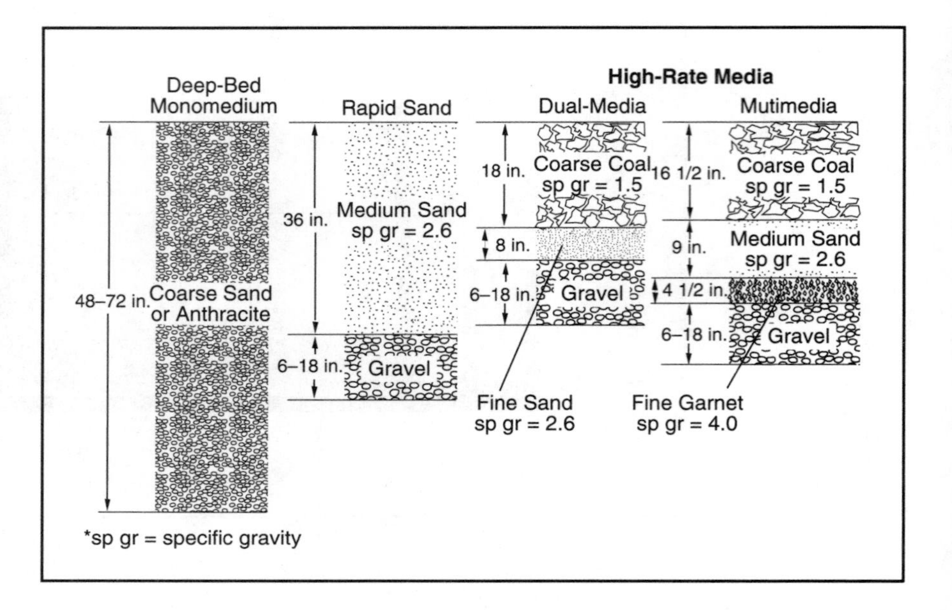

FIGURE 6-8 Comparison of deep-bed, rapid sand, and high-rate filter media

can remain in operation while the other is being backwashed. If there are only two filters, each must be capable of meeting plant design capacity.

High-Rate Filters

As shown in Table 6-1, high-rate filters (i.e., dual-media and multimedia filters) can operate at rates up to four times higher than those of rapid sand filters. As illustrated in Figure 6-8, these filters use a combination of filter media, not just sand.

Dual-media filters usually have a bed of sand covered by a layer of granulated anthracite coal. Multimedia (mixed-media) filters use three or more types of media of varying coarseness and specific gravity (the weight of the material relative to the weight of water). The most common combination is garnet sand on the bottom layer, silica sand in the center, and coarse anthracite on the top.

In dual- and mixed-media filters, the coarsest material also has the lowest specific gravity, so it tends to stay at the top. The heaviest medium is also the finest, and it stays near the bottom. Some mixing of the layers does occur, but as illustrated in Figure 6-9, the medium will approximately maintain their respective positions in the filter bed after backwashing.

In operation, the coarse layer on top removes most of the suspended particles. The particles that do pass through this layer are removed by finer

FIGURE 6-9 Mixing of filter media: left, as laid; right, after backwash

Courtesy of Wheelabrator Engineered Systems – Microfloc

media below. As a result, most of the filter bed is used to remove suspended particles. This allows for longer filter runs and higher filtration rates than for a conventional sand filter, which traps most suspended matter near the sand surface. As a result, head loss does not build up as quickly as with a rapid sand filter. Multimedia filters are now used almost exclusively because they can greatly increase a treatment plant's capacity while maintaining excellent water quality.

The types of filter media used depend on many factors, including general raw-water quality, variations in water quality, and type of chemical treatment. Pilot tests using different types of media are usually conducted to determine which media combination performs best for a particular water source.

Deep-Bed, Monomedium Filters

Deep-bed filters use a single filter medium — sand or anthracite — from 48 to 72 in. (1.2 to 1.8 m) deep. The bed must be deeper than a conventional filter because the medium is coarser. These filters are washed by the relatively gentle concurrent upflow of air and water. This wash causes the medium to be cleaned and mixed, but little or no stratification by size occurs. Monomedium filters can be operated at higher rates than dual-media filters and are well suited for use with direct filtration.

Equipment Associated With Gravity Filters

This section describes equipment associated with gravity filters, including filter tanks, filter media, underdrain systems, wash-water troughs, and filter bed agitation equipment.

Filter Tanks

Filter tanks are generally rectangular and constructed of concrete. However, prefabricated tanks and units for package plants are often made of steel. Several filter tanks are usually constructed side by side on either side of a central pipe gallery to minimize piping. Figure 6-10 illustrates this arrangement (it shows only one row of filters).

Filter Media

Sand or other media used for filtration must be prepared specifically for filtration use. The original design of the conventional sand filters used for many years placed three to five layers of graded gravel between the sand and the underdrain system. The gravel bed served the dual purposes of

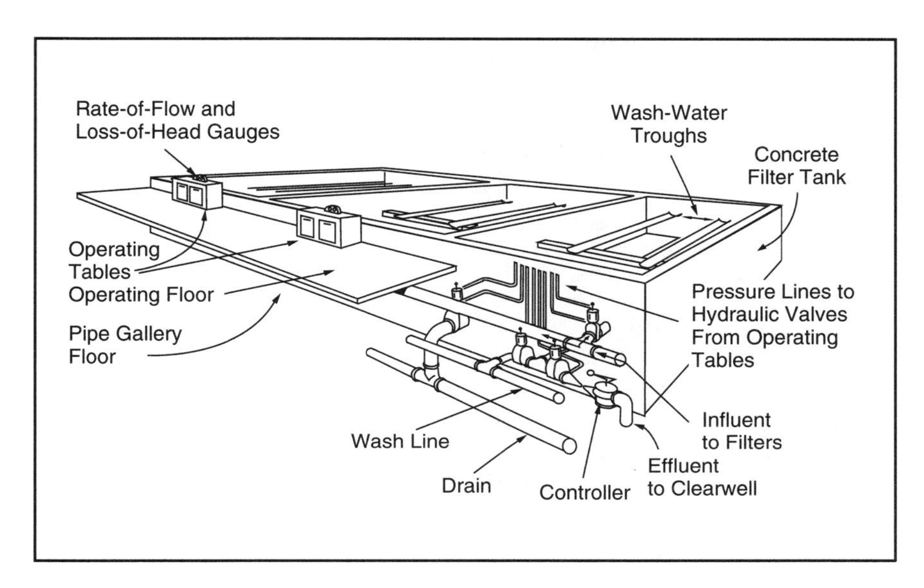

FIGURE 6-10 Filter tank construction

preventing sand from entering the underdrains and of helping distribute the backwash water evenly across the bed. The total gravel bed may be from 6 to 18 in. (150 to 450 mm) thick, depending on the type of underdrain system.

Many types of filter underdrain systems are now available that do not require a gravel layer. They are designed to support the sand or other media directly.

Additional details on the requirements for filtering materials are covered in AWWA Standard B100.

Filter Sand

The sand used for filtration is specifically manufactured for filtration use. If the sand is too fine, it will resist the flow of water and require frequent backwashing. If too coarse, it may not effectively remove turbidity. Included in the specifications for filter sand are the grain size, size distribution, shape, density, hardness, and porosity. Hardness is particularly important because soft sand will break down quickly during the agitation of backwashing.

Grain size has important effects on the efficiency of filtration and on backwashing requirements. Proper size is determined by sieve analysis that uses the American Society for Testing and Materials (ASTM) Standard Test C136 (most recent edition). In the United States, the sand is described in terms of the effective size (ES) and the uniformity coefficient (UC). The ES is that size for which 10 percent of the grains are smaller by weight. The UC is a measure of the size range of the sand.

A conventional rapid sand filter typically uses sand that has a fairly uniform grain size of 0.10 to 0.15 in. (0.4–0.6 mm) in diameter, in a bed 24 to 30 in. (0.6 to 0.75 m) deep. This sand is much coarser than that used for slow sand filters. The sand ordinarily used in conventional rapid sand filters has an ES of 0.11 to 0.13 in. (0.45–0.5 mm) and a UC varying from 1.6 to 1.75.

Anthracite

Crushed and graded anthracite is often used as a filter medium along with sand. It is much lighter than sand, so it always stays on top of the sand during backwashing. It is usually graded to a size slightly larger than sand.

Granular activated carbon (GAC) can also be used instead of anthracite as a filter medium, in which case it plays a dual role by also adsorbing organic compounds in the water. Its principal applications have been for taste-and-odor control, but it can also be used for removing organic compounds suspected of being carcinogenic or causing adverse health effects. Experience has shown that where the organic loading is not too heavy, tastes and odors can successfully be removed for periods of from one to five years. When the effectiveness of the GAC has ceased, it must be removed and regenerated, or replaced with a new medium.

When GAC is used in the retrofit of a conventional rapid sand filter, 15–30 in. (0.38–0.76 m) of GAC is placed over several inches (centimeters) of sand. The effectiveness of GAC as an adsorber in a filter is limited because the water is in contact with it for only a few minutes while passing through the filter.

The use of GAC in filters is covered in more detail in chapter 13.

Garnet Sand

Garnet sand, or ilmenite, is used as the bottom medium in multimedia filters. It is graded to be finer than sand, but because of its greater density, it tends to stay at the bottom of the filter.

Underdrain Systems

Filter underdrains serve two functions. They collect the filtered water uniformly across the bottom of the filter, so that the filtration rate will be uniform across the filter surface. They also distribute the backwash water evenly, so that the sand and gravel beds will expand but not be unduly disturbed by the backwashing. The common types of underdrain systems are

- pipe lateral collectors
- perforated tile bottoms
- "Wheeler" bottoms
- porous plate bottoms
- false-floor underdrain systems

Pipe lateral, perforated tile, and Wheeler filter bottom systems all require a gravel support bed to prevent the sand or anthracite from flowing into the underdrains and to distribute the backwash water evenly. More new systems are now available that allow fine media to be placed directly on the filter bottom so that a gravel layer will not be required.

Pipe Lateral Collectors

Pipe lateral collectors are the first and oldest type of underdrain system; many installations are still in use (Figure 6-11). These systems have a central manifold pipe with perforated lateral pipes on each side. The pipes can be cast iron, asbestos cement, or polyvinyl chloride (PVC). Small holes are usually spaced along the underside of the pipes, where they are least likely to become plugged with sand. This hole placement will also force the backwash water against the floor where it will be evenly distributed and not disrupt the media.

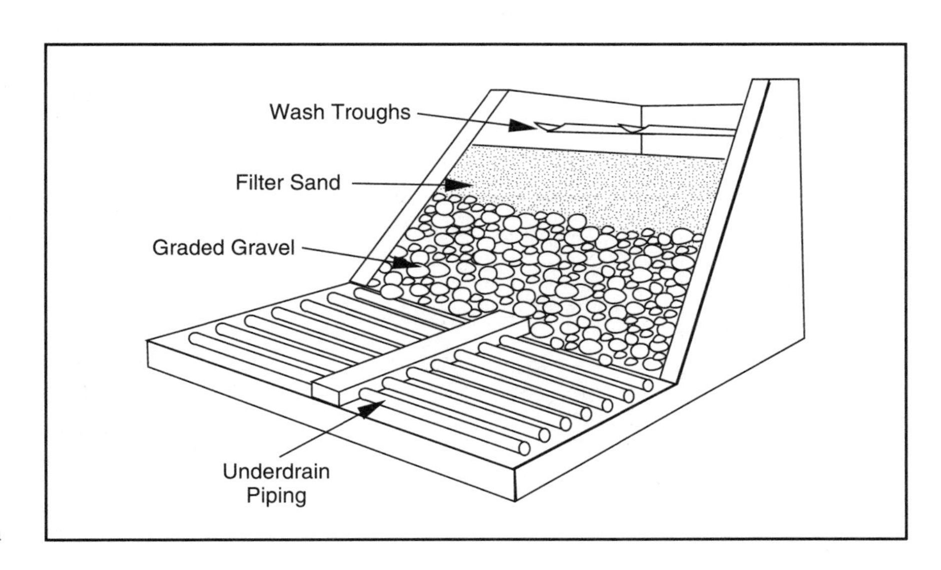

FIGURE 6-11 Pipe lateral collector under a conventional rapid sand system

In some cases, the holes are fitted with brass inserts to prevent plugging caused by corrosion. Plugged holes in the underdrain piping can cause one area to be "dead" during backwashing and can increase velocity through the holes in other areas. This can eventually lead to a serious disruption of the filter media if the problem is not corrected.

Perforated Tile Bottoms

Perforated tile bottoms have been installed in many water filtration plants (Figure 6-12). They consist of perforated, vitrified clay blocks with channels inside to carry and distribute the water. Figure 6-13 shows details of the same type of filter underdrain blocks that are currently available; the blocks are made of high-density polyethylene.

Wheeler Filter Bottoms

Wheeler filter bottoms have also been widely used and consist of conical concrete depressions filled with porcelain spheres (Figure 6-14). Each cone has an opening in the bottom and contains 14 spheres ranging in diameter

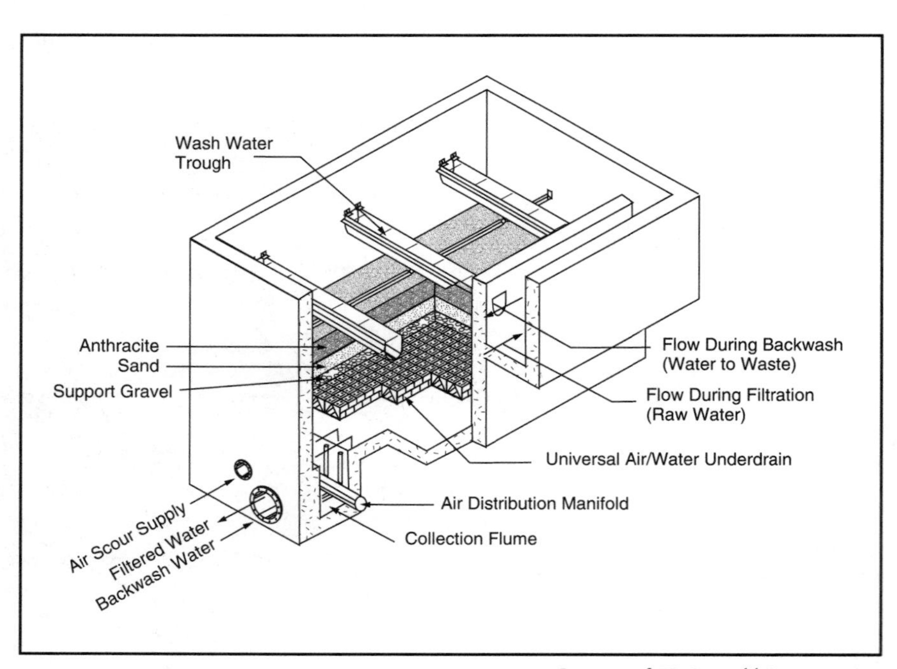

FIGURE 6-12 Conventional rapid sand filter with perforated tile underdrain system

Courtesy of F.B. Leopold Company, Inc.

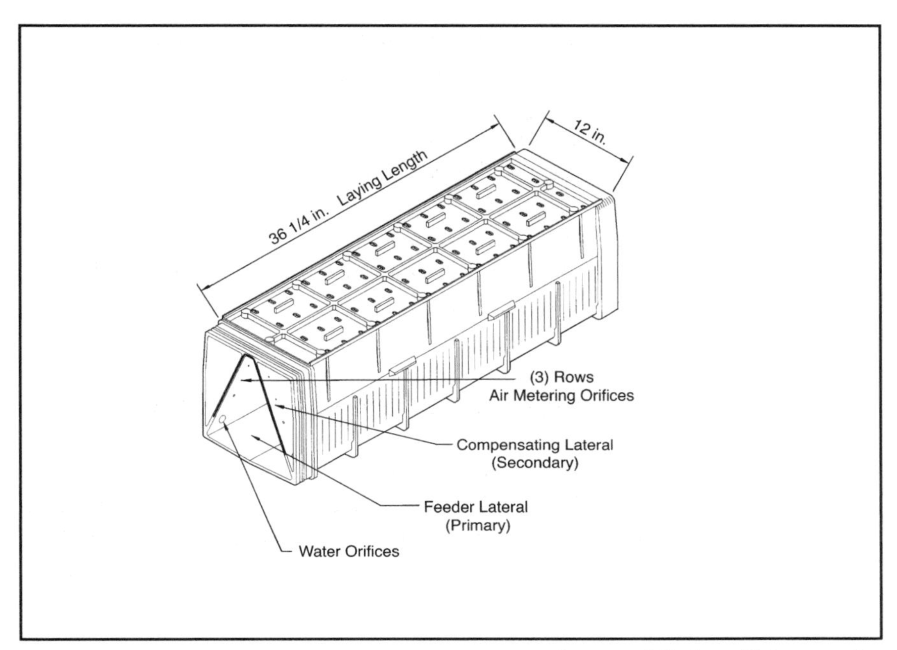

FIGURE 6-13 A section of a perforated underdrain block

Courtesy of F.B. Leopold Company, Inc.

FIGURE 6-14 A Wheeler bottom

Courtesy of Leeds & Northrup, a Division of General Signal

from 1⅜ to 3 in. (30 to 80 cm). The spheres are arranged to lessen the velocity of the wash water and distribute it evenly.

Porous Plate Bottoms

Porous plates can be used to make up an entire filter bottom. The plates are made of a ceramically bonded aluminum oxide and are supported over the bottom of the filter tank by long bolts, steel beams, concrete piers, or fiberglass supports. Typical installations are illustrated in Figure 6-15 and

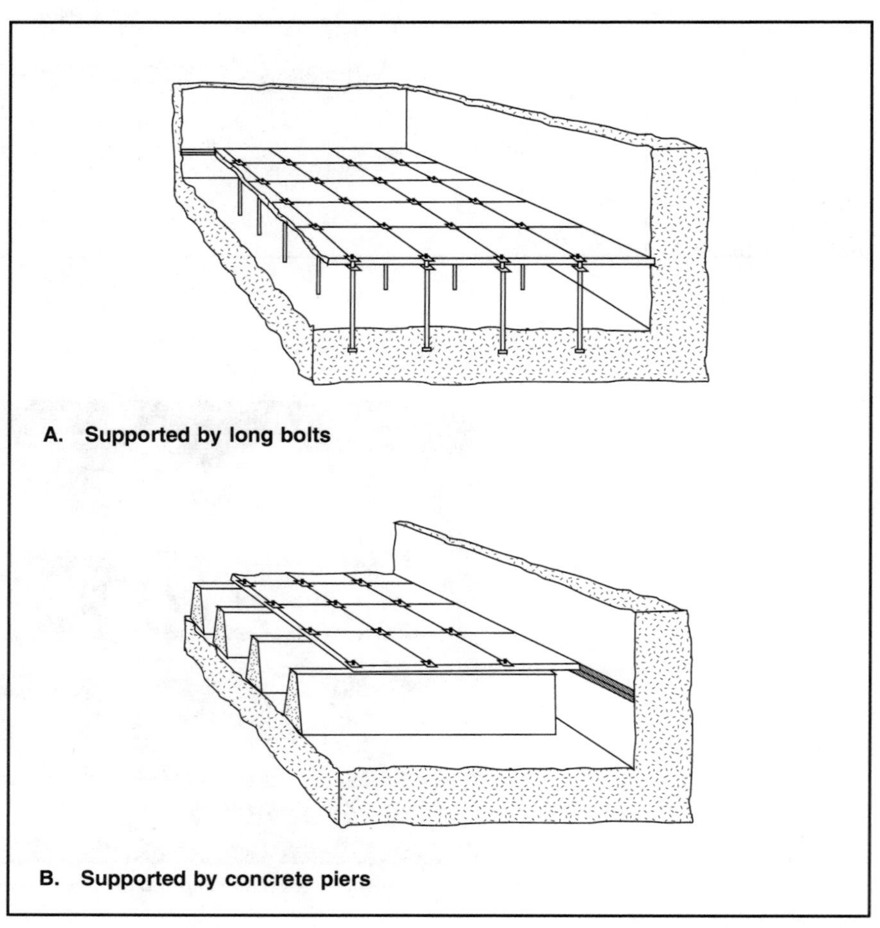

FIGURE 6-15
Porous plate underdrain system

Courtesy of Christy Refractories Co.

FIGURE 6-16 Porous plate installation showing contour gaskets between plates

Courtesy of Christy Refractories Co.

Figure 6-16. Porous plates are not recommended for filtering hard water because a calcium buildup can form in the plates.

False-Floor Underdrain Systems

False-floor underdrain systems are constructed by placing a concrete or steel plate 1–2 ft (0.3–0.6 m) above the filter tank bottom, creating an underdrain plenum below the false floor (Figure 6-17). Various types of nozzles are available for installation in the floor. These nozzles have coarse openings if a gravel layer is to be used or very fine openings if a fine medium is to be placed directly adjacent to them (Figures 6-18 and 6-19).

Other Systems

Several other types of underdrain systems are offered by various manufacturers. Each has some advantages and disadvantages that should be carefully investigated before a new system is installed.

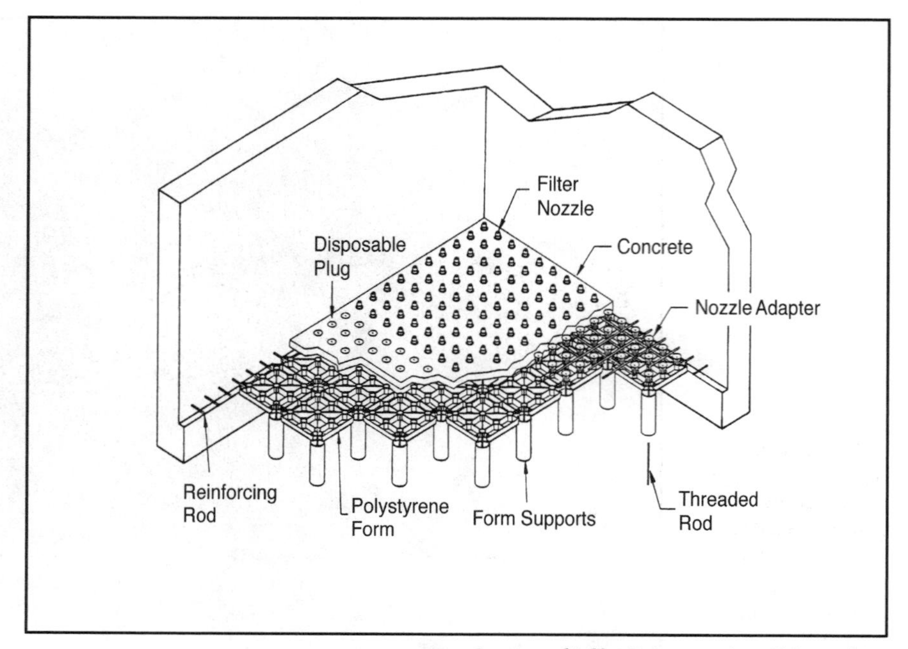

FIGURE 6-17 Typical false-floor filter underdrain system constructed of poured concrete

Courtesy of Infilco Degremont Inc., Richmond, Va.

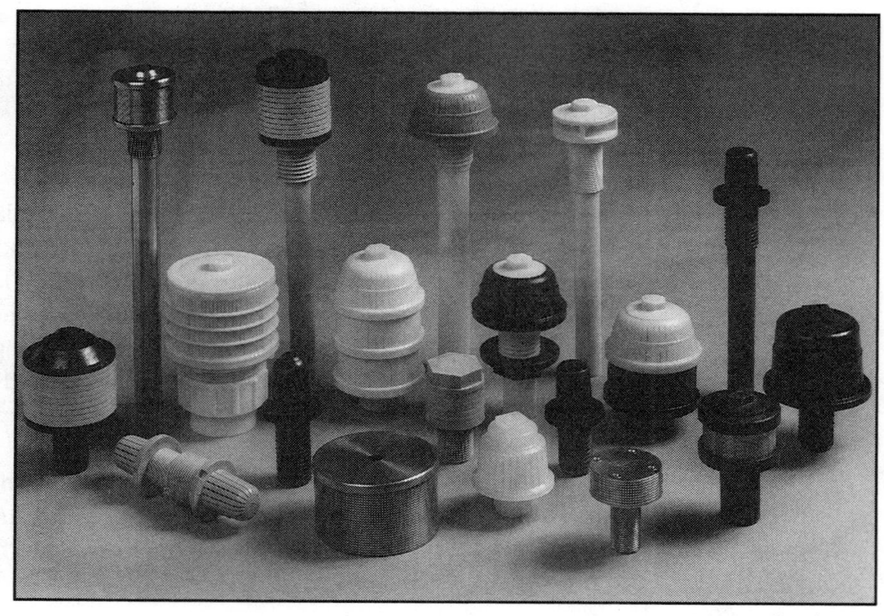

FIGURE 6-18 Typical distribution nozzles used on false-floor underdrain systems

Courtesy of Orthos Inc.

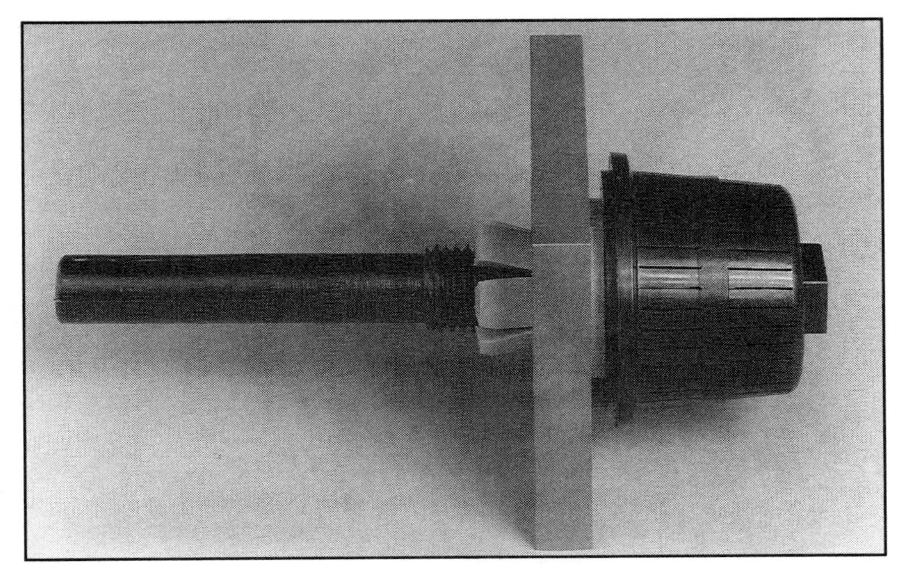

FIGURE 6-19 Typical distribution nozzle installed on a thin plate

Courtesy of ORTHOS Inc.

Wash-Water Troughs

All filters require a trough placed over the filter media to collect the backwash water during a wash and also carry it to waste. Proper placement of the troughs is very important. The media bed must be properly expanded during a wash, but no appreciable quantity of media should be wasted by carrying it over into the troughs. The troughs must also be placed so that equal flow is maintained in all parts of the filter bed and so that the wash will be uniform across the bed. The troughs are usually made of concrete, fiberglass, or steel.

Filter Bed Agitation

Most filters require some type of surface agitation to help release suspended matter trapped in the upper layers of filter media. Normally, this matter is not completely removed by backwashing alone. An additional system is added to provide adequate agitation of the top few inches (centimeters) of media.

Surface wash systems consist of nozzles attached to either a fixed-pipe or rotary-pipe arrangement installed just above the filter media (Figure 6-20). Water or an air–water mixture is pumped through the nozzles, producing high-velocity jets.

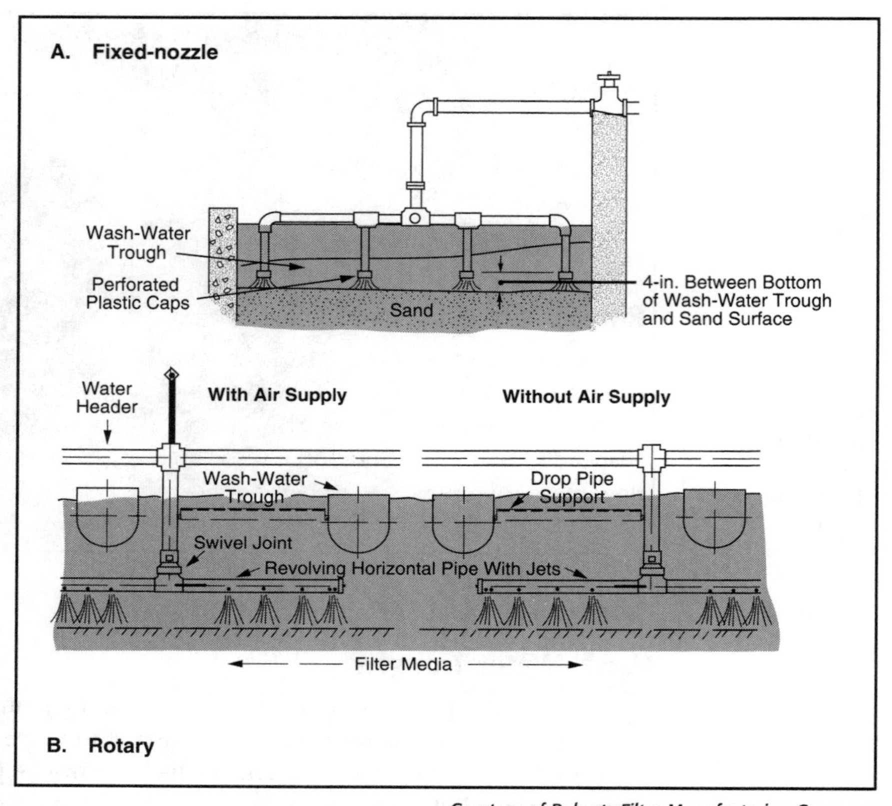

FIGURE 6-20 Filter bed agitators using surface wash systems

Courtesy of Roberts Filter Manufacturing Company

Air scouring systems are sometimes used in place of surface wash systems to ensure complete cleaning of the media. Air is usually applied through the backwash piping or separate outlets located at the bottom of the filter. The air is applied at the beginning of the wash cycle, before the rising water reaches the lip of the wash-water troughs. Air scouring tends to disrupt the media considerably, so it must be followed by a fluidization wash period to allow mixed-media beds to restratify.

When air is used with mixed-media beds, provision must be made to ensure uniform distribution of the air across the filter bottom to prevent undue disruption of the media. The length and rate of air application must also be closely controlled; otherwise, undue amounts of media will be lost over the wash-water troughs. Air scouring works particularly well with monomedium beds because disruption of the bed is not a problem for them.

Filter Control Equipment

Rapid sand and multimedia filters generally require the following control equipment:

- a rate-of-flow controller
- a loss-of-head indicator
- on-line turbidimeters

Rate-of-Flow Controllers

Modulating control valves are usually installed on the effluent discharge of each filter. The controller maintains a fairly constant flow through the filter so that flow surges do not occur. Without a controller, the surges would force suspended particles through the filter. The controller generally consists of a flow-measuring device, a throttling valve, and a means to set the throttling valve automatically or manually to maintain a fixed flow rate. A typical rate-of-flow controller is illustrated in Figure 6-21.

Loss-of-Head Indicators

Indicators are required to monitor the status of resistance to flow in the filter as suspended matter builds up in the media. The head loss should be continuously measured to help determine when the filter should be

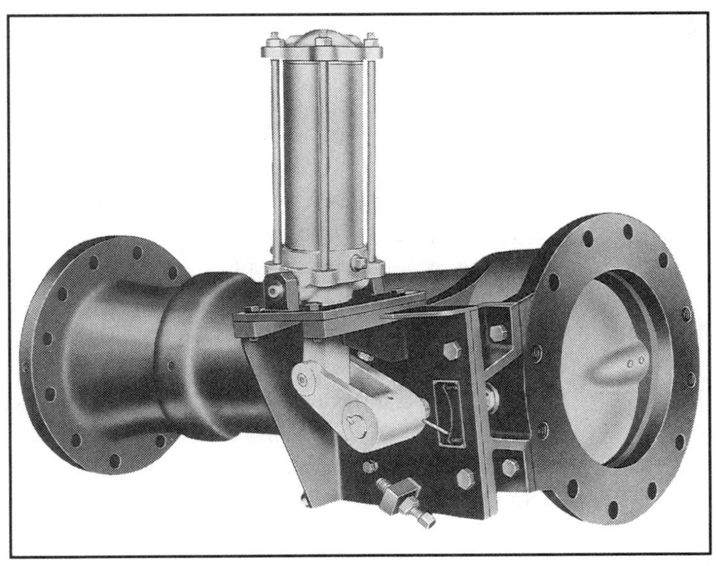

FIGURE 6-21 Rate-of-flow controller

Courtesy of Leeds & Northrup, A Division of General Signal

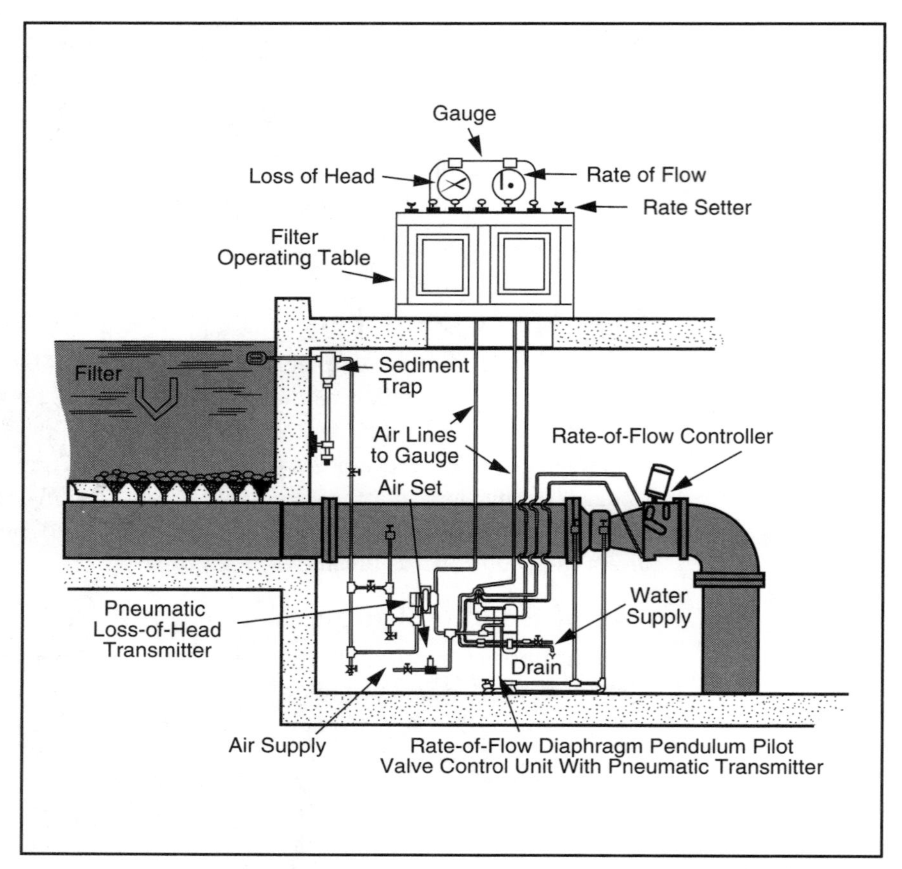

FIGURE 6-22 Configuration to measure head loss based on air pressure data

Courtesy of Hach Company

backwashed. In the simplest form of measurement, a clear plastic hose can be connected to the filter influent and another to the filter effluent. The difference in height between the water levels of the two hoses represents the head loss through the filter.

The head loss can also be measured by devices that use air pressure data (Figure 6-22) or by electronic equipment. Recorders are commonly installed on loss-of-head indicators for maintaining a record of operation. The recorder also provides a visible indication of the rate at which loss of head is increasing, so that a projection can be made of when it will be necessary to backwash a filter.

On-Line Turbidimeters

After a filter has been operating for a period of time, the suspended material will begin to break through the filter bed. This will cause turbidity to increase in the filtered water. If the filter effluent is continuously monitored, the filter can be backwashed as soon as the breakthrough starts, preventing excessive turbidity from passing into the distribution system. A typical on-line turbidimeter is illustrated in Figure 6-23.

If it is necessary to analyze the size and density of particles in the filter effluent at levels below the range of a turbidimeter, a particle counter must be used. (See discussion in chapter 4.)

Operation of Gravity Filters

Filtration involves three procedures: filtering, backwashing, and filtering to waste. Figure 6-24 indicates the key valves used for filtration and their positions during the filtration procedures.

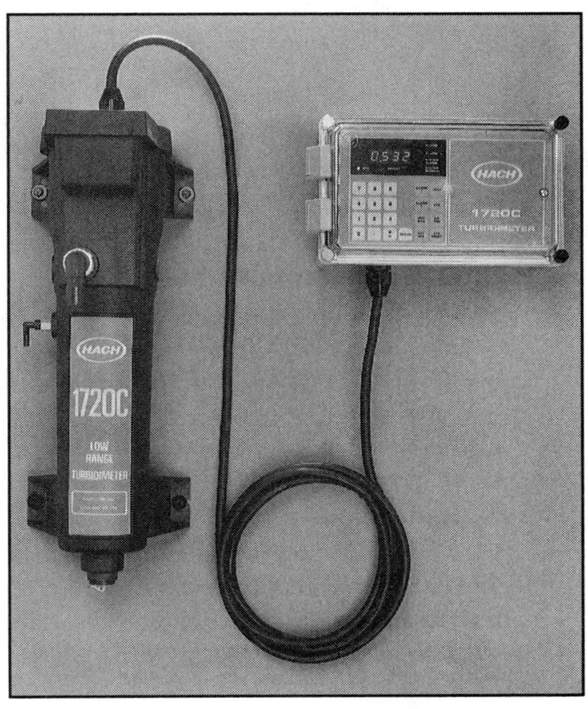

FIGURE 6-23
On-line turbidimeter

Courtesy of Hach Company

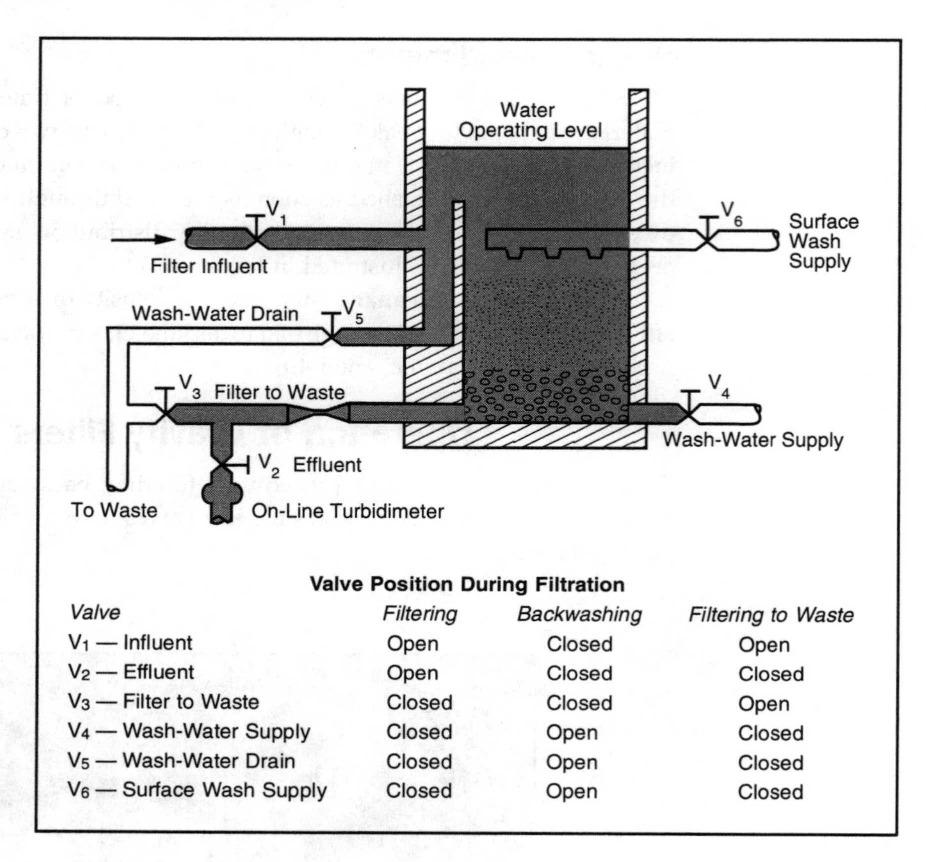

Valve Position During Filtration

Valve	*Filtering*	*Backwashing*	*Filtering to Waste*
V_1 — Influent	Open	Closed	Open
V_2 — Effluent	Open	Closed	Closed
V_3 — Filter to Waste	Closed	Closed	Open
V_4 — Wash-Water Supply	Closed	Open	Closed
V_5 — Wash-Water Drain	Closed	Open	Closed
V_6 — Surface Wash Supply	Closed	Open	Closed

FIGURE 6-24
Key valves used in filter operation

Filter Operation Methods

The flow rate through filters can be controlled by a rate-of-flow controller or it may proceed at a variable declining rate. Regardless of the control method used, as filtration progresses, suspended matter builds up within the filter bed. At some time, usually after 15 to 36 hours of operation, the filter must be cleaned by backwashing.

Controlled Rate of Flow

During filtering, water is applied to the filter to maintain a constant depth of 4–5 ft (1–1.5 m) over the media. The media bed is initially clean, and head loss is very low. The filtration rate can be kept at the desired level by using a modulating control valve called a rate-of-flow controller. The controller is also important in preventing harmful surges that can disturb the media and force floc through the filter.

The controller is used to maintain a constant desired filtration rate — usually 2 gpm/ft^2 (1.4 mm/s) for rapid sand filters and 4–6 gpm/ft^2 (2.8–4 mm/s) for high-rate filters. At the beginning of the filter run, the flow controller valve is almost closed, which produces the necessary head loss and maintains the desired flow rate.

As filtration continues and suspended material builds up in the bed, head loss increases. To compensate for this increase, the controller valve is gradually opened. When the valve is fully opened, the filter run must be ended because further head loss cannot be compensated for and the filter rate will drop sharply. Rate-of-flow controllers must be carefully maintained because a malfunctioning controller can damage the filter bed and degrade water quality by allowing sudden changes in the filtration rate.

Variable Declining-Rate Filtration

Another commonly used control method is variable declining-rate filtration. As shown in Figure 6-25, the filtration rate is not kept constant for this method. The rate for a particular filter starts high and gradually decreases as the filter gets dirty and head loss increases. One advantage of this system is that it does not require a rate-of-flow controller or constant attention by the operator. In addition, harmful rate changes cannot occur.

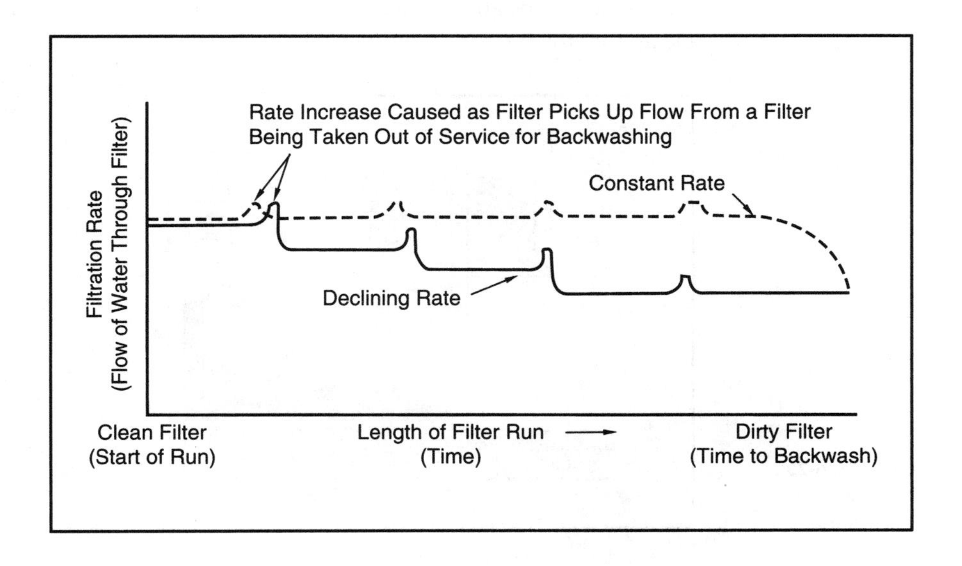

FIGURE 6-25
Declining-rate filtration

With this method, each filter accepts the proportion of total flow that its bed condition can handle. As a filter gets dirty, the flow through it decreases. Flow then redistributes to cleaner filters, and total plant capacity does not decrease. To prevent excessive flow rates from occurring in clean filters, a flow-restricting orifice plate is placed in the effluent line of each filter.

Backwashing Procedures

The Importance of Careful Backwashing

Backwashing is a critical step in the filtration process. Inadequate backwashing causes most of the operating problems associated with filtration. For a filter to operate efficiently, it must be cleaned thoroughly before the next filter run begins. In addition, properly backwashed filters require far less maintenance.

Treated water is always used for backwashing so that the bed will not be contaminated. This treated water can be delivered from elevated storage tanks designed for this purpose or pumped in from the clearwell.

During filtration, the voids between the media grains fill with filtered material (floc). The grains also become coated with the floc and become very sticky, making the filter bed difficult to clean. To clean the filter bed, the media grains must be agitated violently and rubbed against each other to dislodge the sticky coating. Therefore, the backwash rate must be high enough to completely suspend the filter media in the water.

The backwash causes the filter bed to expand, as shown in Figure 6-26. The expansion, however, should not be so large that the media flow into the wash-water troughs. Because normal backwash rates are not sufficient to clean the media thoroughly, auxiliary scour (surface wash) equipment is

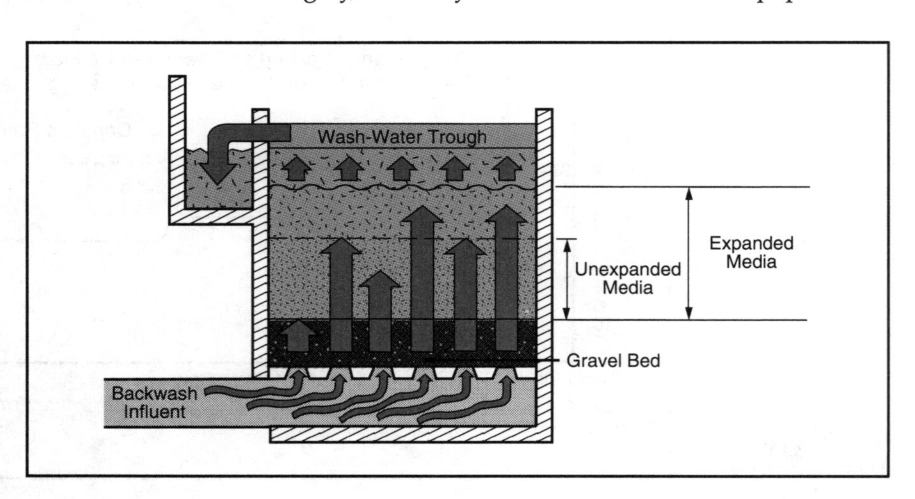

FIGURE 6-26 Filter bed expansion

recommended to provide the extra agitation needed before backwashing begins. Auxiliary scouring is a must for high-rate filters because the filtered material penetrates much deeper into the bed.

Factors That Determine Backwash Frequency

Head loss, filter effluent turbidity, and length of filter run must all be considered while deciding when a filter needs backwashing. Generally, a filter should be backwashed when

- head loss is so high that the filter no longer produces water at the desired rate. This is known as terminal head loss, and is usually about 8 ft (2.4 m).
- floc starts to break through the filter bed, causing the filter effluent turbidity to increase.
- a filter run reaches a time limit established for the plant. Many systems set a limit of 36 hours.

Figure 6-27 shows the relationship of these three factors in a typical filter run. However, the decision to backwash should not be based on only one of these three factors — that can lead to operating problems. For example, if a filter is not washed until terminal head loss is reached, a serious increase in filtered-water turbidity could occur well before the filter is backwashed. Likewise, initiating backwash based only on effluent turbidity could result in a drop in filter bed pressure below atmospheric, which occurs if maximum head loss is reached before turbidity breakthrough.

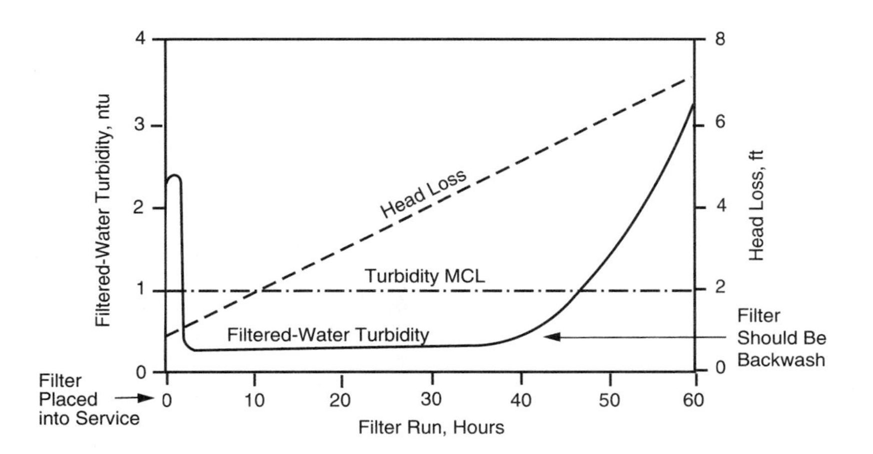

FIGURE 6-27 Typical filter run

A filter must be backwashed when the effluent turbidity begins to increase, as illustrated in Figure 6-27. The turbidity should never be allowed to increase to 1 ntu before backwashing (Figure 6-27). Most surface systems must now maintain a turbidity of less than 0.5 ntu entering the distribution system according to the Safe Drinking Water Act Surface Water Treatment Rule. In fact, filtration tests have shown that microorganisms start passing through the filter rapidly once breakthrough begins, even though the turbidity may be well below 1 ntu.

Length of Filter Runs

If high-quality water is applied to the filter, then filter runs, based on head loss or effluent turbidity, can be very long. In some cases, filters may have to be washed only once a week based on these criteria. It must be remembered that long filter runs can cause a gradual buildup of organic materials and bacterial populations within the filter bed. This, in turn, can lead to tastes and odors in the treated water, as well as slime growths within the filter that are difficult to remove.

However, short filter runs decrease finished-water production because of the time the filters are out of production and the increased quantity of backwash water needed. In general, the amount of backwash water used should not exceed 4 percent of the amount of water treated for rapid sand filters, and 6 percent for dual-media and multimedia filters. Filter runs in most treatment plants range between 15 and 24 hours.

Backwash Sequence

A typical backwash sequence is as follows:

1. The water in the filter is drained down to a level about 6 in. (150 mm) above the media.
2. The surface washers are turned on and allowed to operate for 1 to 2 minutes. This allows the high-velocity water jets to break up any surface layers of filtered material.
3. The backwash valve is partly opened to allow the bed to expand to just above the level of the washers. This provides violent scrubbing of the top portion of the media, which has the greatest accumulation of filtered material. Intense scrubbing is particularly important for rapid sand filters because the top 8 in. (200 mm) of media removes most of the suspended solids.

4. After a few minutes, the backwash valve is fully opened to allow a filter bed expansion of 20 to 30 percent. The actual amount of expansion that is best for a filter depends on how much agitation is needed to suspend the coarsest grains of media in the bed. With multimedia filters, the bed must be expanded so that the surface washers can scrub the area between the anthracite and sand layers (known as the interface), where most of the filtered material has penetrated. A backwash rate of 15–20 gpm/ft^2 (10–14 mm/s) is usually sufficient to provide the expansion needed.
5. The surface washers are usually turned off about 1 minute before the backwash flow is stopped. This allows the bed to restratify into layers, which is particularly important for multimedia filters.
6. The expanded bed is washed for 5 to 15 minutes, depending on how dirty the filter is. The clarity of the wash water as it passes into the wash-water troughs can be used as an indicator of when to stop washing.

If surface wash equipment is not available, a two-stage wash should be used. The initial wash velocity should be just enough to expand the top portion of the bed slightly, generally about 10 gpm/ft^2 (7 mm/s). Although not as effective as surface washing, this method will provide some scrubbing action to clean the surface media grains. After the surface has been cleaned, the full backwash rate is applied.

Turning on the backwash too quickly can severely damage the underdrain system, as well as heave the gravel bed and media to a point where they will not restratify. The time from starting backwash flow to reaching the desired backwash flow rate should be 30 to 40 seconds. To prevent accidents the backwash valve controls should be set to open slowly.

Disposal of Backwash Water

To avoid water pollution, backwash water must not be returned directly to streams or lakes. The water is usually routed to a lagoon or basin for settling. After settling, the water may be recycled to the treatment plant, as shown in Figure 6-28. The settled solids are combined with the sludge from the sedimentation basins and disposed of as discussed in chapter 5.

Because backwash water does not usually contain a very high concentration of suspended solids, some treatment plant operators have found that they can send the wash water directly to the intake well to be blended with the incoming raw water.

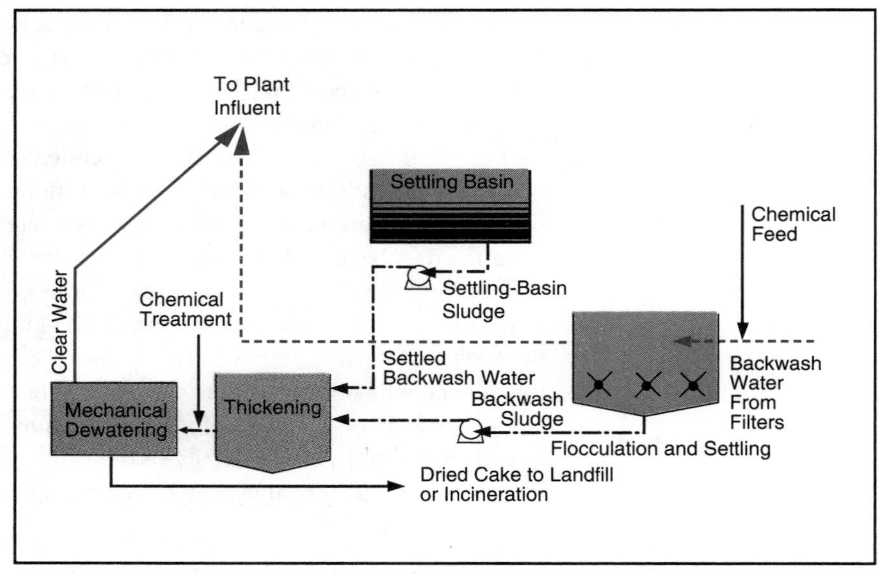

FIGURE 6-28 Disposal of backwash water

Courtesy of Wheelabrator Engineered Systems – Microfloc

Optimizing Filter Effluent Quality

Filtering to Waste

Once backwashing is completed, the water applied to the filter should be filtered to waste until the turbidity of the effluent drops to an acceptable level. As shown in Figure 6-27, the effluent turbidity remains high for a certain time period at the beginning of the filter run. Depending on the type of filter and treatment processes used before filtering, this period may be from 2 to 20 minutes. This is probably due, in part, to filtered material that remains in the bed after backwashing. In addition, the initial high filtration rates continue to wash fine material through the filter until the media grains become sticky and more effectively adsorb the suspended material.

If filtering to waste cannot be done, a slower filtration rate can be used for the first 15 to 30 minutes of each filter run to minimize breakthrough of filtered material. The material that passes through the filter at this time can include large quantities of microorganisms, so it is extremely important that breakthrough be prevented.

Filter Aids

As water passes through the filter, the floc can be torn apart, resulting in very small particles that can penetrate the filter. In such instances, turbidity

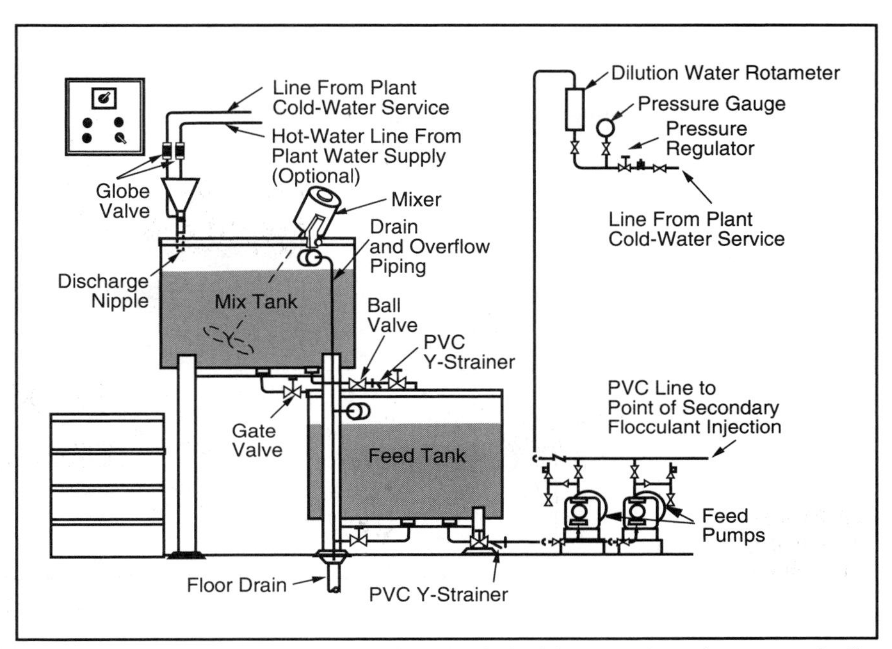

FIGURE 6-29 Filter aid feed system for use with dry polymers

Courtesy of Wheelabrator Engineered Systems – Microfloc

breakthrough occurs well before the terminal head loss is reached, resulting in short filter runs and greater use of backwash water. This is particularly true of high-rate filters because their loadings, in terms of suspended material and the filter rate, are much higher than with rapid sand filters.

To help solve this problem, polymers (also called polyelectrolytes) can be used as filter aids. Polymers are water-soluble, organic compounds that come in either dry or liquid form. The dry form is more difficult to use because it does not dissolve easily. Mixing and feeding equipment, as illustrated in Figure 6-29, is required.

Polymers have high molecular weights and may be nonionic, cationic, or anionic. As discussed in chapter 4, they are also used as primary coagulants or coagulant aids. Nonionic or slightly anionic polymers are normally used as filter aids.

When used as a filter aid, the polymer strengthens the bonds between the filtered particles and coats the media grains to improve adsorption. The floc then holds together better, adheres to the media better, and resists the shearing forces exerted by the water flowing through the filter. For best results, the polymer should be added to the water just upstream of the filters.

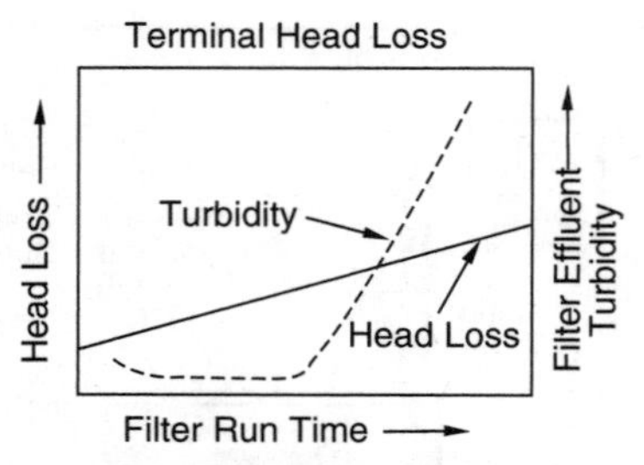

A. Inadequate Polymer Dosage

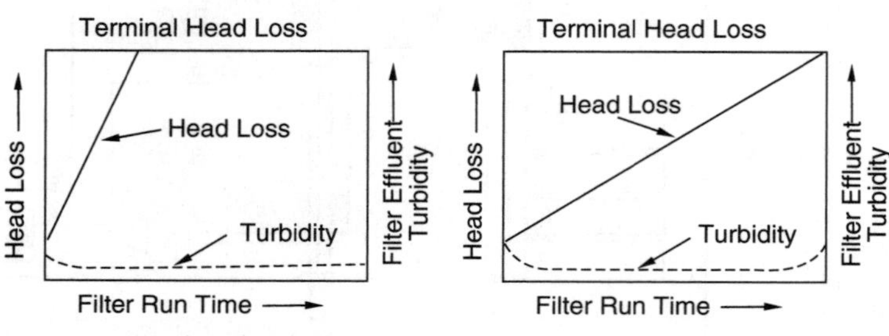

B. Excessive Polymer Dosage C. Optimal Polymer Dosage

Courtesy of Hach Company

FIGURE 6-30 Effect of polymer dosage on turbidity and head loss

In order to select the proper polymer for a particular treatment plant, several kinds should be tested and their performance and relative cost compared.

The required dosage is normally less than 0.1 mg/L, but the proper dosage must be determined through actual use. Continuous turbidity monitoring of the filtered water is essential to ensure that the proper dose is being applied. Figure 6-30A illustrates the effect of an inadequate polymer dosage: turbidity breakthrough occurs well before the terminal head loss is reached. Figure 6-30B illustrates the effect of an excessive polymer dosage. In this case, too much polymer makes the filtered material stick in the upper few inches (centimeters) of the filtered bed, creating a rapid head loss. The optimal dosage causes maximum head loss to be reached just before turbidity breakthrough occurs, as illustrated in Figure 6-30C.

Monitoring Filter Operation

Head Loss Monitors

Head loss must be monitored and recorded continuously to provide the operator with the most important information concerning filter operation. If head loss is allowed to become excessive, the operating efficiency of the filter

FIGURE 6-31 Nephelometric turbidimeter

Courtesy of Hach Company

will decrease, effluent turbidity will increase, and turbidity will be driven deeper than desired into the bed, where it may be hard to remove.

Effluent Turbidity Monitors

The turbidity of filter effluent should be monitored with either an on-line turbidimeter (Figure 6-23) or a laboratory turbidimeter to meet the requirements of the Surface Water Treatment Rule (Figure 6-31). If continuous monitors are used, they should be checked periodically for accuracy against a laboratory unit. A sudden increase in effluent turbidity indicates filter breakthrough. Turbidity measurements are discussed further in *Water Quality*, also part of this series.

Filterability Tests and Zeta Potential Tests

As described in chapter 4, filterability and zeta potential tests can be used to monitor the coagulation process to ensure the optimal coagulant dosage. If the direct filtration process is being used, one of these types of tests is essential to properly control the chemical dosage in the short time span between application of the chemicals and the point when the water reaches the filters.

Visual Inspection

Filter beds should periodically be visually inspected by observing the backwash and the filter's surface. Media loss is usually apparent from the undue amounts of media grains flowing into the backwash troughs. The top of the media surface should also be measured periodically to see if media grains are being lost. A serious decline in the media level may indicate an excessive wash-water rate.

A bed in good condition and with even backwash distribution should appear very uniform, with the media moving laterally on the surface. Violent upswelling or boils of water indicate problems. If some areas do not appear to clear up as quickly as others, an uneven distribution of backwash flow is indicated. If backwash is uneven, the problem will not fix itself and will probably get worse.

When the filter is drained, its surface should appear smooth. Cracks, mudballs, or ridges also indicate problems with backwashing. Backwash problems could mean that some of the underdrain system is blocked or broken, depending on the type of system. This indicates that the media will have to be removed for repair; at the same time, the media will probably have to be regraded, or new media purchased for replacement. This is a difficult and expensive task, and expert advice should be sought before these repairs are attempted.

Mudball Volume Determination

The volume of mudballs in the filter sand can be determined easily without expensive equipment. The procedure is described in *Basic Science Concepts and Applications* (part of this series), Mathematics Section.

Filter Operating Problems

Most filtration problems occur in the following major areas:

- chemical treatment before the filter
- control of filter flow rate
- backwashing the filter

If these three procedures are not performed effectively, the quality of filtered water will suffer and additional maintenance problems will occur.

Chemical Treatment Before the Filter

The importance of proper coagulation and flocculation and the advantages of using a filter aid have already been discussed. Successful adjustments to these processes are possible only if the filtration process is closely monitored. Because many raw-water characteristics such as turbidity and temperature are not constant, dosage changes during filtration are usually necessary. Consequently, instruments that continuously record turbidity, head loss, and flow rate are very important.

If short filter runs occur because of turbidity breakthrough, perhaps more coagulant, better mixing, or less filter aid is needed. If short runs are due to a rapid buildup of head loss, perhaps less coagulant or less filter aid is required. It is the operator's job to recognize these types of problems and choose the proper solution.

Control of Filter Flow Rate

Increases or rapid fluctuations in the flow rate can force previously deposited filtered material through the media. The dirtier the filter, the more problems rate fluctuations can cause. These fluctuations may be due to an increase in total plant flow, a malfunctioning rate-of-flow controller, a flow increase when a filter is taken out of service for backwashing, or operator error.

Obviously, as demand increases, filter rates may have to be increased to meet the demand, particularly if there is inadequate treated-water storage. If an increase is necessary, it should be made gradually over a 10-minute period to minimize the impact on the filters. Filter aids can also reduce the harmful effects of a rate increase.

When a filter is backwashed, the filters remaining in operation must pick up the nonoperating filter's load. This can create an abrupt surge through the filters, particularly if rate-of-flow controllers are used. This problem can be avoided if a clean filter is kept in reserve. When a filter is taken out of service for backwashing, the clean filter is placed in service to pick up the extra load. This will probably not work for plants with fewer than four filters, because all filters are generally needed to keep up with demand.

In plants that operate for only part of a day, the filtered material remaining on the filters can be shaken loose by the momentary surge that occurs when filtration is again started. These problems can be avoided if the filters are backwashed before they are placed into service.

If rate-of-flow controllers are used, they should be well maintained so that they function smoothly. Malfunctioning controllers will "hunt" for the proper valve position, causing harmful rate fluctuations.

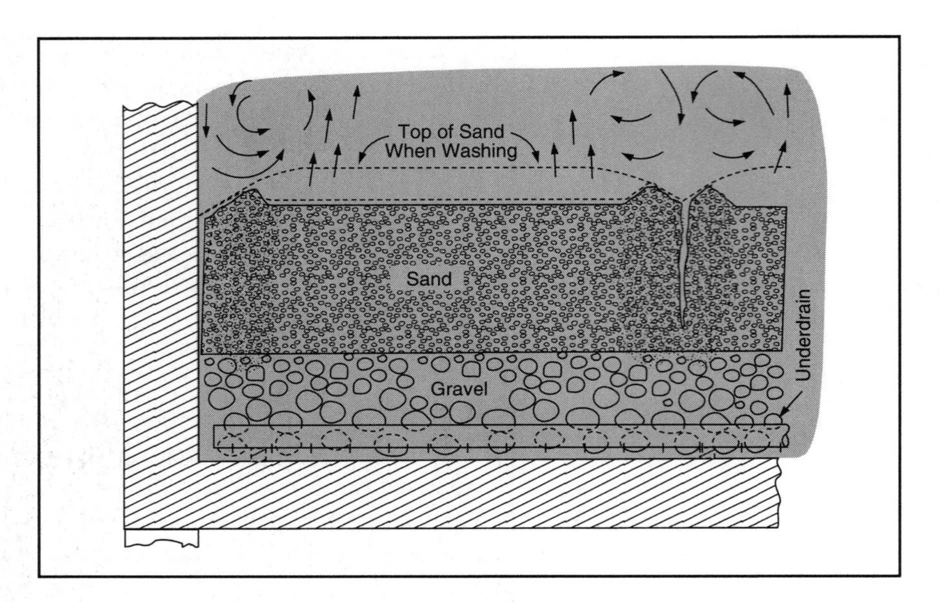

FIGURE 6-32
Clogged filter bed

Results of Ineffective Backwashing

Effective backwashing is essential for the consistent production of high-quality water. Ineffective backwashing can cause many problems.

Mudball formation. During filtration, grains of filter media become covered with sticky floc material. Unless backwashing removes this material, the grains clump together and form mudballs. As the mudballs become larger, they can sink into the filter bed during backwashing and clog those areas where they settle. These areas then become inactive, causing higher-than-optimal filtration rates in the remaining active areas and unequal distribution of backwash water (Figure 6-32).

Additional problems, such as cracking and separation of the media from the filter walls, may also result. Mudballs are usually seen on the surface of the filter after backwashing, particularly if the problem is severe, as shown in Figure 6-33.

A periodic check for mudballs should be made. Mudballs can be prevented by adequate backwash flow rates and surface agitation. Filter agitation is essential for dual-media and multimedia filters because mudballs can form deep within the bed.

Filter bed shrinkage. Bed shrinkage, or compaction, can result from ineffective backwashing. Clean media grains rest directly against each other, with little compaction even at terminal head loss. However, dirty media

FIGURE 6-33 Mudballs on filter surface

grains are kept apart by the layers of soft filtered material. As the head loss increases, the bed compresses and shrinks, resulting in cracks (Figure 6-34) and separation of the media from the filter walls. The water then passes rapidly through the cracks and receives little or no filtration.

Gravel displacement. If the backwash valve is opened too quickly, the supporting gravel bed can be washed into the overlying filter media. This can also happen over a period of time if a section of the underdrain system is clogged, causing unequal distribution of the backwash flow. Eventually, the increased velocities displace the gravel and create a sand boil, as shown in Figure 6-35. When this occurs, there is little or no media over the gravel at the boil, so water passing through the filter at that point receives very little filtration. In addition, media may wash down at this point and start flowing into the underdrain system.

Because some gravel movement always occurs, filters should be probed at least once a year to locate the gravel bed. This can be done with a metal rod ¼ in. (6 mm) in diameter while the filter is out of service. By probing the

FIGURE 6-34
Cracks in the filter bed

bed on a grid system and keeping track of the depths at which the gravel is located, one can determine if serious displacement has occurred.

If displacement has occurred, the media must be removed and the gravel regraded or replaced. Future displacement can be minimized by placing a 3-in. (80-mm) layer of coarse garnet between the gravel and the media and by not using excessive backwash rates.

Additional Operating Problems

Air binding and media loss are common problems for gravity filters.

Air binding. If a filter is operated so that the pressure in the bed is less than atmospheric (a condition known as negative head), the air dissolved in the water will come out of solution and form bubbles within the filter bed. This process, known as air binding, creates resistance to flow through the filter and leads to very short filter runs.

Upon backwashing, the release of the trapped air causes violent agitation, which can cause a loss of media into the wash-water troughs. Negative head typically occurs in filters with less than 5 ft (1.5 m) of water depth over the unexpanded filter bed. If a filter has been designed for a shallower water depth, filter runs may have to be terminated at a head loss of about 4.5 ft (1.4 m) to prevent negative head.

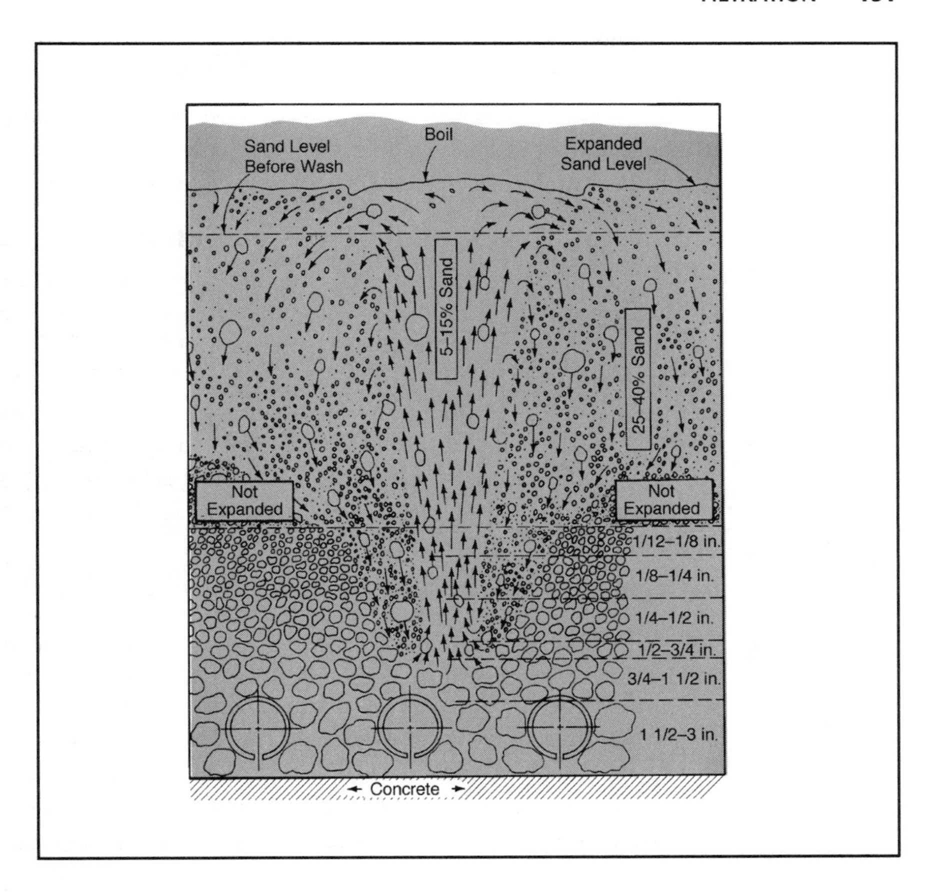

FIGURE 6-35 A sand boil in a filter bed

Air binding can also occur when cold water that is supersaturated with air warms up, typically in the spring. Unfortunately, not much can be done in this case except to keep the water at maximum levels over the filter bed and to backwash frequently.

Media loss. Some media are always lost during backwashing. This is especially true if surface washers are used. However, if considerable quantities of media are being lost, backwashing procedures should be examined. Because the bed is usually completely fluidized at 20 percent expansion, further expansion may not be needed. Turning the surface washers off about 1 to 2 minutes before the end of the main backwash will also help reduce media loss. If serious loss continues, the only solution is to raise the level of the wash-water troughs.

Pressure Filtration

Pressure filters have been used extensively to filter water for swimming pools, and more small water systems are now using them because installation and operating costs are low.

Sand or Mixed-Media Pressure Filters

The operating principle of pressure filters is similar to that of gravity filters, but the filtration process takes place in a cylindrical steel tank. Another difference is that water entering the filter is under pressure, so it is forced through the filter. As with gravity filters, sand or a combination of media is used. Filtration rates are about the same as for gravity filters.

If these filters are not to be backwashed very often, the valves may be manually operated. Systems with automatic backwashing that uses mechanically operated valves are also available. Because the water is under pressure, air binding will not occur.

The principal disadvantage of pressure filters is that the filter bed cannot be observed during operation or backwashing. The operator consequently cannot observe the backwash process or the condition of the filter bed. Horizontal and vertical pressure filter tanks are shown in Figures 6-36 and 6-37.

Diatomaceous Earth Filters

Diatomaceous earth (DE) filters were developed by the US military during World War II to remove the organisms causing amoebic dysentery. Today they are used extensively to filter swimming pool water, and they are used to some extent by small public water systems.

DE filters may be installed in either horizontal or vertical tanks and can be operated as either pressure or vacuum filters. A vacuum filter has a pump on the filter discharge that pulls the water through the filter. Regardless of the design, the components and the operation are similar. The filter medium is diatomaceous earth — the skeletal remains of microscopic aquatic plants called diatoms. Prehistoric deposits of diatoms are mined and processed to produce the diatomaceous earth medium.

A DE filter is put into operation by first applying a precoat of diatomaceous earth to the filter element (also known as the septum), which provides a filter surface. Figure 6-38 shows the components of a DE filter. Figure 6-39 shows the basic operation of a DE filter, and Figure 6-40 shows a precoat tank used in the process. The precoat normally has a thickness of about ⅛ in. (3 mm). Only previously filtered water should be used for

FIGURE 6-36 Horizontal pressure filter tanks

Courtesy of Infilco Degremont, Inc., Richmond, Va.

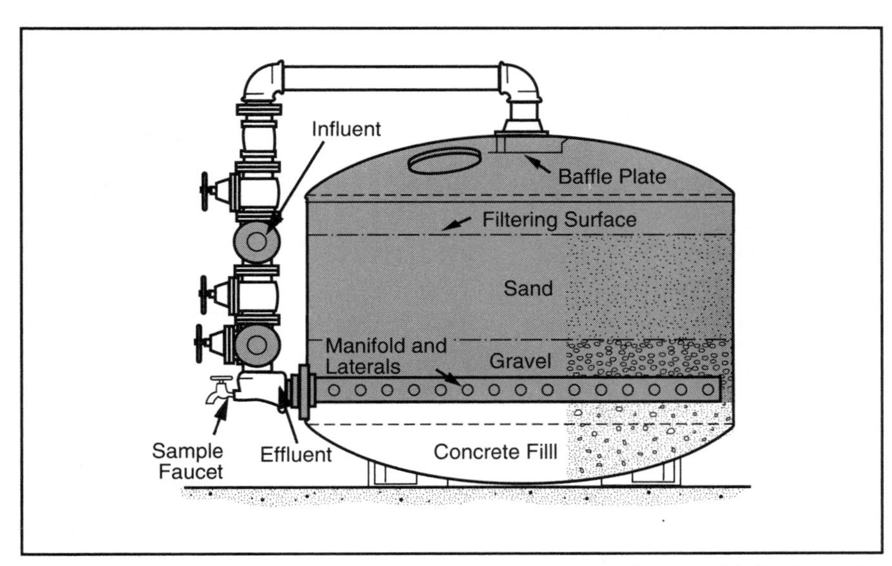

FIGURE 6-37 Vertical sand pressure filter

Courtesy of Celite Corporation

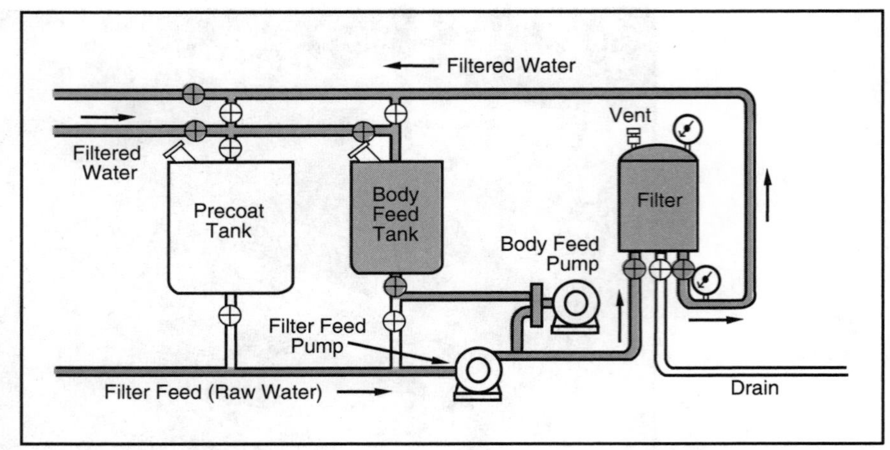

FIGURE 6-38 Components of a diatomaceous earth filter

Courtesy of Celite Corporation

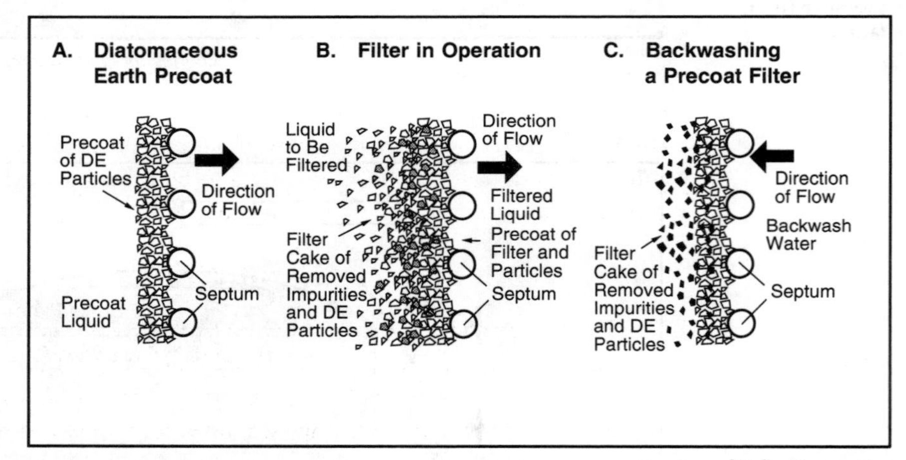

FIGURE 6-39 Operation of a diatomaceous earth filter

Courtesy of Celite Corporation

applying the precoat. Untreated water can then be applied to the filter, but to prevent possible cracking and clogging of the precoat, a body feed of diatomaceous earth must be added continuously while the filter is in operation. When the filter run is completed, the filter cake is washed from the septum by a reversal of the water flow; it is usually discharged to a lagoon for disposal.

The use of DE filters for potable water treatment has been limited because of the difficulty in maintaining an effective filter cake at all times. If

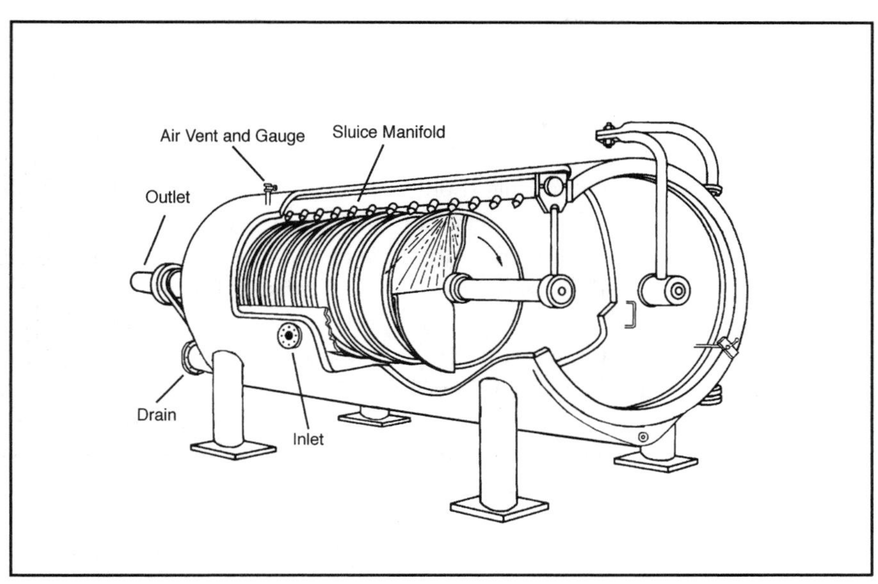

FIGURE 6-40 Precoat tank of rating leaf type

Courtesy of Celite Corporation

there is a defect in the cake at any time, microorganisms could easily pass into the water system. If DE filters are to be used for treating surface water, the filtered water must be monitored continuously.

Regulations

The Surface Water Treatment Rule (SWTR), enacted by the US Environmental Protection Agency (USEPA) in June 1989, affects the operation of essentially every public water system in the United States that uses surface water as a source. It also imposes new requirements on systems using groundwater that might become contaminated by surface water. These systems are described as having "groundwater under the direct influence of surface water," or GWUI.

In essence, the rule requires most systems using surface water to practice filtration treatment. Only those systems having a water source that is consistently very low in turbidity may be allowed to use disinfection without filtration. Many systems using groundwater will also have to add filtration treatment if they are classified as GWUI by the state. Specific filtration requirements include design and operation standards, as well as maximum allowable effluent turbidity requirements. More details on the requirements of the SWTR are included in *Water Quality*, also part of this series.

Safety Precautions

Guardrails should be installed to prevent falls into the filters. Operators should work in pairs if at all possible, and life rings or poles should be located near the filter area to allow quick rescue in case of an accident.

Pipe galleries are ideal places for accidents. They should be well ventilated, well lit, and provided with good drainage to reduce the possibility of slipping. Any ramps or stairs in this area should be equipped with nonskid treads. Painting pipes also makes them easier to see and maintain.

Polymers used for filter aids are particularly slippery, so spills should be cleaned up immediately. If the containers or tanks are leaking, they should be repaired or discarded.

Record Keeping

Good record keeping can identify problems and indicate the proper steps to be followed. The type of filtration records maintained will depend on the treatment process being used, but all records should include the following:

- rate of flow, in million gallons per day (or megaliters per day)
- head loss, in feet (or meters)
- length of filter runs, in hours
- backwash water rate, in gallons per minute (or liters/min)
- volume of wash water used, in gallons (or liters)
- volume of water filtered, in gallons (or liters)
- length of backwash, in minutes
- length of surface wash, in minutes
- filter aid dosage, in milligrams per liter

Figures 6-41 and 6-42 are examples of typical filtration record-keeping forms.

Daily Filter Record
Filter Plant No. 2

Filter No. ______ Prev. Run ________ Hours Date: _____/_____/ 19____

Oper.	Time	Rate of Flow, mgd	Loss of Head, ft	Surface Wash, min	Rate Water Wash, mgd	Water Wash, min	Polymer Feed		Wash Water Used	Water Filtered, mil gal
							On	Off		
	12 mid									
	1 a.m.									
	2									
	3									
	4									
	5									
	6									
	7									
	8									
	9									
	10									
	11									
	Noon									
	1									
	2									
	3									
	4									
	5									
	6									
	7									
	8									
	9									
	10									
	11									
	12 mid									

FIGURE 6-41 Typical daily filter record

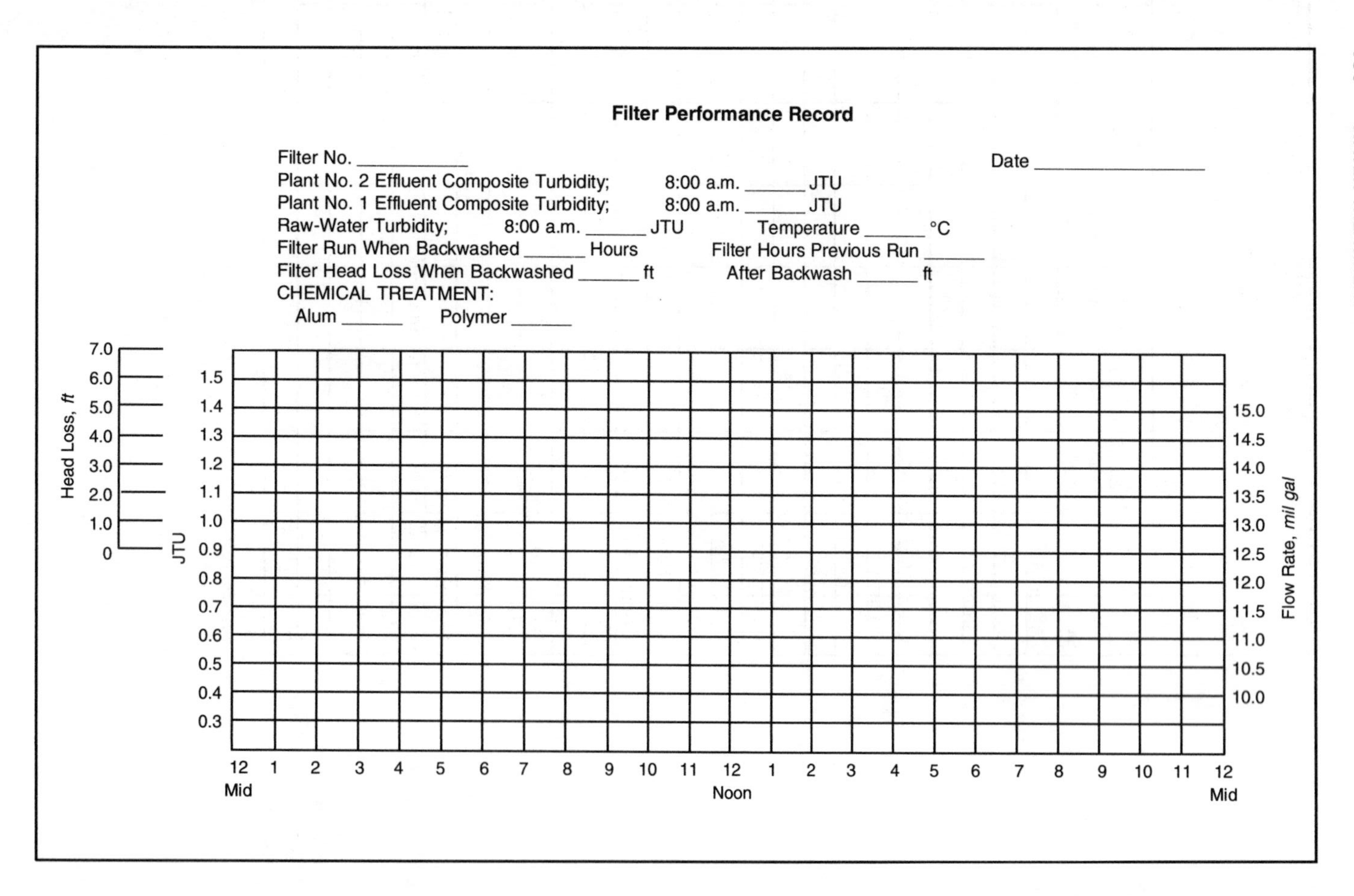

Filter Performance Record

Filter No. ______

Date ______

Plant No. 2 Effluent Composite Turbidity; 8:00 a.m. ______ JTU

Plant No. 1 Effluent Composite Turbidity; 8:00 a.m. ______ JTU

Raw-Water Turbidity; 8:00 a.m. ______ JTU Temperature ______ °C

Filter Run When Backwashed ______ Hours Filter Hours Previous Run ______

Filter Head Loss When Backwashed ______ ft After Backwash ______ ft

CHEMICAL TREATMENT:

Alum ______ Polymer ______

FIGURE 6-42 Typical filter performance record

Selected Supplementary Readings

Amirtharajah, A., N. McNelly, G. Page, and J. McLeod. 1991. ***Optimum Backwash of Dual Media Filters and GAC Filter-Adsorbers With Air Scour.*** Denver, Colo.: American Water Works Association Research Foundation and American Water Works Association.

AWWA Standard for Filtering Material, ANSI/AWWA B100. (latest edition). Denver, Colo.: American Water Works Association.

Back to Basics Guide to Slow Sand Filtration. 1993. Denver, Colo.: American Water Works Association.

Back to Basics Guide to Surface Water Treatment. 1992. Denver, Colo.: American Water Works Association.

Barrett, J.M., et al. 1991. ***Manual of Design for Slow Sand Filtration.*** Denver, Colo.: American Water Works Association Research Foundation and American Water Works Association.

Cleasby, J.L., A.H. Dharmarajah, G.L. Sindt, and E.R. Baumann. 1989. ***Design and Operation Guidelines for Optimization of the High-Rate Filtration Process: Plant Survey Results.*** Denver, Colo.: American Water Works Association Research Foundation and American Water Works Association.

Collins, M.R., and N.J.D. Graham, eds. 1994. ***Slow Sand Filtration: An International Compilation of Recent Scientific and Operational Developments.*** Denver, Colo.: American Water Works Association.

Collins, M.R., T.T. Eighmy, and J.P. Malley Jr. 1991. Evaluating Modifications to Slow Sand Filters. ***Jour. AWWA,*** 83(9):62.

Cornwell, D.A., M.M. Bishop, T.R. Bishop, N.E. McTigue, A.T. Rolan, and T. Bailey. 1991. ***Full-Scale Evaluation of Declining and Constant Rate Filtration.*** 1991. Denver, Colo.: American Water Works Association Research Foundation and American Water Works Association.

Eighmy, T., M.R. Collins, J.O. Matley Jr., J. Royce, and D. Morgan. 1993. ***Biologically Enhanced Slow Sand Filtration for Removal of Natural Organic Matter.*** 1993. Denver, Colo.: American Water Works Association Research Foundation and American Water Works Association.

Graham, N.J.D., et al. 1992. Evaluating the Removal of Color From Water Using Direct Filtration and Dual Coagulants. ***Jour. AWWA,*** 84(5):105.

Guidance Manual for Compliance With the Filtration and Disinfection Requirements for Public Water Systems Using Surface Water Sources. 1990. Denver, Colo.: American Water Works Association.

Hendricks, D.W. 1988. ***Filtration of* Giardia *Cysts and Other Particles Under Treatment Plant Conditions.*** Denver, Colo.: American Water Works Association Research Foundation and American Water Works Association.

LeChevallier, M.W., et al. 1992. Performance Evaluation of Biologically Active Rapid Filters. ***Jour. AWWA,*** 84(8):36.

Leland, D.E. 1990. Slow Sand Filtration in Small Systems in Oregon. ***Jour. AWWA,*** 82(6):50.

Letterman, R.D. 1991. ***Filtration Strategies to Meet the Surface Water Treatment Rule.*** Denver, Colo.: American Water Works Association.

Manem, J.A., and B.E. Rittmann. 1992. Removing Trace Level Pollutants in a Biological Filter. ***Jour. AWWA,*** 84(4):152.

Manual M30, Precoat Filtration. 1995. Denver, Colo.: American Water Works Association.

Manual M37, Operational Control of Coagulation and Filtration Processes. 1992. Denver, Colo.: American Water Works Association.

Manual of Instruction for Water Treatment Plant Operators. 1975. Albany, N.Y.: New York State Department of Health.

Manual of Water Utility Operations. 8th ed. 1988. Austin, Texas: Texas Water Utilities Association.

Reckhow, D.A., J.K. Edzwald, and J.E. Tobiason. 1993. ***Ozone as an Aid to Coagulation and Filtration.*** Denver, Colo.: American Water Works Association Research Foundation and American Water Works Association.

Recommended Standards for Water Works. 1992. Albany, N.Y.: Health Education Services.

Schuler, P.F., and M.M. Ghosh. 1990. Diatomaceous Earth Filtration of Cysts and Their Particulates Using Chemical Additives. ***Jour. AWWA,*** 82(12):67.

Tanner, S.A. 1990. Evaluation of Slow Sand Filters in Northern Idaho. ***Jour. AWWA,*** 82(12):51.

Tobiason, J.E., et al. 1992. Pilot Study of the Effect of Ozone and PEROXONE on In-Line Direct Filtration. ***Jour. AWWA,*** 84(12):72.

Uluatam, S.S. 1991. Assessing Perlite as a Sand Substitute. ***Jour. AWWA,*** 83(6):70.

Visscher, J.T. 1990. Slow Sand Filtration: Design, Operation, and Maintenance. ***Jour. AWWA,*** 82(6):67.

Von Huben, H. 1991. ***Surface Water Treatment: The New Rules.*** Denver, Colo.: American Water Works Association.

Water Quality and Treatment. 4th ed. 1990. New York: McGraw-Hill and American Water Works Association (available from AWWA).

CHAPTER 7

Disinfection

Disinfection is the treatment process used to destroy or inactivate disease-causing (pathogenic) organisms. Diseases caused by pathogenic organisms in water are called waterborne diseases; the more common ones are summarized in Table 7-1. The consequences of waterborne disease range from mild illness to death.

Disinfection should not be confused with sterilization. Sterilization is the destruction of *all* living microorganisms. To sterilize drinking water completely would require the application of much higher doses of chemical disinfectants, which would greatly increase operating costs. These larger doses of chemicals would create tastes that, in most cases, would be very objectionable to the public. In addition, excessive application of disinfectants usually generates excessive levels of unwanted disinfection by-products. It has been found that treatment for turbidity removal and subsequent disinfection to the extent necessary to eliminate known disease-causing organisms is sufficient to protect public health.

Destroying Pathogens in Water

Most pathogens are accustomed to living in the temperatures and conditions found in the bodies of humans and warm-blooded animals. In general, they do not survive outside of this environment, but there are some significant exceptions. Of those that do not survive for very long, significant numbers can still survive in water long enough to cause sickness, or even death, if ingested in drinking water. In addition, certain viruses and protozoans that form cysts can survive for surprisingly long periods, even under adverse conditions. Some pathogenic organisms also tend to be

TABLE 7-1 Common waterborne diseases

Waterborne Disease	Causative Organism	Source of Organism in Water	Symptom
Gastroenteritis	*Salmonella* (bacteria)	Animal or human feces	Acute diarrhea and vomiting
Typhoid	*Salmonella typhosa* (bacteria)	Human feces	Inflamed intestine, enlarged spleen, high temperature — fatal
Dysentery	*Shigella* (bacteria)	Human feces	Diarrhea — rarely fatal
Cholera	*Vibrio comma* (bacteria)	Human feces	Vomiting, severe diarrhea, rapid dehydration, mineral loss — high mortality
Infectious hepatitis	Virus	Human feces, shellfish grown in polluted waters	Yellowed skin, enlarged liver, abdominal pain — low mortality, lasts up to 4 months
Amoebic dysentery	*Entamoeba* histolytica (protozoan)	Human feces	Mild diarrhea, chronic dysentery
Giardiasis	*Giardia lamblia* (protozoan)	Animal or human feces	Diarrhea, cramps, nausea, and general weakness — not fatal, lasts 1 week to 30 weeks
Cryptosporidiosis	*Cryptosporidium* (protozoa)	Human and animal feces	Acute diarrhea, abdominal pain, vomiting, low-grade fever
Legionellosis	*Legionella pneomophila* and related bacteria		Acute respiratory illness

somewhat resistant to disinfection processes, so disinfection alone cannot always be assumed to ensure safe drinking water.

Some pathogens can be destroyed by simply storing water in open tanks for extended periods of time. Some pathogens are removed by sedimentation in those tanks, and others experience natural die-off. This is not usually a practical treatment method because of the large investment required for the storage facilities. In addition, other nuisance organisms, such as algae, can actually multiply in the water while it is in storage.

A significant number of pathogens are removed during the processes of coagulation, flocculation, sedimentation, and filtration. As a result, these processes are normally required in addition to disinfection if the source water turbidity and pathogen loading are significant. Table 7-2 lists the percentage of pathogen reduction from various treatment processes.

Detecting Pathogens in Water

Relatively simple, inexpensive tests are available for detecting the presence of coliform bacteria in water. The presence of these bacteria may indicate the presence of actual pathogens. All public water systems are required by federal and state regulations to collect representative samples from the distribution system periodically for coliform analysis. (Details of coliform sampling are included in *Water Quality*, part of this series).

However, these tests indicate only the likelihood that water is contaminated by feces from a warm-blooded animal. They do not indicate the presence of specific, harmful organisms. Unfortunately, no tests are simple and inexpensive enough to be used routinely for indicating the presence or absence of pathogens, including *Giardia*, viruses, *Legionella*, and *Cryptosporidium*.

This inability to conduct routine tests for the presence of specific disease-causing microorganisms has been recognized in the federal Surface Water Treatment Rule, which is discussed later in this chapter. In essence, the rule requires a "treatment technique" for all systems using surface water sources. The technique must consist of one or more methods of treatment that will ensure almost complete removal and/or inactivation of the most

TABLE 7-2 Pathogen reduction from various treatment processes

Unit Process	Percent Reduction
Storage*	Significant amounts
Sedimentation*	0–99
Coagulation*	Significant amounts
Filtration*	0–99
Chlorination	99

*These methods do not, in themselves, provide adequate pathogen reduction. However, their use prior to disinfection may significantly lower the costs associated with disinfection.

resistant pathogenic organisms presently known to be a threat to public health. In other words, establishing a maximum contaminant level (MCL) for pathogenic organisms is not practical because of the lack of practical tests for their presence, so compliance with regulations is based on properly operating the treatment process.

Disinfection Methods

Although chlorination is the most common disinfection method, other methods are available and can be used in various situations. The three general types of disinfection are

- heat treatment
- radiation treatment
- chemical treatment

Heat Treatment

A method of disinfection first discovered many years ago is to boil the water. It is still a good emergency procedure for small quantities of water. When contamination of a public water supply is suspected, a "boil order" should be issued to the public, suggesting that all water for consumption be boiled before use until there is further notice that the water is proven safe. Campers should also take the precaution of boiling water from surface sources or other sources not known to be safe.

For proper disinfection, the water should be maintained at a rolling boil for at least 5 minutes to ensure inactivation of the most resistant organisms. At higher altitudes, a longer boiling time should be used because water boils at a lower temperature. Boiling is obviously not well suited for large-scale use because of the high cost of energy required.

Radiation Treatment

Ultraviolet (UV) radiation can be used to disinfect water. Specially designed lamps produce UV light radiation for disinfecting water. Because UV light is readily adsorbed and scattered by turbidity in water, the water must pass very close to the lamp for the process to be effective. There are no residual effects with radiation treatment; in other words, UV light does not remain in the water to continue disinfection after its initial application. All of the killing action must occur during the brief exposure. UV light disinfection is normally used only for treating very small quantities of water.

Chemical Treatment

Although the primary use of chemical oxidants is for disinfection, these chemicals can serve other purposes during the disinfection process. In some cases, the choice of chemicals used in a treatment system is dictated by the ability of the chemicals to perform these secondary functions, which include

- control of biological growth in pipelines and basins
- control of tastes and odors
- removal of color
- reduction of some organic compounds, particularly those that are precursors to the formation of disinfection by-products
- aids to flocculation
- oxidation of iron and manganese so that they can subsequently be removed by precipitation

Chemicals used for treating potable water include the following:

- bromine
- iodine
- ozone, alone or in combination with other chemicals
- potassium permanganate
- chlorine dioxide
- chlorine and chlorine compounds
- oxygen

The general effectiveness of commonly used oxidants in treating various water problems is summarized in Table 7-3. Bromine and iodine are not covered in this table because they are rarely used for public water system treatment. Advantages and disadvantages of the principal water treatment oxidants are summarized in Table 7-4.

Bromine. Bromine is a dark reddish-brown liquid. It vaporizes at room temperature and has a penetrating, suffocating odor. The vapor is extremely irritating to the eyes, nose, and throat, and it is very corrosive to most metals. If splashed onto the skin, bromine causes painful burns that are slow to heal.

The residual formed when bromine is added to water is as effective a disinfectant as chlorine, but not as stable. Consequently, depending on the constituents in the water being treated, it may be necessary to add bromine at two or three times the concentration required for chlorine. Because of the higher cost of bromine and its handling hazards, liquid bromine is not used to disinfect public water supplies. Bromine is sometimes used in a safer, but

TABLE 7-3 General effectiveness of water treatment oxidants

Purpose	Chlorine	Chlora-mines	Ozone	Chlorine dioxide	Potassium perman-ganate	Oxygen
Iron removal	E	N	E	E	E	E
Manganese removal*	S	N	E	E	E	N
Sulfide removal	E	N	S	S	S	E†
Taste-and-odor control	S	N	E	E	S‡	S‡
Color removal	E	N	E	E	S	N
Flocculation aid	E	N	E	U	S§	N
Trihalomethane formation potential control	N	N	E**	E	S	N
Synthetic organics removal	S††,‡‡	N	S††	S††	S††	N
Biological growth control	E	S	N§§	E	S	N

Source: Water Quality and Treatment. *4th ed. 1990.*

NOTES: E = effective, S = somewhat effective, N = not effective, U = unknown.

*Above pH 7.

†By stripping.

‡Except earthy–musty odor-causing compounds.

§May involve adsorption on MnO_2.

**May increase problem at low doses.

††Depending on compound.

‡‡May form chlorinated by-products.

§§Except with dual-stage ozonation.

more costly, solid "stick form" (organobromine compound) to disinfect swimming pools.

Iodine. Iodine is a lustrous, blue-black solid that is about five times the density of water and has a peculiar chlorine-like odor. The solid can quickly change to a gas, releasing a characteristic violet vapor. Iodine has been used extensively for medicinal purposes.

Because of its possible adverse health effects, long-term consumption of iodine is not recommended. For this reason, it is not used as a disinfectant for water supplies serving permanent populations. However, it is occasionally used for water disinfection at campgrounds and other locations where use by most persons is limited to about two weeks. It can also be used for emergency disinfection of water and is available in small tablets for

TABLE 7-4 Advantages and disadvantages of water treatment oxidants

Oxidant	Advantages	Disadvantages
Chlorine	Strong oxidant Simple feeding Persistent residual Long history of use	Chlorinated by-products Taste-and-odor problems possible pH influences effectiveness
Chloramines	No trihalomethane formation Persistent residual Simple feeding Long history of use	Weak oxidant Some total organic halide formation pH influences effectiveness Taste, odor, and growth problems possible
Ozone	Strong oxidant Usually no trihalomethane or total organic halide formation No taste-or-odor problems Some by-products biodegradable Little pH effect Coagulant aid	Short half-life On-site generation required Energy intensive Some by-products biodegradable Complex generation and feeding Corrosive
Chlorine dioxide	Strong oxidant Relatively persistent residual No trihalomethane formation No pH effect	Total organic halide formation ClO_3 and ClO_2 by-products On-site generation required Hydrocarbon odors possible
Potassium permanganate	Easy to feed No trihalomethane formation	Moderately strong oxidant Pink H_2O By-products unknown Causes precipitation
Oxygen	Simple feed No by-products Companion stripping Nontoxic	Weak oxidant Corrosion and scaling

disinfecting small quantities of water. Crystalline iodine is available for use in saturator-type feeders.

Ozone. Ozone (O_3) cannot be stored, so it must be manufactured on-site as needed. It is created by passing a high voltage through the air between two electrodes. It can also be formed photochemically in the atmosphere and is one of the constituents of smog. Ozone is a bluish, toxic gas with a pungent odor. It is considered hazardous to health at relatively low concentrations in air. The threshold odor level is 0.05 ppm and the 8-hour Occupational Safety and Health Administration (OSHA) standard is 0.1 ppm.

Ozone has been used widely in Europe and on a limited basis in the United States for drinking water treatment. Federal and state rules in the United States now require that a disinfectant residual be maintained in the distribution system. As a result, water systems using ozone as a primary disinfectant must still add a small dose of chlorine or chloramine as the water enters the distribution system.

Ozone is a powerful oxidizing agent when used in water treatment. It is effective not only as a disinfectant, but also for controlling color, taste, and odor. The gas is chemically unstable and disappears in just a few minutes, leaving no residual disinfectant to continue further disinfecting action. Because of the lack of a long-lasting residual, it is difficult to determine whether enough ozone has been added to destroy all pathogens.

The electrical energy required to generate ozone is quite large, but the greater expense of generating ozone can often be offset by its effectiveness in controlling taste, odor, and color and in oxidizing organic substances. Because of increasing concern and regulation of the by-products of chlorine (trihalomethanes and possibly other disinfection by-products [DBPs] in the future), the use of ozone is gaining greater acceptance in the United States. Another advantage is that ozone often improves coagulation, which has the effect of lowering the cost of coagulant chemicals, reducing sludge disposal costs, and lengthening filter runs. Research is still ongoing as to whether there are any adverse health effects of the DBPs formed during the ozonation process.

An emerging technology concerns the use of ozone along with other oxidants. One recent development uses ozone along with hydrogen peroxide; this is called PEROXONE. Tests have shown that, in some cases, PEROXONE may be more effective than ozone alone both as a disinfectant and in reducing DBPs.

Potassium permanganate. Potassium permanganate ($KMnO_4$) is an oxidant that was first used in 1910 for water treatment in London. However, widespread use of potassium permanganate did not occur until the 1960s, when its effectiveness for controlling tastes and odors had become recognized.

When added to water, permanganate turns the water purple until it finally dissipates after the completion of the oxidizing action. For this reason, it should be fed as early in the treatment process as possible to allow completion of the reaction before the water enters the distribution system. Concentrations as low as 0.05 ppm may still have some noticeable color.

Permanganate is frequently used as the initial chemical fed into surface water systems because it will control many taste-and-odor-causing substances in the raw water. In particular, it will eliminate a number of taste-and-odor conditions that will not be controlled or may be accentuated by chlorine. It also works well in removing hydrogen sulfide, iron, and manganese. When permanganate alone is not completely successful in controlling tastes and odors, it may be used in combination with activated carbon.

Permanganate is also being used for trihalomethane (THM) control. The main cause of THMs is chlorination of precursors, primarily humic and fulvic acids, found in raw water. Feeding permanganate as the initial oxidant allows chlorine to be applied later in the treatment process, when the precursors have been reduced. THMs can usually be significantly reduced by this process.

Permanganate is also widely used by groundwater systems to oxidize iron and manganese (see chapter 10). It has also been found effective for controlling zebra mussels and algae in reservoirs (see chapter 2).

Although potassium permanganate is reported by many water systems to reduce coliforms, it is not registered with the US Environmental Protection Agency (USEPA) as a disinfectant. Some laboratory tests have shown its effectiveness against certain microorganisms, but its effectiveness against *Giardia lamblia* and viruses is still under study. Surface water systems electing to use permanganate as a disinfectant should contact their state drinking water agency for advice on the $C \times T$ (concentration multiplied by time) credit allowed under the Surface Water Treatment Rule.

Chlorine dioxide. Chlorine dioxide (ClO_2) is a powerful oxidant that is generally prepared on-site. It is used by some water systems as an initial oxidant for water having high humic and fulvic substance levels, in order to reduce the formation of THMs. In some cases, it is used only when the THM precursor level in the raw water is high.

During the formation of ClO_2, a small amount of chlorate and chlorite is formed. There is evidence that these substances can cause adverse health effects in some people. Water systems considering the use of chlorine dioxide should check with their state public water supply control agency for current requirements or restrictions on use.

Chlorine dioxide is capable of oxidizing iron and manganese, removing color, and lowering THM formation potential. It also oxidizes many organic and sulfurous compounds that cause taste-and-odor problems. Chlorine dioxide does add a specific taste to water that is objectionable to some people. The maximum residual that does not cause tastes or odors is usually about 0.4–0.5 mg/L as ClO_2.

All chemicals used for disinfection must conform to standards set by the American Water Works Association and NSF International. See appendix A for details.

Oxygen. Oxygen is not effective in pathogen reduction, so it is not used for this purpose. It is frequently used to oxidize various contaminants for removal. The introduction of oxygen into water by aeration is covered in chapter 14.

Chlorine and Chlorine Compounds

Chlorination, the addition of chlorine to water, is the most common form of disinfection currently practiced in the United States. When properly understood and correctly operated, the chlorination process is a safe, practical, and effective way to destroy disease-causing organisms.

Several secondary benefits are gained from using chlorine as the disinfectant for treated water, and chlorine may also be used as part of other treatment processes. Chlorine is useful for disinfecting storage tanks and pipelines; for oxidizing iron, manganese, and hydrogen sulfide; and for controlling tastes, odors, algae, and slime. These uses are discussed in greater detail in subsequent chapters.

Chlorine is available in gaseous, liquid, and solid forms. The chemicals and equipment used for chlorination depend primarily on the type of chlorine used.

Chlorine Chemicals

Hypochlorous acid (HOCl) and hypochlorite ion (OCl^-) are the most effective residuals. They can be derived from the following three chemicals:

- chlorine, Cl_2
- calcium hypochlorite, $Ca(OCl)_2$
- sodium hypochlorite, NaOCl

A comparison of the chlorine content for these chemicals is given in Table 7-5.

TABLE 7-5 Chlorine content of common disinfectants

Compound	Percentage Cl	Amount of Compound Needed to Yield 1 lb of Pure Cl
Chlorine gas or liquid (Cl_2)	100	1 lb (0.454 kg)
Sodium hypochlorite (NaOCl)*	15	0.8 gal (3 L)
	12.5	1.0 gal (3.8 L)
	5	2.4 gal (9.1 L)
	1	12.0 gal (45.4 L)
Calcium hypochlorite [$Ca(OCl)_2$]	65	1.54 lb (0.7 kg)

*Sodium hypochlorite is available in four standard concentrations of available chlorine. Ordinary household bleach contains 5% chlorine.

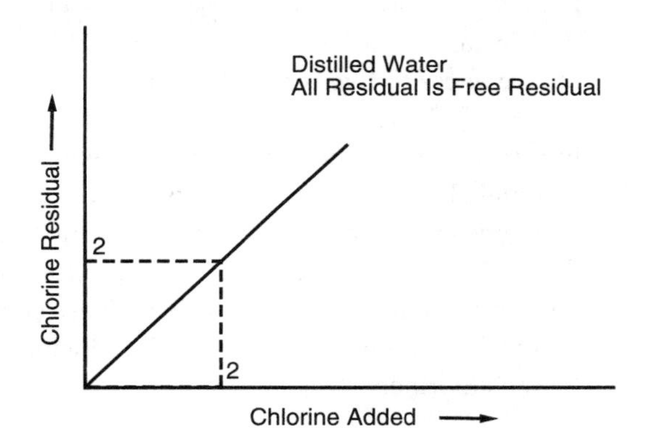

FIGURE 7-1 The reaction of chlorine in distilled water

Chemistry of Chlorination

To understand reactions of chlorine in natural water, consider the reaction of chlorine in distilled water. As shown in Figure 7-1, the amount of free chlorine residual is directly related to the amount (dose) of chlorine added. For example, if 2 mg/L of chlorine is added, 2 mg/L of free residual is produced. The reactions that occur are as follows (the chemistry of chlorination is discussed in detail in *Basic Science Concepts and Applications*, part of this series):

$$\underset{\text{chlorine}}{Cl_2} + \underset{\text{water}}{H_2O} \rightarrow \underset{\text{hypochlorous acid}}{HOCl} + \underset{\text{hydrochloric acid}}{HCl} \quad (7\text{-}1)$$

The products are weak compounds that dissociate as follows:

$HOCl$	$\rightarrow$	H^+	+	OCl^-	(7-2)
hypochlorous acid		hydrogen ion		hypochlorite ion	

HCl	$\rightarrow$	H^+	+	Cl^-	(7-3)
hydrochloric acid		hydrogen ion		chlorine ion	

Hypochlorous acid, one of two forms of free chlorine residual, is the most effective disinfectant available. When it dissociates as in Eq 7-2, the hypochlorite ion (the second form of free chlorine residual) is formed. The hypochlorite ion is only 1 percent as effective as hypochlorous acid as a disinfectant. This is indicated in Table 7-6, which lists the estimated effectiveness of five types of residuals.

Natural water is not pure, and the reaction of chlorine with the impurities in the water interferes with the formation of a free chlorine residual. For example, if the water contains organic matter, nitrites, iron, manganese, and ammonia, then the chlorine added will react as shown in Figure 7-2. Between points 1 and 2, added chlorine combines immediately with iron, manganese, and nitrites. These chemicals are reducing agents, and no residual can be formed until all reducing agents are completely destroyed by the chlorine.

TABLE 7-6 Estimated effectiveness of types of residual chlorine

Type	Chemical Abbreviation	Estimated Effectiveness Compared With HOCl
Hypochlorous acid	$HOCl$	1
Hypochlorite ion	OCl^-	1/100
Trichloramine*	NCl_3	†
Dichloramine	$NHCl_2$	1/80
Monochloramine	NH_2Cl	1/150

*Commonly called nitrogen trichloride.
†No estimate; possibly more effective than dichloramine.

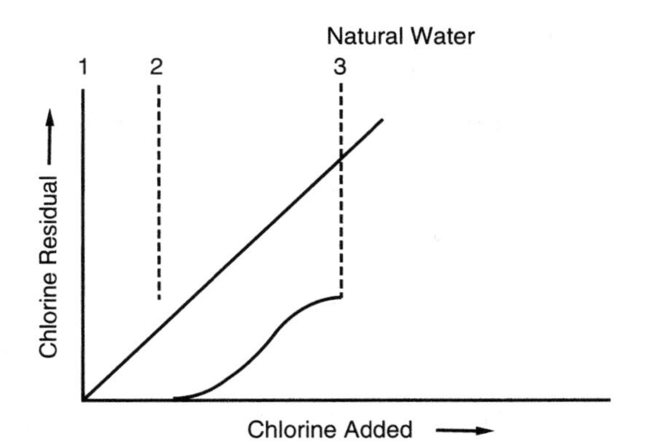

FIGURE 7-2 The reaction of chlorine with impurities in natural water

As more chlorine is added between points 2 and 3, the chlorine begins to react with ammonia and organic matter to form chloramines and chloroorganic compounds. These are called combined chlorine residuals. Because the chlorine is combined with other compounds, this residual is not as effective as a free chlorine residual.

Between points 2 and 3, the combined residual is primarily monochloramine, that is,

$$\underset{\text{ammonia}}{NH_3} + \underset{\text{hypochlorous acid}}{HOCl} \rightarrow \underset{\text{monochloramine}}{NH_2Cl} + \underset{\text{water}}{H_2O} \quad (7\text{-}4)$$

Adding more chlorine to the water actually decreases the residual (Figure 7-3). The decrease (shown from point 3 to point 4) results because the additional chlorine oxidizes some of the chloroorganic compounds and ammonia. The additional chlorine also changes some of the monochloramine to dichloramine and trichloramine.

$$\underset{\text{monochloramine}}{NH_2Cl} + \underset{\text{hypochlorous acid}}{HOCl} \rightarrow \underset{\text{dichloramine}}{NHCl_2} + \underset{\text{water}}{H_2O} \quad (7\text{-}5)$$

$$\underset{\text{dichloramine}}{NHCl_2} + \underset{\text{hypochlorous acid}}{HOCl} \rightarrow \underset{\text{trichloramine}}{NCl_3} + \underset{\text{water}}{H_2O} \quad (7\text{-}6)$$

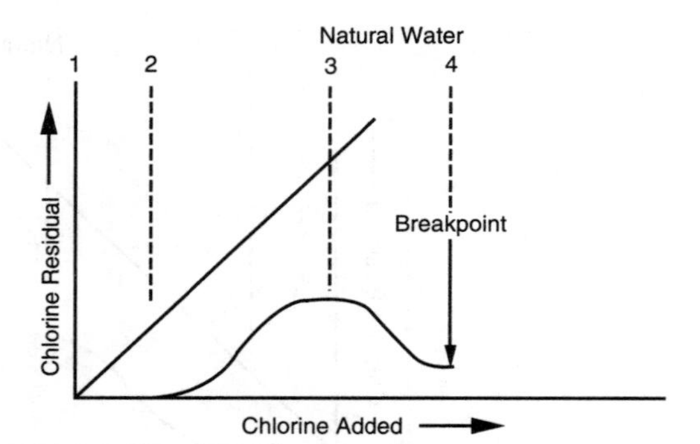

FIGURE 7-3
Decrease of chlorine residual

As additional chlorine is added between points 3 and 4, the amount of chloramine reaches a minimum value. Beyond this minimum point, the addition of more chlorine produces free residual chlorine. The point at which this occurs (point 4 in Figure 7-4) is known as the *breakpoint*.

Past the breakpoint, an increase in the chlorine dose will usually produce a proportionate increase in the free chlorine residual; the free chlorine residual should be 85–90 percent of the total chlorine residual. The remaining percentage is combined residual consisting of dichloramines, trichloramines, and chloroorganic compounds. One group of the chloroorganic compounds — trihalomethanes — is discussed later in this chapter.

Principle of Disinfection by Chlorination

The five factors important to the success of chlorination are

- concentration of chlorine
- contact time between the chlorine and water
- temperature of the water
- pH of the water
- foreign substances in the water

Concentration and contact time. The effectiveness of chlorination depends primarily on two factors, concentration C and contact time T. The destruction of organisms, often referred to as the "kill," is directly related to these two factors as follows:

$$\text{kill is proportional to } C \times T$$

This means that if the chlorine concentration is decreased, then the contact time — the length of time the chlorine and the organisms are in

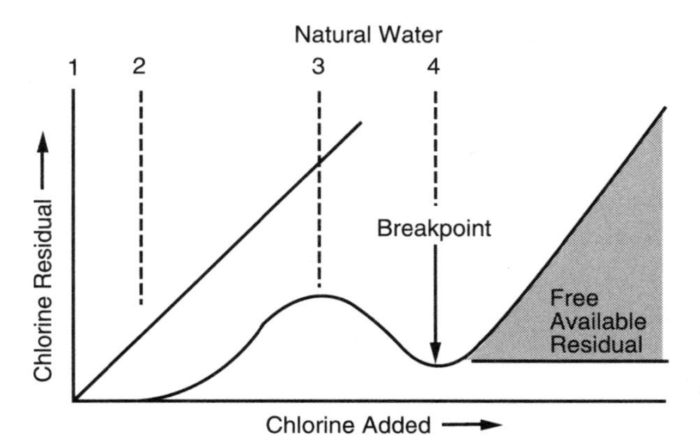

FIGURE 7-4 The chlorine breakpoint

physical contact — must be increased to ensure that the kill remains the same. Similarly, as the chlorine concentration increases, the contact time needed for a given kill decreases.

A combined chlorine residual, which is a weak disinfectant, requires a greater concentration, acting over a longer period of time, than is required for a free chlorine residual. Therefore, when the contact time between the point of chlorine application and the consumption of water by customers is short, only a free residual will provide effective disinfection. It is important to know the contact time and type of residual chlorine available so that the proper concentration can be provided. Figure 7-5, which shows how many minutes are needed by different residual concentrations to achieve 99 percent destruction of *E. coli* at 2–6°C, illustrates this point. In general, a minimum free chlorine residual of 0.2 mg/L should be maintained at the extremities of the distribution system.

Temperature. The effectiveness of chlorination is also related to the temperature of the water. At lower temperatures, bacterial kill tends to be slower. However, chlorine is more stable in cold water, and the residual will remain for a longer period of time, compensating to some extent for the lower rate of disinfection. Other factors being equal, chlorination is more effective at higher water temperatures.

It is important for the operator to maintain a record of water temperatures. As temperatures change seasonally, the chlorine dosage will also need to be adjusted. The effectiveness of combined chlorine residuals is influenced more by low temperatures than that of free chlorine residuals.

pH. The pH of the water affects the disinfecting action of chlorine because it determines the ratio of $HOCl$ to OCl^-. In other words, depending

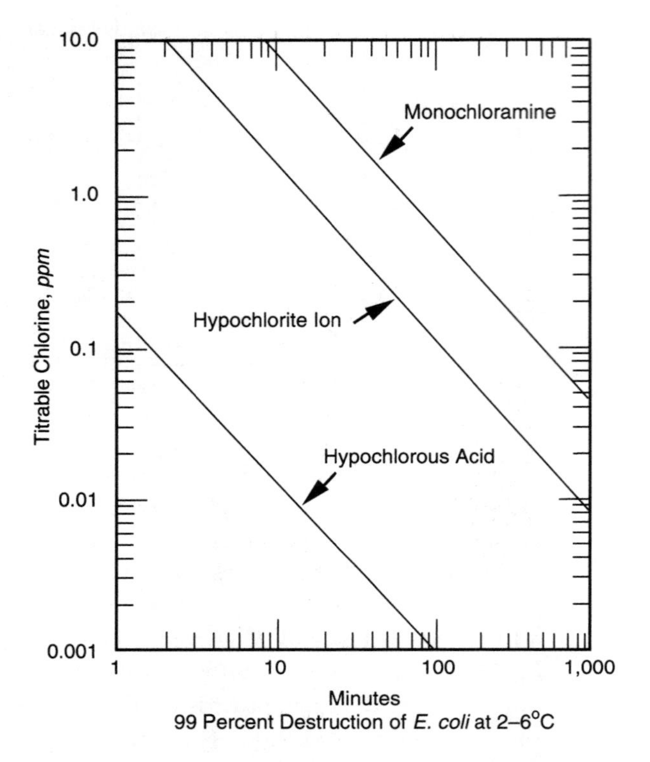

FIGURE 7-5 Efficiency of hypochlorous acid, hypochlorite ion, and monochloramine as disinfectants

From Handbook of Chlorination *by Geo. Clifford White, copyright © 1972 by Van Nostrand Company. Reprinted by permission of the publisher.*

on the pH, either more hypochlorite ion or more hypochlorous acid could be present. As shown in Figure 7-6, the ratio of the ions will shift as the pH changes.

Hypochlorous acid dissociates poorly at low pH levels. The dominant residual is then HOCl. However, HOCl will dissociate almost completely at high pH levels, leaving OCl^- as the dominant residual. Note in Figure 7-6 that temperature has very little effect on the dissociation at various pH levels.

Figure 7-7 summarizes the effects of pH on free and combined residuals. It is essential that the operator understand and use these relationships in order to obtain the most effective disinfectant. The pH of the water should be checked routinely. This is particularly important if the pH of the water is being raised to control corrosion, because the chlorine dosage will have to be raised to maintain an effective level. Addition of chlorine gas lowers the pH of the water. The use of hypochlorites raises the pH slightly.

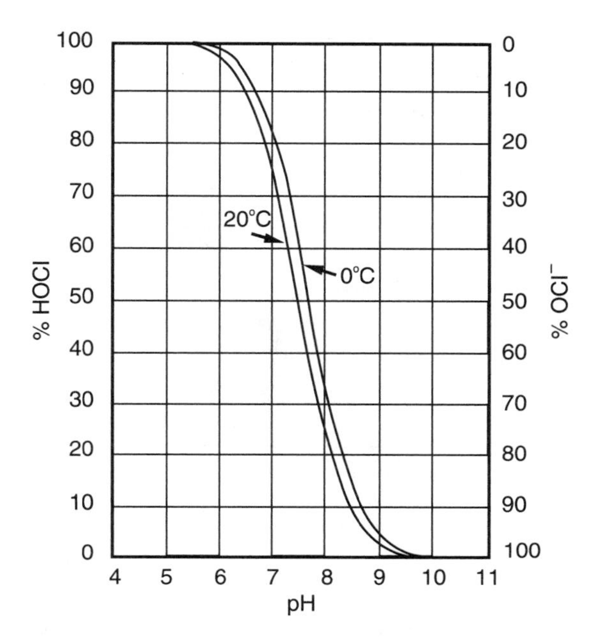

FIGURE 7-6 Distribution of HOCl and OCl^- in water at indicated pH levels

Interference Substances

Chlorine acts as an effective disinfectant only if it comes in contact with the organisms to be killed. Turbidity, caused by tiny particles of dirt and other impurities suspended in the water, can prevent good contact and protect the pathogens. Therefore, for chlorination to be effective, turbidity must be reduced as much as possible through the use of coagulation, flocculation, and filtration.

As discussed earlier, chlorine reacts with other substances in water, such as organic matter and ammonia. Because these compounds result in the formation of the less-effective combined residuals, their concentrations are an important factor in determining chlorine dosages.

Superchlorination and Dechlorination

The process of superchlorination has generally been used in the treatment of poor-quality water, including water with high ammonia concentrations or severe taste-and-odor problems. In these cases, chlorine is added beyond the breakpoint, which oxidizes the ammonia nitrogen present.

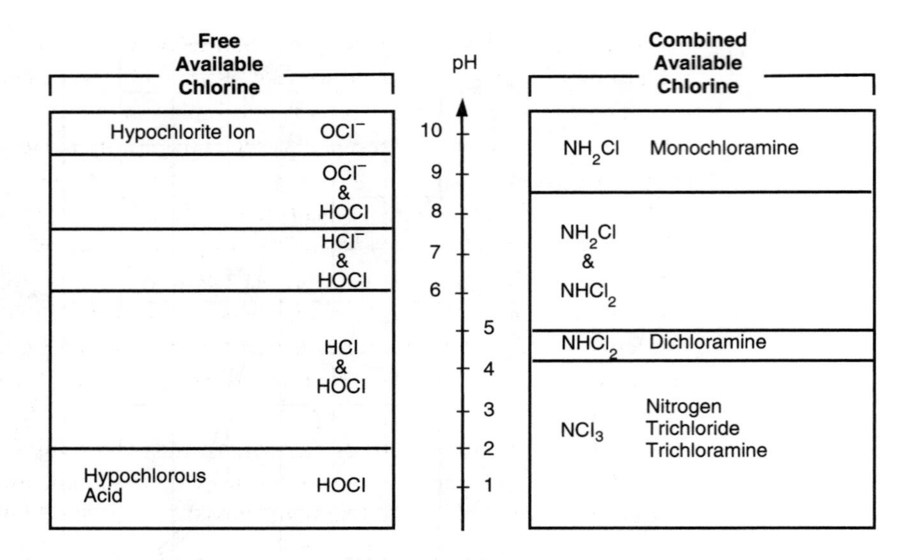

FIGURE 7-7
Effects of pH on free and combined chlorine residual

The residual chlorine present at this point is usually higher than desired for the distribution systems, so the residual may have to be decreased before the water leaves the treatment plant. The chlorine residual can be reduced by dosing the water with a substance that reacts with the residual chlorine or accelerates its rate of decomposition. Compounds that can perform this function include thiosulfate, hydrogen peroxide, and ammonia. However, the chemicals most commonly used for plant-scale dechlorination are sodium bisulfite, sodium sulfite, and sulfur dioxide. Dechlorination can also be accomplished by passing the water through a bed of activated carbon.

Chloramination

Chloramine can be formed as a result of the reaction between applied chlorine and ammonia present in raw water. It can also be formed in the treatment process by the addition of ammonia to react with chlorine so that a combined chlorine residual will be formed. This practice has been used by many water systems for over 70 years. In practice, the ammonia can be added before, at the same time as, or after the chlorine feed.

Chloramine has been used principally in systems requiring a reduction in tastes and odors, particularly where the raw water may contain phenol. Free chlorine normally reacts with phenol to form chlorophenol, which has a very disagreeable taste and odor. Many systems have also found that chloramine can be used to reduce the THM level in their water.

The primary disadvantage of chloramine is that it is a much weaker disinfectant than free chlorine, chlorine dioxide, or ozone. It is particularly weak for inactivating certain viruses. In most water systems that use chloramine as the principal disinfectant, the ammonia is added at a point downstream from the initial chlorine application so that microorganisms, including viruses, will be exposed to the free chlorine for a short period before the chloramine is formed. In order to meet SWTR requirements, systems using chloramine must carry considerably higher residuals or provide a longer contact time than would be necessary if they were using free chlorine.

If a system changes from free chlorine to chloramine for disinfection, hospitals and kidney dialysis centers must be alerted. Cases of chloramine-induced hemolytic anemia in patients have been reported when their dialysis water was not appropriately treated.

Disinfectant Application Points

Disinfectants are commonly applied at two points: where the raw water enters a treatment plant and again after treatment has been completed. A growing number of treatment plants have also found it advantageous to add disinfectants at intermediate points in the treatment process.

Source Water Chlorination

As illustrated in Figure 7-8, most surface water systems apply chlorine (or alternative disinfectants) at two points. Source water chlorination (prechlorination) is performed to

- begin the process of killing and/or inactivating pathogenic organisms.
- minimize operational problems and tastes and odors that could be caused by biological growths on filters, pipes, and basins.
- oxidize hydrogen sulfide, iron, and manganese that may be in the raw water.
- oxidize various organic substances in the raw water.

Historically, the most common point for prechlorinating surface water is the intake well or rapid-mix basin. Because the formation of THMs and other DBPs is a concern, many systems are moving the application point to later in the treatment process and/or feeding a different chemical oxidant as the water enters the treatment plant.

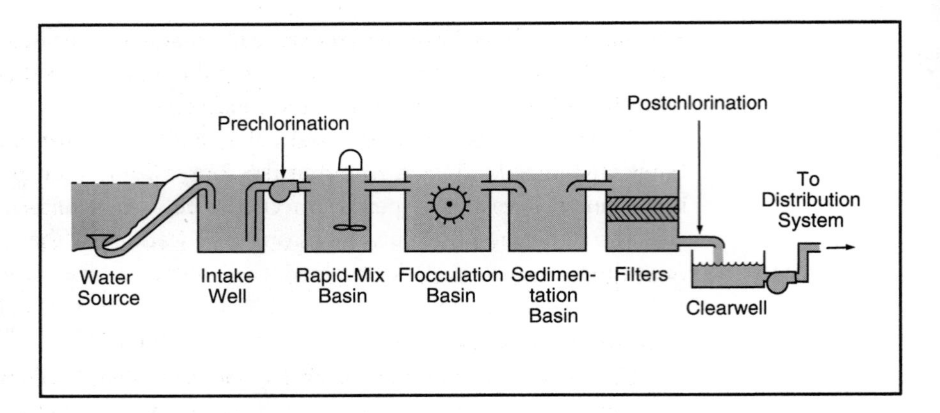

FIGURE 7-8 Common chlorination points in a conventional filtration plant

Postchlorination

Terminal disinfection (postchlorination) is the application of chlorine to treated water. This is necessary to meet federal and state requirements for maintaining a minimum chlorine residual in the water, both at the entry to the distribution system and at points throughout the distribution system. The presence of a residual indicates that a more than adequate amount of disinfectant has been added to complete the reaction. The residual is also considered to be a safeguard against contamination that could be introduced into the distribution system or customer plumbing systems.

Postchlorination is usually performed immediately before the clearwell or immediately before the sand filters. Although the clearwell is intended to provide some contact time with the chlorine to ensure adequate inactivation of pathogens, most clearwells do not have baffles, so there could be a short-circuiting of flow. To meet the $C \times T$ requirements of the SWTR, some systems may find it necessary to install baffles in order to ensure adequate contact time before the water enters the distribution system.

Systems having difficulty in maintaining the required chlorine residual in the distribution system may find it necessary to install booster chlorination facilities. A particularly good location for adding more chlorine is on the discharge from a storage reservoir.

Additional Application Points

In the process of balancing the multiple requirements of minimizing tastes and odors, reducing DBPs, and meeting the SWTR requirements, many systems have found it advantageous to use two or more types of disinfectants and to feed them at different points during the treatment process.

Figure 7-9 illustrates changes typically made to a surface water system to reduce THMs; permanganate is applied at the intake, and the prechlorination point is moved to a later stage of the treatment process.

Figure 7-10 illustrates a system that has raw-water storage tanks. An initial disinfectant dose is added at the source, another application is added before treatment, and a third is applied as the water enters the distribution system. This situation is often ideal for meeting *C×T* requirements for poor-quality water while still minimizing DBPs.

Groundwater Systems

Unless special treatment such as iron removal or softening is required, groundwater systems generally pump water directly from wells either to the

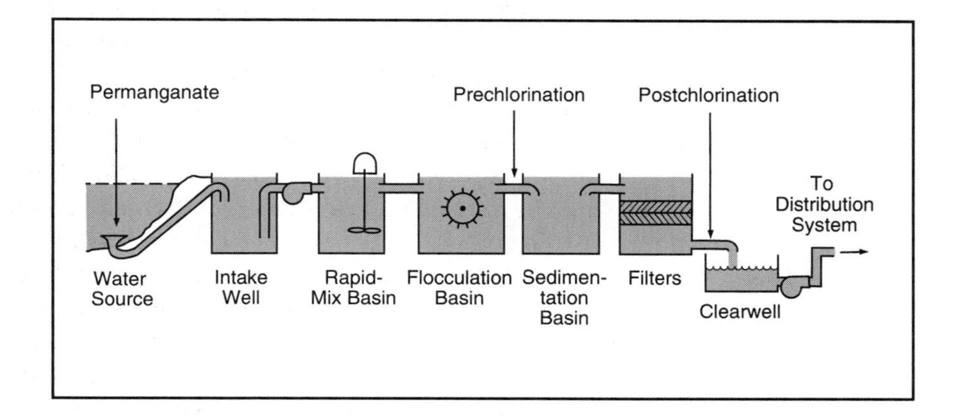

FIGURE 7-9 Use of multiple oxidants to minimize disinfection by-products

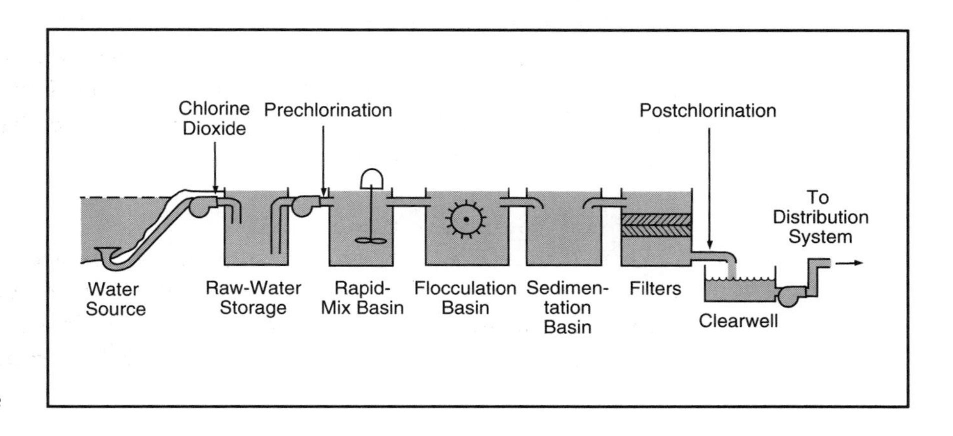

FIGURE 7-10 Initial oxidant feed before raw-water storage

distribution system or into a storage reservoir that pumps to the system. In either case, the chlorination point is generally located immediately past the wellhead.

Historically, most groundwater systems that required no special treatment did not chlorinate. In recent years, many states have enacted a requirement for all community public water supplies to chlorinate. Planned federal regulations will, within a few years, require chlorination of all public water supplies; there will be some allowance for the states to provide an exemption to some systems under certain conditions.

Use of Multiple Disinfectants

There is a growing tendency for water systems to use multiple disinfectants. Many systems feed ozone, chlorine dioxide, or potassium permanganate as the initial oxidant and then apply chlorine later in the treatment process.

Gas Chlorination Facilities

Chlorine gas, Cl_2, is about 2.5 times as dense as air. It has a pungent, noxious odor and a greenish-yellow color, although it is visible only at a very high concentration. The gas is very irritating to the eyes, nasal passages, and the respiratory tract, and it can kill a person in a few breaths at concentrations as low as 0.1 percent (1,000 ppm) by volume. Its odor can be detected at concentrations above 0.3 ppm.

Chlorine liquid is created by compressing chlorine gas. The liquid, which is about 99.5 percent pure chlorine, is amber in color and about 1.5 times as dense as water. It can be purchased in cylinders, containers, tank trucks, and railroad cars (Figures 7-11 through 7-14).

Liquid chlorine changes easily to a gas at room temperatures and pressures. One volume of liquid chlorine will expand to about 460 volumes of gas. Dry chlorine gas will not corrode steel or other metals, but it is extremely corrosive to most metals in the presence of moisture.

Chlorine will not burn. But, like oxygen, *it will support combustion* — that is, it takes the place of oxygen in the burning of combustible materials. Chlorine is not explosive, but it will react violently with greases, turpentine, ammonia, hydrocarbons, metal filings, and other flammable materials. Chlorine will not conduct electricity, but the gas can be very corrosive to exposed electrical equipment. Because of the inherent hazards involved, chlorine requires special care in storage and handling.

Handling and Storing Chlorine Gas

Safe handling and storage of chlorine are vitally important to the operator and to the communities immediately surrounding a treatment plant. An error or accident in chlorine handling can cause serious injuries or even fatalities.

The containers commonly used to supply chlorine in smaller water treatment plants are 150-lb (68-kg) cylinders. Larger plants find it more

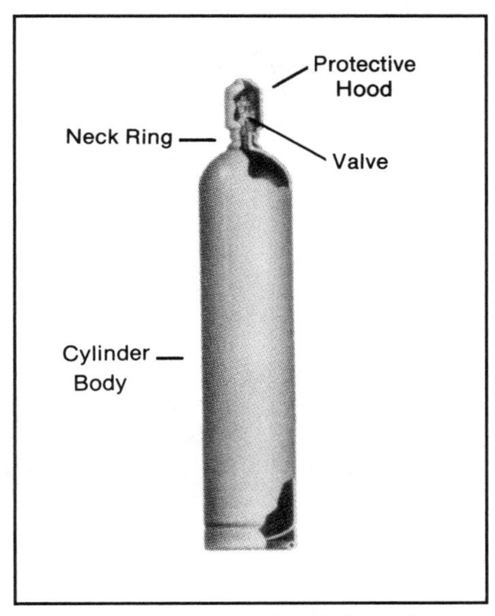

FIGURE 7-11 Chlorine cylinder

Courtesy of the Chlorine Institute, Inc.

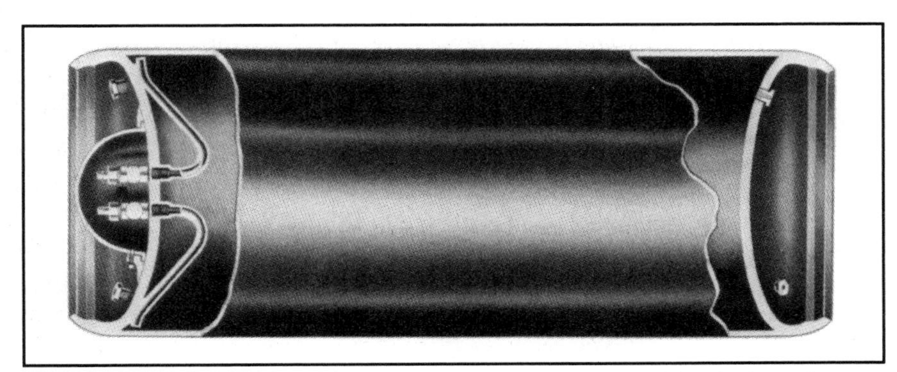

FIGURE 7-12 Chlorine ton container

Courtesy of the Chlorine Institute, Inc.

FIGURE 7-13 Chlorine ton container truck

Courtesy of PPG Industries, Inc.

FIGURE 7-14 Chlorine tank car

Courtesy of the Chlorine Institute, Inc.

economical to use ton containers. Some very large plants are equipped to draw chlorine directly from tank cars.

The decision of whether to use cylinders or ton containers should be based on cost and capacity. The cost per pound (kilogram) of chlorine in cylinders is usually substantially more than that of chlorine in ton containers. If a plant's needs for chlorine are lower than 50 lb/d (23 kg/d), cylinders should usually be selected. For systems that use large amounts, ton containers will probably be more economical.

Cylinders

Chlorine cylinders hold 150 lb (68 kg) of chlorine and have a total filled weight of 250–285 lb (110–130 kg). They are about 10.5 in. (270 mm) in diameter and 56 in. (1.42 m) high. As illustrated in Figure 7-11, each cylinder is equipped with a hood that protects the cylinder valve from damage during shipping and handling. The hood should be properly screwed in place whenever a cylinder is handled, and should be removed only during use.

Cylinders are usually delivered by truck. Each cylinder should be unloaded to a dock at truck-bed height if possible. If a hydraulic tailgate is used, the cylinders should be secure to keep them from falling. Cylinders must never be dropped, including "empty" cylinders, which actually still contain some chlorine.

The easiest and safest way to move cylinders in the plant is with a hand truck. As shown in Figure 7-15, the hand truck should be equipped with a restraining chain that fastens snugly around the cylinder about two-thirds of the way up. Slings should never be used to lift cylinders, and a cylinder should never be lifted by the protective hood because the hood is not designed to support the weight of the cylinder.

Cylinders should not be rolled to move them about a plant. Tipping the cylinders over and standing them up can lead to employee injury. In addition, the rolled cylinders might strike something that could break off the valve.

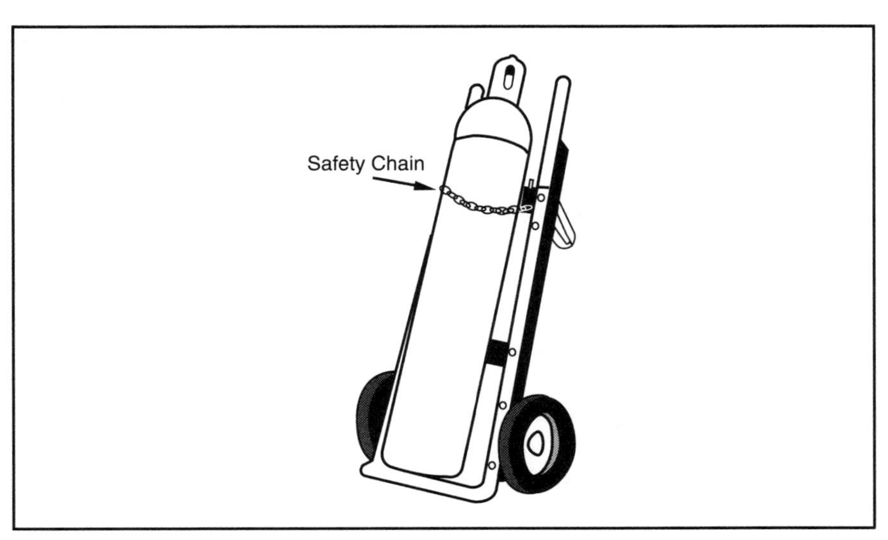

FIGURE 7-15 Hand truck for moving chlorine cylinders

Courtesy of the Chlorine Institute, Inc.

Cylinders can be stored indoors or outdoors. If cylinders are stored indoors, the building should be fire resistant, have multiple exits with outward-opening doors, and be adequately ventilated. Outdoor storage areas must be fenced and protected from direct sunlight, and they should be protected from vehicles or falling objects that might strike the cylinders. If standing water accumulates in an outdoor storage area, the cylinders should be stored on elevated racks. Avoiding contact with water will help minimize cylinder corrosion.

Some operators find it convenient to hang "full" or "empty" identification tags on cylinders in storage, so that the status of the chlorine inventory can be quickly determined. Other plants maintain separate storage areas for full and empty cylinders, but all cylinders, full or empty, should receive the same high level of care. In addition, protective hoods should be placed on empty and full cylinders in storage. Even when a cylinder no longer has sufficient chlorine for plant use, a small amount of gas remains and could escape if the cylinder or valve were damaged. Both full and empty cylinders should always be stored upright and secured with a chain to prevent them from tipping over.

Ton Containers

The ton container is a reusable, welded tank that holds 2,000 lb (910 kg) of chlorine. Containers weigh about 3,700 lb (1,700 kg) when full and are generally 30 in. (0.76 m) in diameter and 80 in. (2.03 m) long. As shown in Figure 7-12, the ends are concave. The container is crimped around the perimeter of the ends, forming good gripping edges for the hoists used to lift and move them. The ton container is designed to rest horizontally both in shipping and in use. It is equipped with two valves that provide the option of withdrawing either liquid or gaseous chlorine. The upper valve will draw gas, and the lower valve will draw liquid.

Handling the heavy containers is, by necessity, far more mechanized than handling cylinders. Containers are loaded or unloaded by a lifting beam in combination with a manual or motor-operated hoist mounted on a monorail that has a capacity of at least 2 tons (1,815 kg) (Figure 7-16). To prevent accidental rolling, containers are stored on trunnions, as illustrated in Figure 7-17. The trunnions allow the container to be rotated so that it can be positioned correctly for connection to the chlorine supply line.

Ton containers can be stored indoors or outdoors, and require the same precautions as chlorine cylinders. The bowl-shaped hood that covers the two

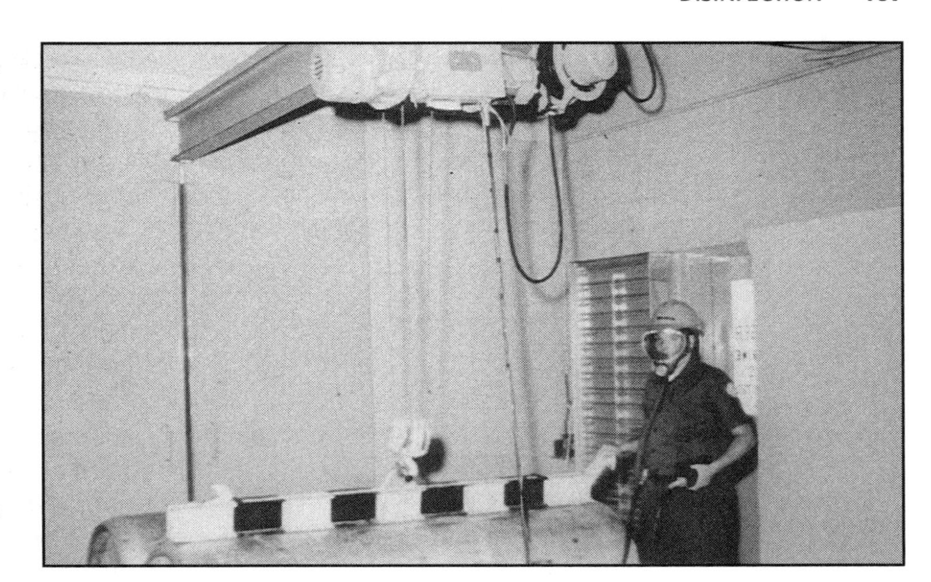

FIGURE 7-16 Lifting beam with motorized hoist for ton containers

FIGURE 7-17 Ton containers stored on trunnions

valve assemblies when the tank is delivered should be replaced each time the container is handled, as well as right after it has been emptied.

The chlorine storage area should provide space for a 30- to 60-day supply of chlorine. Some systems feed chlorine directly from this storage area. When ton containers are used, the chlorination feed equipment is usually housed in a separate room (Figure 7-18).

Feeding Chlorine Gas

Chlorine feeding begins where the cylinder or ton container connects to the manifold that leads to the chlorinator. The feed system ends at the point where the chlorine solution mixes into the water being disinfected. The main components of the system are

- weighing scale
- valves and piping
- chlorinator
- injector or diffuser

Weighing Scales

It is important that an accurate record be kept of the amount of chlorine used and the amount of chlorine remaining in a cylinder or container. A simple way to do this is to place the cylinders or containers on weigh scales.

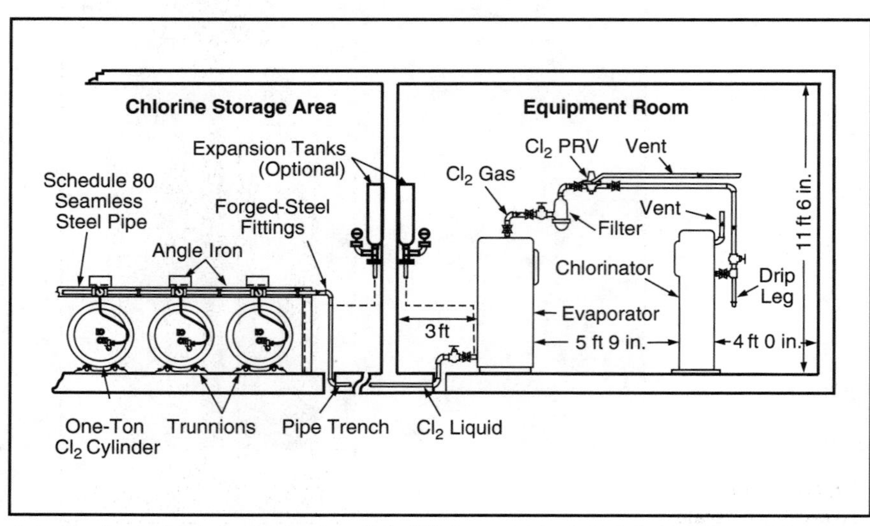

FIGURE 7-18 Chlorination feed equipment located in a separate room

From Handbook of Chlorination *by Geo. Clifford White, copyright © 1972 by Van Nostrand Company. Reprinted by permission of the publisher.*

The scales can be calibrated to display either the amount used or the amount remaining. By recording weight readings at regular intervals, the operator can develop a record of chlorine-use rates. Figure 7-19 shows a common type of two-cylinder scale. Figure 7-20 shows a portable beam scale. Figure 7-21 shows a combination trunnion and scale for a ton container; this scale operates hydraulically and has a dial readout.

Valves and Piping

Chlorine cylinders and ton containers are equipped with valves as shown in Figures 7-22 and 7-23. The valves must comply with standards set by the Chlorine Institute (an organization listed in appendix C).

It is standard practice for an auxiliary tank valve to be connected directly to the cylinder or container valve, as illustrated in Figure 7-24. The connection is made with either a union-type or a yoke-type connector. The auxiliary valve can be used to close off all downstream piping, thus minimizing gas leakage during container changes. The auxiliary tank valve will also serve as an emergency shutoff if the container valve fails. If a direct-mounted chlorinator is used, an auxiliary tank valve is not required (Figure 7-25).

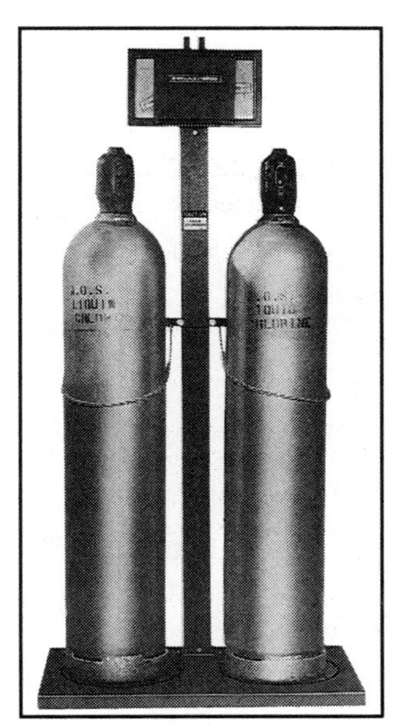

FIGURE 7-19
Two-cylinder scale

Courtesy of Wallace & Tiernan, Inc.

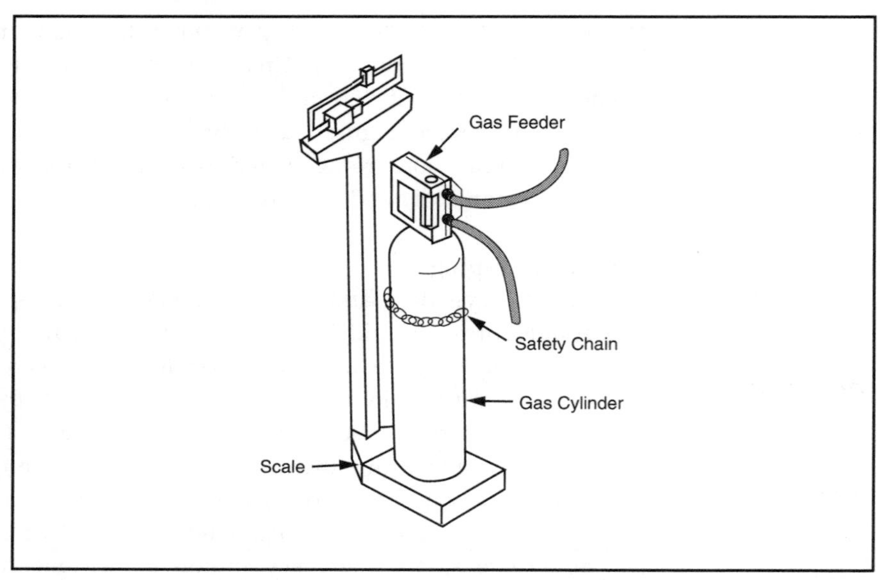

FIGURE 7-20 Portable beam scale

Courtesy of Capital Controls Company, Inc.

FIGURE 7-21 Combination trunnion and scale for a ton container

Courtesy of Force Flow Equipment

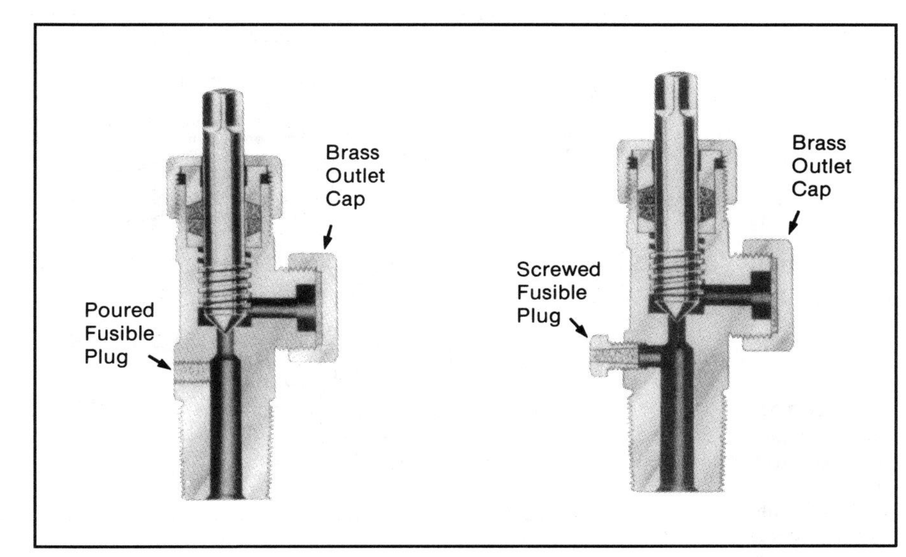

FIGURE 7-22 Standard cylinder valves: poured-type fusible plug (left) and screw-type fusible plug (right)

Courtesy of the Chlorine Institute, Inc.

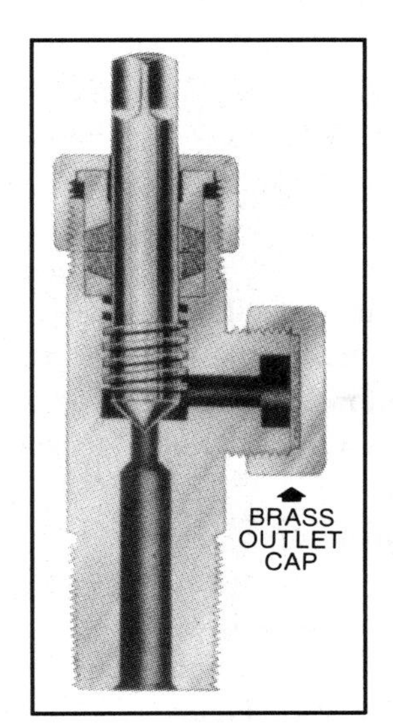

FIGURE 7-23 Standard ton container valve

Courtesy of the Chlorine Institute, Inc.

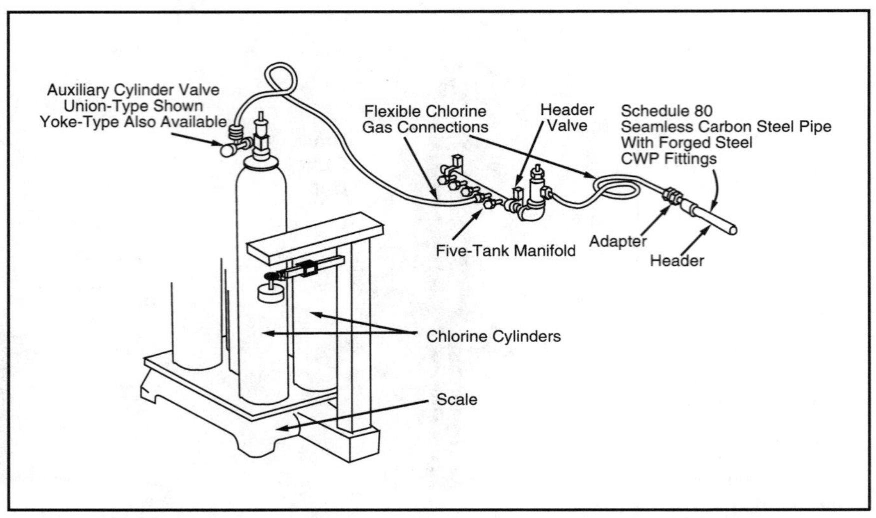

FIGURE 7-24 Auxiliary tank valve connected directly to container valve

Courtesy of Wallace & Tiernan, Inc.

The diagram in Figure 7-24 is of a typical valve assembly. The figure shows that the assembly is connected to the chlorine-supply piping by flexible tubing, which is usually ⅜-in. (10-mm) copper rated at 500 psig (3,500 kPa).

When more than one container is connected, a manifold must be used, as shown in Figure 7-24. The manifold channels the flow of chlorine from two or more containers into the chlorine-supply piping. The manifold and supply piping must meet the specifications of the Chlorine Institute. Manifolds may have from 2 to 10 connecting points. Each point is a union nut suitable for receiving flexible connections. Notice in Figure 7-24 that the header valve is connected at the manifold discharge end, providing another shutoff point. Additional valves are used along the chlorine supply line for shutoff and isolation in the event of a leak.

Chlorinators

The chlorinator can be a simple direct-mounted unit on a cylinder or ton container, as shown in Figure 7-25. This type of chlorinator feeds chlorine gas directly to the water being treated. A free-standing cabinet-type chlorinator is illustrated in Figure 7-26. Cabinet-type chlorinators, which operate on the same principle as cylinder-mounted units, have a sturdier mounting, and are capable of higher feed rates. Schematic diagrams of two typical chlorinators are shown in Figures 7-27 and 7-28.

FIGURE 7-25
Direct-mounted chlorinator

Courtesy of Fischer & Porter Co.

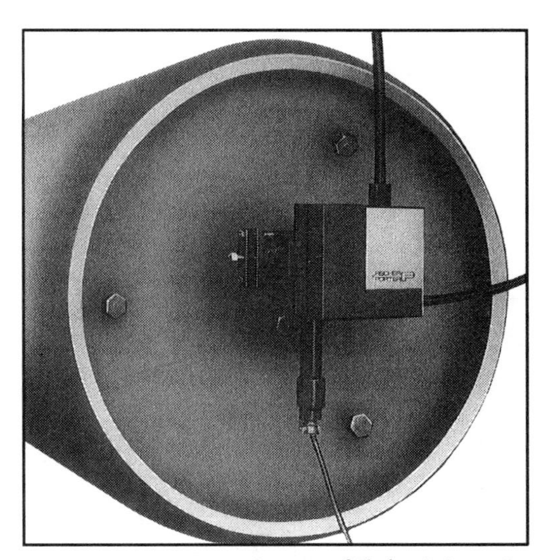

FIGURE 7-26
Free-standing chlorinator cabinet

Courtesy of Fischer & Porter Co.

The purpose of the chlorinator is to meter chlorine gas safely and accurately from the cylinder or container and then accurately deliver the set dosage. To do this, a chlorinator is equipped with pressure and vacuum regulators that are actuated by diaphragms and orifices for reducing the gas pressure. The reduced pressure allows a uniform gas flow, accurately metered by the rotameter (feed rate indicator). In addition, a vacuum is

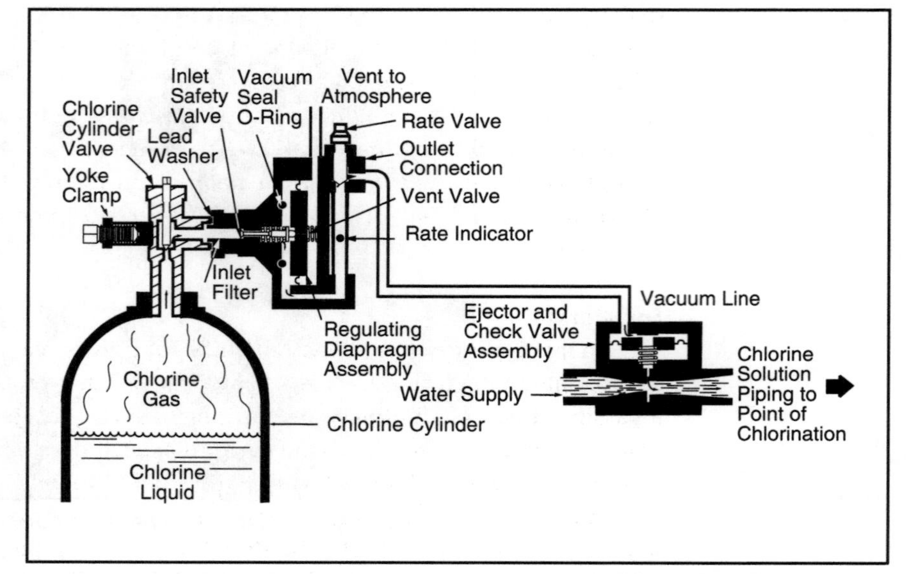

FIGURE 7-27 Schematic of direct-mounted gas chlorinator

Courtesy of Capital Controls Company, Inc.

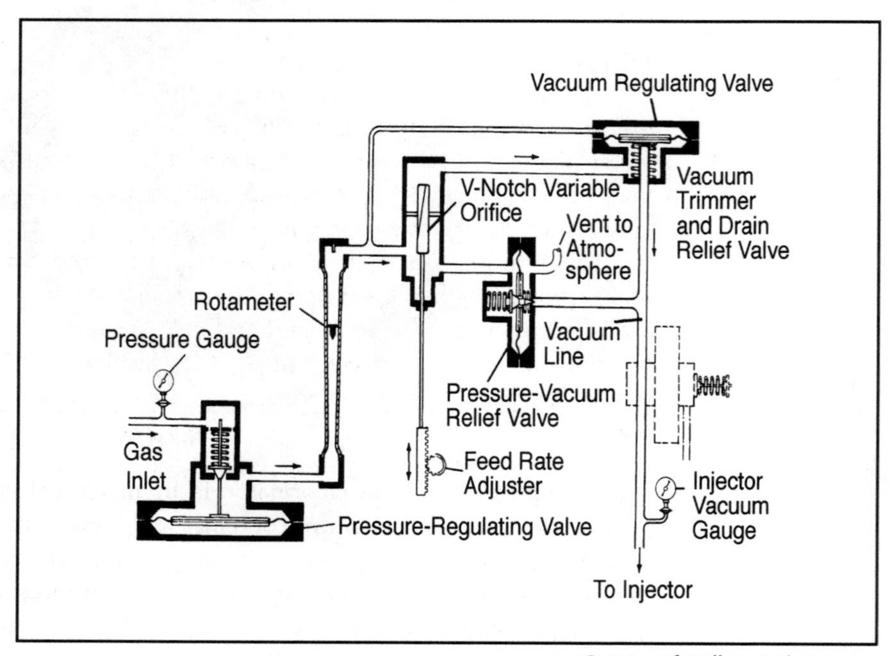

FIGURE 7-28 Schematic of cabinet-style chlorinator

Courtesy of Wallace & Tiernan, Inc.

maintained in the line to the injector for safety purposes. If a leak develops in the vacuum line, air will enter the atmospheric vent, causing the vacuum relief valve to close and stopping the flow of chlorine gas. To vary the chlorine dosage, the operator manually adjusts the setting of the rotameter.

It is normally required that each treatment plant have at least one standby chlorinator ready for immediate use in the event that the primary chlorinator should fail. Automatic switchover equipment is also strongly recommended.

Injectors

An injector (or ejector) is located within or downstream of the chlorinator as illustrated in Figure 7-27. It is a venturi device that pulls chlorine gas into a passing stream of dilution water, forming a strong solution of chlorine and water. The injector also creates the vacuum needed to operate the chlorinator. The highly corrosive chlorine solution (pH about 2 to 4) is carried to the point of application in a corrosion-resistant pipeline. The type of pipe generally used is polyvinyl chloride (PVC), fiberglass, or steel pipe lined with PVC or rubber. A strainer should be installed on the water line upstream of the injector. This prevents any grit, rust, or other material from entering and blocking the injector or causing wear of the injector throat.

Diffusers

A diffuser is one or more short lengths of pipe, usually perforated, that quickly and uniformly disperse the chlorine solution into the main flow of water. There are two types of diffusers: those used in pipelines and those used in open channels and tanks. A properly designed and operated diffuser is necessary for the complete mixing needed for effective disinfection.

The diffuser used in pipelines less than 3 ft (0.9 m) in diameter is simply a pipe protruding into the center of the pipeline. Figure 7-29 shows a diffuser made from Schedule 80 PVC, and Figure 7-30 shows how the turbulence of the flowing water completely mixes the chlorine solution throughout the water. Complete mixing should occur downstream at a distance of 10 pipe diameters.

Figure 7-31 shows a perforated diffuser for use in larger pipelines. A similar design is used to introduce chlorine solution into a tank or open channel, as shown in Figure 7-32. (During normal operations, the diffuser would be completely submerged, but in the figure, the water level has been dropped, for illustrative purposes only, to show the chlorine solution passing out of each perforation.)

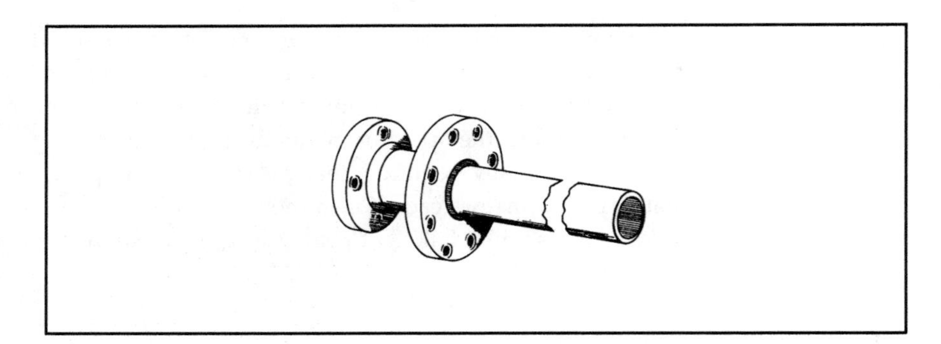

FIGURE 7-29
Diffuser made from Schedule 80 PVC

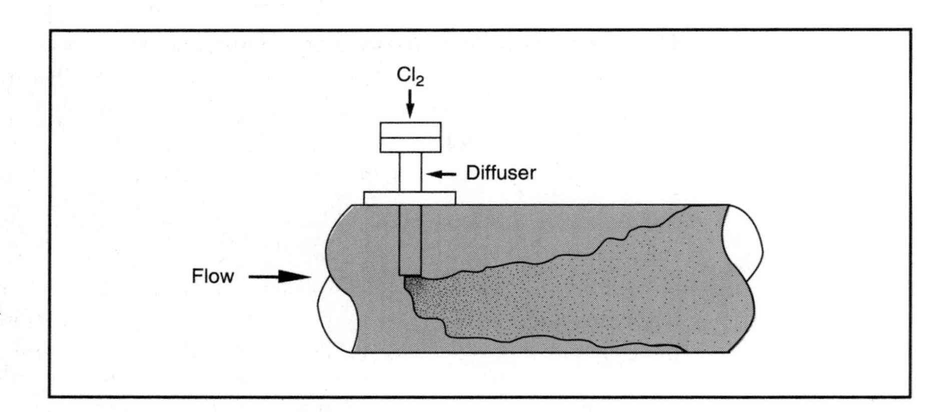

FIGURE 7-30
Chlorine solution mixing in a large-diameter pipeline

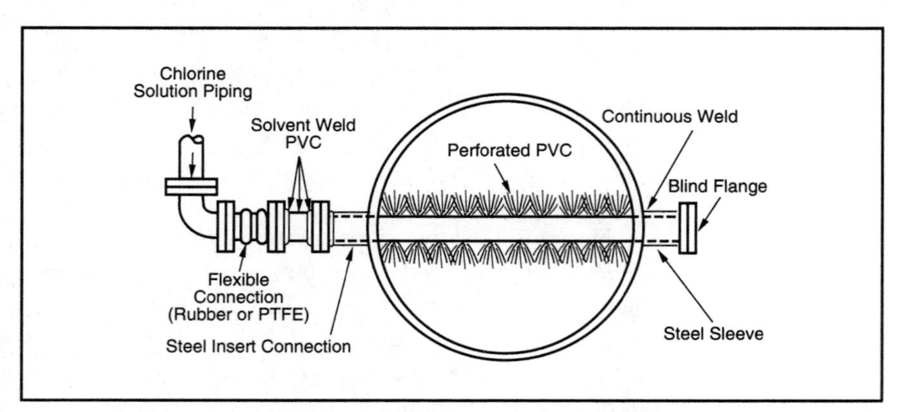

FIGURE 7-31
Perforated diffuser for pipelines larger than 3 ft (0.9 m) in diameter

From Handbook of Chlorination *by Geo. Clifford White, copyright © 1972 by Van Nostrand Company. Reprinted by permission of the publisher.*

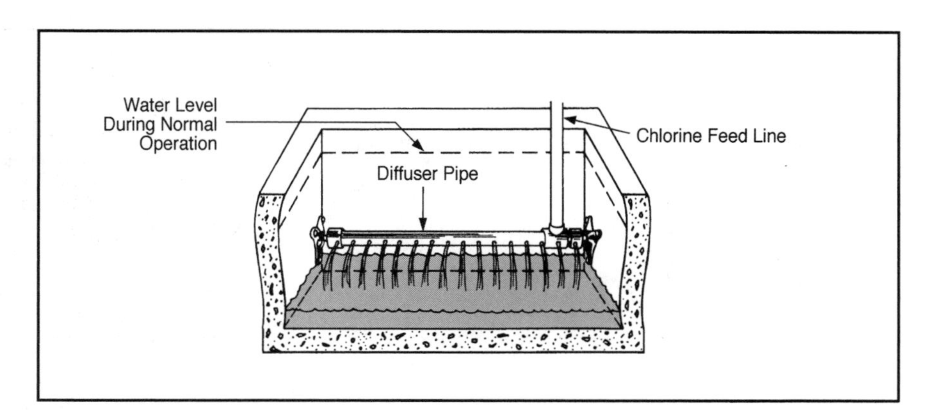

FIGURE 7-32
Open-channel diffuser

Gas Chlorination Auxiliary Equipment

A variety of auxiliary equipment is used for chlorination. The following discussion describes the functions of the more commonly used items.

Booster Pumps

A booster pump (Figure 7-33) is usually needed to provide the water pressure necessary to make the injector operate properly. The booster pump is usually a low-head, high-capacity centrifugal type. It must be sized to overcome the pressure in the line that carries the main flow of water being treated, and it must be rugged enough to withstand continuous use.

Automatic Controls

If a chlorination system is to be manually operated, adjustments must be made each time the flow rate or the chlorine demand changes. For constant or near-constant flow rate situations, a manual system is suitable.

However, when flow rate or chlorine demand is continually changing, the operator is required to change the rotameter settings frequently. In these situations, automatic controls are valuable. Although many automatic control arrangements are possible, there are two common types: flow proportional control and residual flow control.

Flow proportional control. If chlorine demand rarely changes and it is necessary to compensate only for changes in the pumping rate, flow proportional control works well. It will automatically increase or decrease the chlorine feed rate as the water flow rate increases or decreases. The required equipment includes a flowmeter for the treated water, a transmitter to sense the flow rate and send a signal to the chlorinator, and a receiver at the

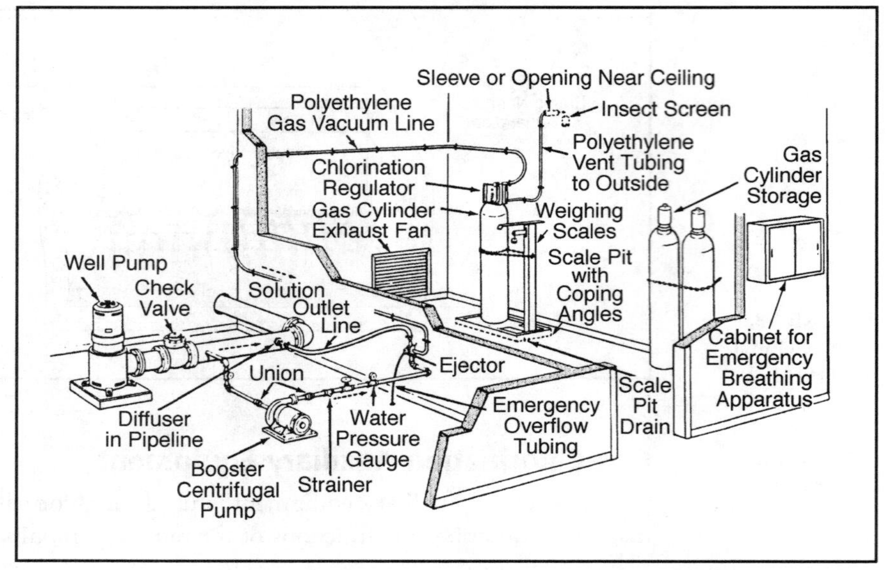

FIGURE 7-33 Typical chlorinator deep-well installation showing booster pump

Courtesy of Fischer & Porter Co.

chlorinator. The receiver responds to the transmitted signal by opening or closing the chlorine flow rate valve.

Residual flow control. If the chlorine demand of the water changes periodically, it is necessary to make corresponding changes in the rate of feed to provide adequate disinfection. Residual flow control, also called compound loop control, automatically maintains a constant chlorine residual, regardless of chlorine demand or flow rate changes. The system uses an automatic chlorine residual analyzer (Figure 7-34) in addition to the signal from a meter measuring the flow rate. The analyzer uses an electrode to determine the chlorine residual in the treated water. Signals from the residual analyzer and flow element are sent to a receiver in the chlorinator, where they are combined to adjust the chlorine feed rate to maintain a constant residual in the treated water.

Evaporators

An evaporator is a heating device used to convert liquid chlorine to chlorine gas. Ton containers are equipped with valves that will draw either liquid or gas. At 70°F (21°C), the maximum gas withdrawal rate from a ton container is 400 lb/d (180 kg/d). If higher withdrawal is required, the liquid feed connection is used and connected to an evaporator. The evaporator

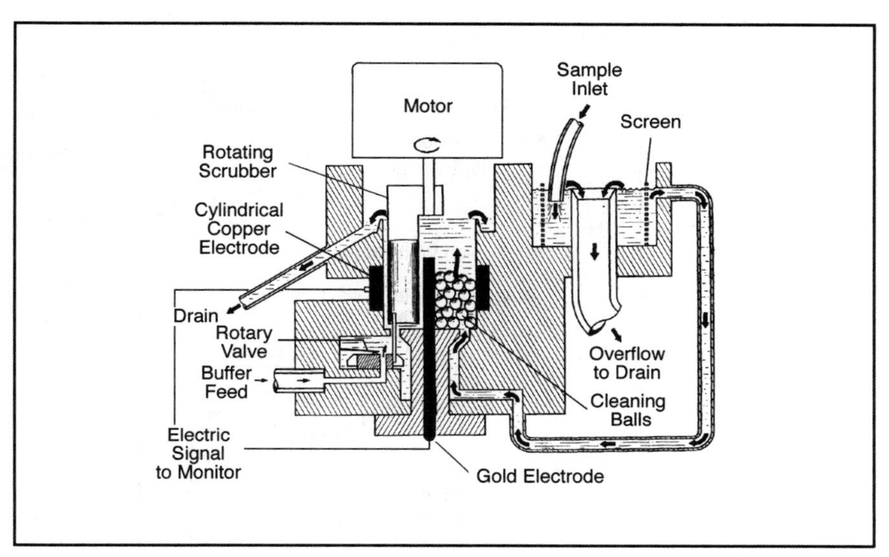

FIGURE 7-34 Automatic chlorine residual analyzer

Courtesy of Capital Controls Company, Inc.

accelerates the evaporation of liquid chlorine to gas, so that withdrawal rates up to 9,600 lb/d (4,400 kg/d) can be obtained.

An evaporator (Figure 7-35) is a water bath heated by electric immersion heaters to a temperature of 170–180°F (77–82°C). The pipes carrying the liquid chlorine pass through the water bath, and liquid chlorine is converted to gas by the heat.

Automatic Switchover Systems

For many small water systems, it is either impossible or uneconomical to have an operator available to monitor operation of the chlorination system at all times. An automatic switchover system provides switchover to a new chlorine supply when the on-line supply runs out. The switchover is either pressure or vacuum activated. The vacuum type of installation is shown in Figure 7-36. The automatic changeover mechanism has two inlets and one outlet. As the on-line supply is exhausted, the vacuum increases, causing the changeover mechanism to close on the exhausted supply and open the new chlorine supply. The unit can also send a signal to notify operating personnel that the one tank is empty and should be replaced. Figure 7-37 shows a typical installation. This system is ideal for remote locations to ensure uninterrupted chlorine feeding.

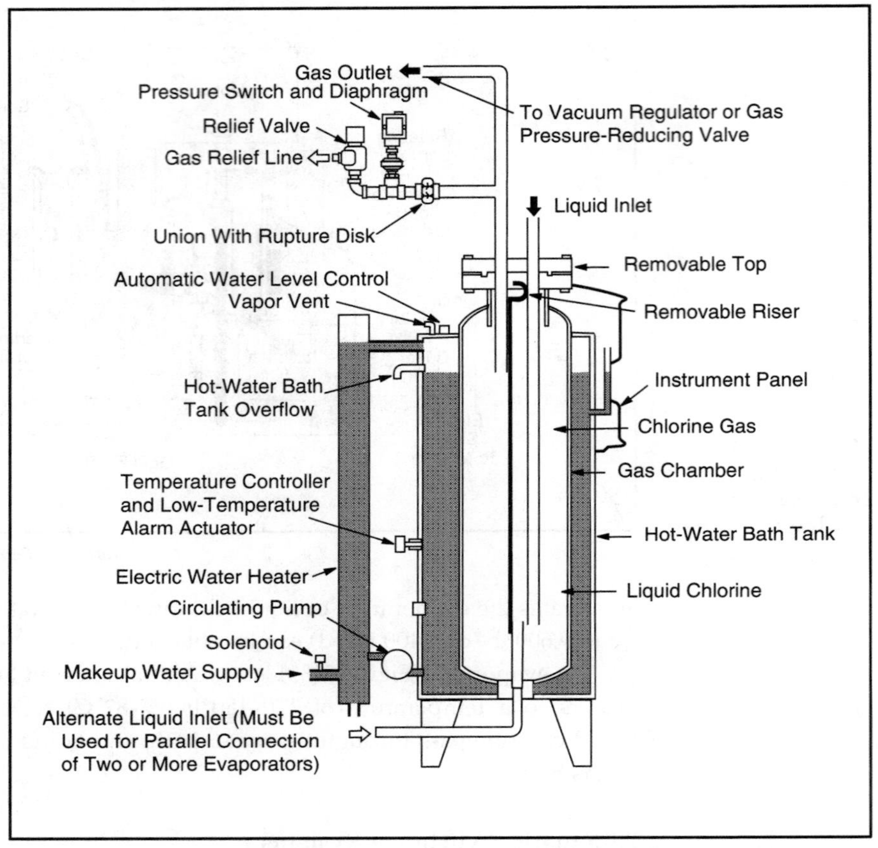

FIGURE 7-35 Chlorine evaporator

Courtesy of Wallace & Tiernan, Inc.

Chlorine Alarms

Chlorinators are often equipped with a vacuum switch that triggers an alarm when it senses an abnormally low or high vacuum. A low-vacuum condition can mean an injector failure, vacuum line break, or booster pump failure. A high-vacuum condition can be caused by a plugged chlorine supply line or by empty chlorine tanks.

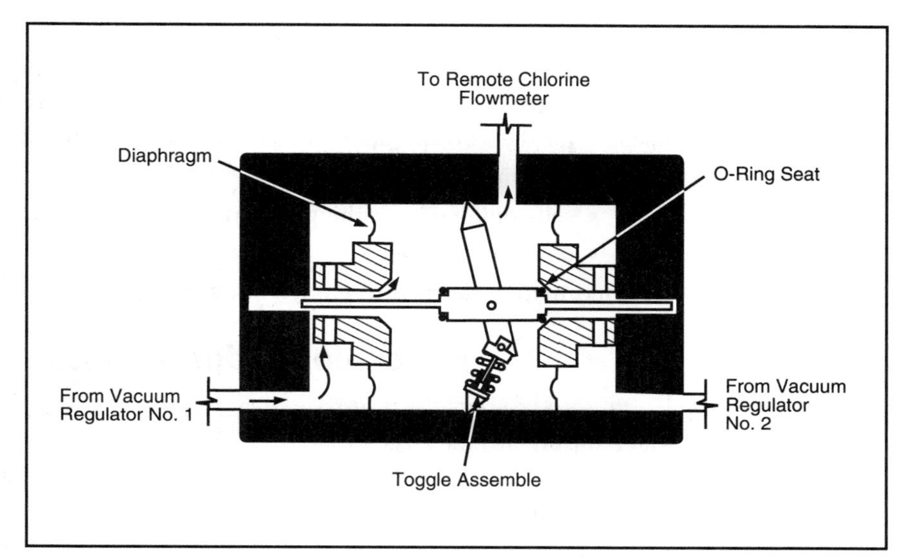

FIGURE 7-36 Automatic switchover unit

Courtesy of Capital Controls Company, Inc.

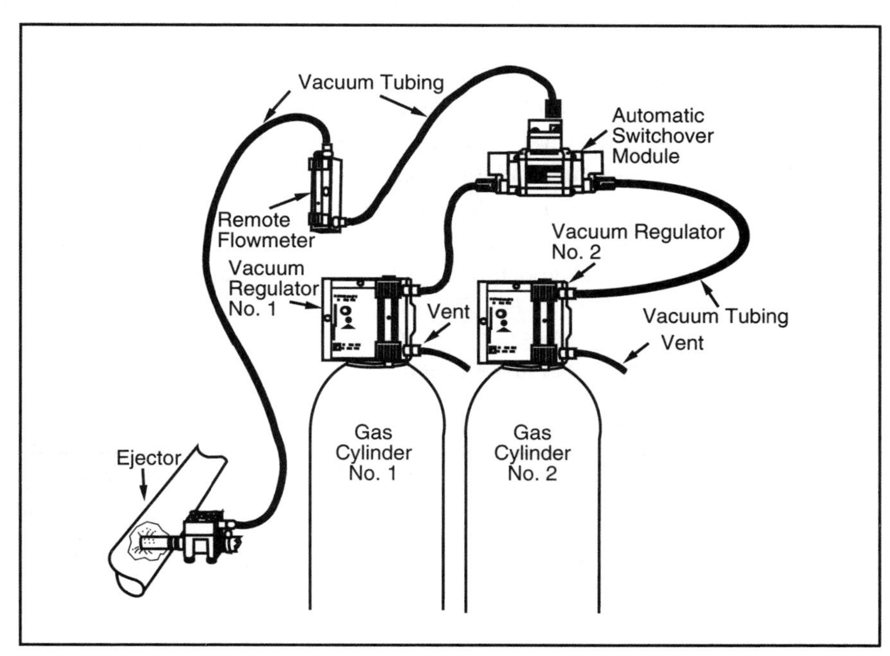

FIGURE 7-37 Typical installation of switchover system

Courtesy of Capital Controls Company, Inc.

Safety Equipment

Safety in and around the gas chlorination process is important to prevent serious accidents and equipment damage. Certain items of equipment are essential for the safe operation of a chlorination facility, including

- chlorine detectors
- self-contained breathing apparatus
- emergency repair kits

Chlorination safety is discussed in detail later in this chapter.

Hypochlorination Facilities

Hypochlorination is a chlorination method commonly used by smaller water supply facilities. As a rule, plants using less than 3 lb/d (1.4 kg/d) of available chlorine use hypochlorinators. However, hypochlorinators can be used by larger systems under some circumstances.

Hypochlorite Compounds

The two most commonly used compounds are calcium hypochlorite and sodium hypochlorite. Table 7-7 lists the properties of both compounds.

Calcium Hypochlorite

Calcium hypochlorite, $Ca(OCl)_2$, is a dry, white or yellow-white, granular material. It is also available in compressed tablets. It generally contains 65 percent available chlorine by weight. This means that when 1 lb (0.5 kg) of the powder is added to water, only 0.65 lb (0.3 kg) of pure chlorine is being added. Conversely, if 1 lb (0.5 kg) of chlorine is added, 1.5 lb (0.7 kg) of calcium hypochlorite must be added (Table 7-5).

TABLE 7-7 Properties of hypochlorites

Property	Sodium Hypochlorite	Calcium Hypochlorite
Symbol	$NaOCl$	$Ca(OCl)_2$
Form	Liquid	Dry granules, powder, or tablets
Strength	Up to 15% available chlorine	65–70% available chlorine, depending on form

Calcium hypochlorite requires special storage to avoid contact with organic material. Its reaction with any organic substances can generate enough heat and oxygen to start and support a fire. When calcium hypochlorite is mixed with water, heat is given off. To provide adequate dissipation of the heat, the dry chemical should be added to the water — the water should *not* be added to the chemical.

Sodium Hypochlorite

Sodium hypochlorite, NaOCl, is a clear, light-yellow liquid commonly used for bleach. Ordinary household bleach contains 5 percent available chlorine. Industrial bleaches are stronger, containing from 9 to 15 percent.

The sodium hypochlorite solution is alkaline, with a pH of 9 to 11, depending on the available chlorine content. For common strengths, Table 7-5 shows the amount of solution needed to supply 1 lb (0.5 kg) of pure chlorine. Large systems can purchase the liquid chemical in carboys, drums, and railroad tank cars. Very small water systems often purchase it in 1-gal (3.8-L) plastic jugs.

There is no fire hazard in storing sodium hypochlorite, but the chemical is quite corrosive and should be kept away from equipment susceptible to corrosion damage. Sodium hypochlorite solution can lose 2 to 4 percent of its available chlorine content per month at room temperature. It is therefore recommended that it not be stored for more than 60 to 90 days.

Common Equipment

Plants using calcium hypochlorite should be equipped with a cool, dry storage area to stockpile the compound in the shipping containers. A variable-speed chemical feed pump (hypochlorinator), such as a diaphragm pump, is all that is required for feeding the chemical to the water. A mix tank and a day tank (Figure 7-38) are also required. After calcium hypochlorite is mixed with water, impurities and undissolved chemicals settle to the bottom of the mix tank. The clear solution is then transferred to the day tank for feeding. This prevents any of the solids from reaching and plugging the hypochlorinator or rupturing the diaphragm.

Because sodium hypochlorite is a liquid, it is simpler to use than calcium hypochlorite. It can be fed full strength or it can be diluted with water to a 1 percent solution. In either case, only one feed tank is required.

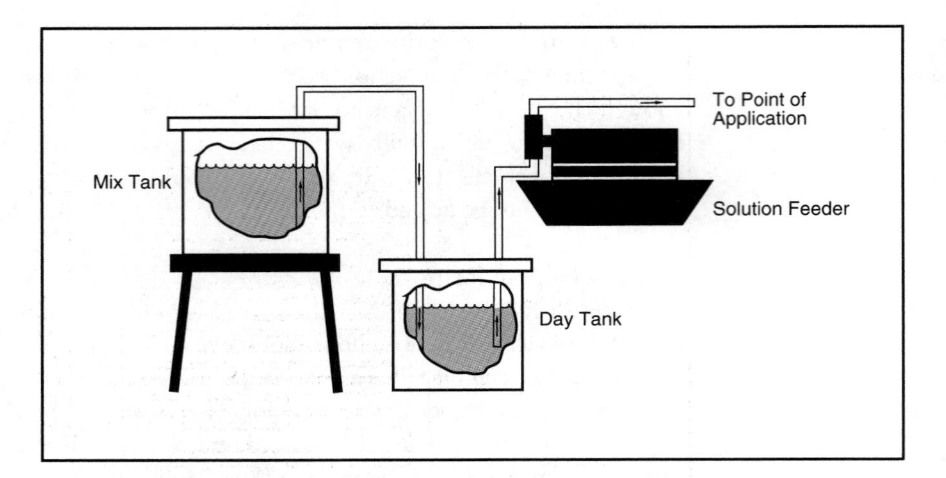

FIGURE 7-38
Mix tank and day tank

Facilities for Other Oxidants

This section discusses the facilities associated with the use of ozone, potassium permanganate, and chlorine dioxide.

Ozone Equipment and Facilities

Because ozone cannot be stored, it must be generated on-site as it is needed. It is generated by passing an electrical current through air or pure oxygen. An ozonation system consists of the following components:

- an air compression and drying unit, or oxygen source
- an ozonator, or ozone generator
- a contactor to introduce the ozone into the water
- a residual ozone destruction unit

Air Preparation

When air is used as the feed gas for an ozone generator, it must be extremely dry. Special compressors and drying systems must be installed to treat the air properly before it is used in a generator.

When oxygen is used as the feed gas, it is simply piped to the generator from a storage tank of compressed oxygen.

Ozone Generators

Most ozone generators for water treatment use one of two designs. The most common type used in large plants is a bank of glass-tube generators, as

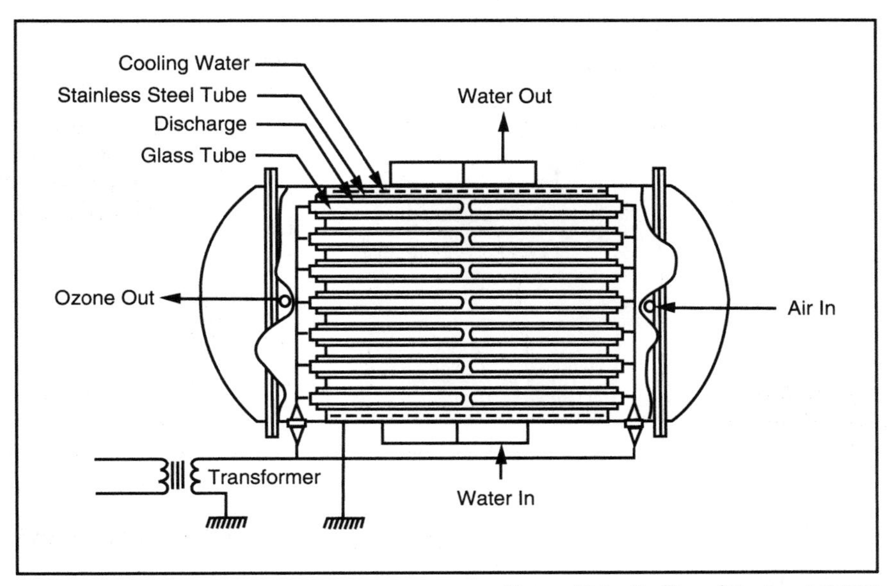

Source: Water Quality and Treatment (1990)

FIGURE 7-39 Large-scale, tube-type ozone generator

illustrated in Figure 7-39. Smaller plants may use either this type or a plate-type generator in which the ozone is generated between ceramic plates.

Ozone generators can use pure oxygen, oxygen-enriched air, or air as the feed gas. If air is used, the product is about 2 percent ozone, whereas a generator using oxygen can produce 5–7 percent ozone. Figure 7-40 is a typical flow diagram for a treatment plant using oxygen-enriched air for ozone generation.

Ozone Contactors

Ozone is only sparingly soluble in water. It has a very short life, generally from a few seconds to a few minutes, depending on a number of variables. It must therefore be efficiently introduced into the water to be treated. This is critical to the success of the system. A common type of contactor is illustrated in Figure 7-41. The gas containing ozone is applied under pressure to a porous stone at the bottom of the unit, where it forms small bubbles. As the bubbles rise in the tank, ozone is transferred into the water. Another design applies the ozone through a venturi injector.

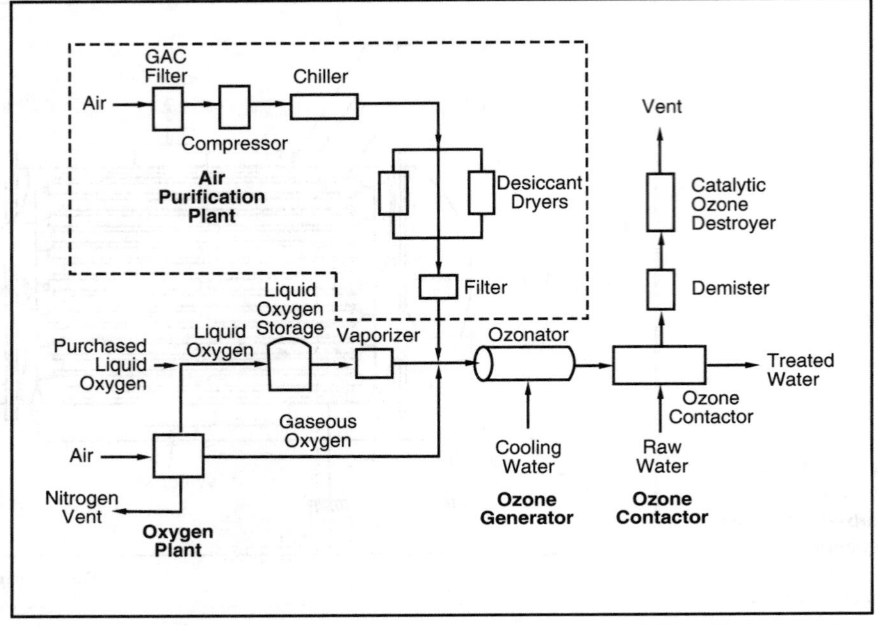

FIGURE 7-40
Flow diagram for air and oxygen purification for ozone production

Source: Water Quality and Treatment *(1990)*

Residual Ozone Destruction

Ozone contactors must have a system to collect ozone off-gas. Ozone is toxic and must be kept within allowable limits — set by the Occupational Safety and Health Administration (OSHA) — in the treatment plant and in the area surrounding the plant. In some areas of the United States, ozone discharge from treatment plants may be further regulated.

As a result, ozone-generating installations must include a thermal or catalytic ozone destroyer. In some installations, the off-gas is also reused by piping it to the air compressor.

Ozone is so highly corrosive that only certain materials can be used in constructing treatment plant equipment. Suppliers of ozone generators should be consulted about appropriate materials.

Potassium Permanganate Equipment and Facilities

Permanganate is shipped as crystals in pails, drums, returnable bins, and bulk truckloads. In water plants using less than 25 lb/d (11.3 kg/d) of permanganate, the dry chemical is usually dissolved in batches in a solution tank. The facility normally includes two solution tanks and at least two

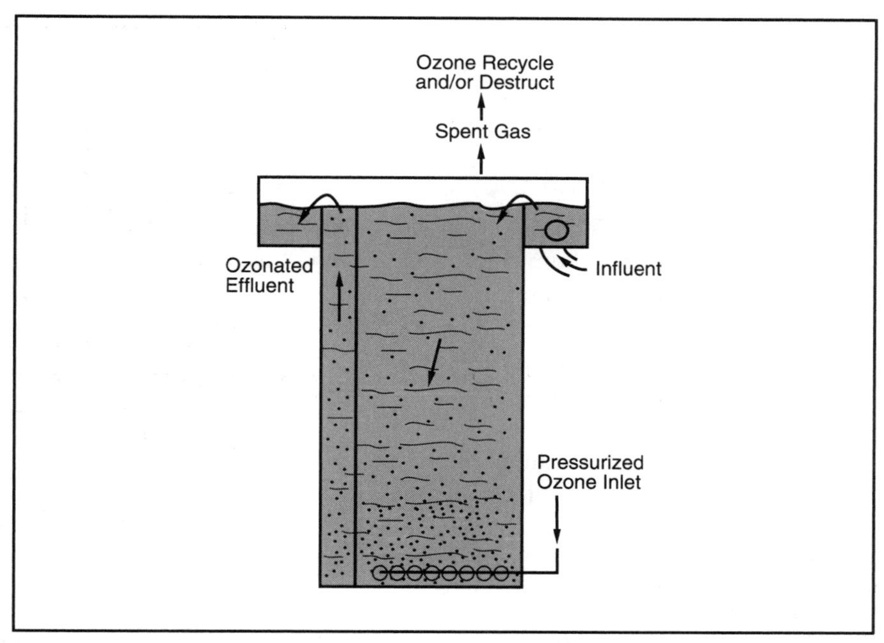

FIGURE 7-41 Typical countercurrent ozone contactor

Source: Water Quality and Treatment *(1990)*

chemical feed pumps. Each pump should be capable of meeting peak-day needs, and each tank should have a capacity for a three-day supply.

For larger plants, dry volumetric feeders with hoppers and dust collectors are normally used. A float-controlled solution tank with a mixer is usually located under the feeders, and injectors are used for powering the flow to the feed point. Permanganate dosages of from 0.5 to 2.5 mg/L are usually sufficient to control most oxidizable taste-and-odor-causing materials in raw water.

As the purple color of permanganate in water dissipates, a residual of manganese dioxide is left in the water that can give it a yellow-brown color. For this reason, permanganate must be fed only as a pretreatment step so that the manganese dioxide can be removed by clarification. If the water is still purple when it reaches clarification treatment, too much permanganate has been applied. The ideal application point is the raw-water intake. This allows the maximum reaction time for oxidation and complete coagulation of the manganese dioxide.

If possible, permanganate should be fed ahead of chlorine. The oxidation of organics before chlorination will reduce both the amount of chlorine needed and the formation of total trihalomethanes (TTHMs). If powdered activated carbon is used in addition to permanganate, it should, if possible, be fed after the permanganate has completed its oxidizing action. If both are fed together, the permanganate consumption increases substantially.

Chlorine Dioxide Equipment and Facilities

Chlorine dioxide is usually generated on-site by the reaction of chlorine and sodium chlorite. One method of preparation is to introduce the chlorine to the water and then add the water to a solution containing hydrochloric acid and sodium chlorate. This method is likely to leave substantial amounts of chlorine in the solution, which may lead to the formation of chlorinated by-products and undesirable amounts of chlorate and chlorite. Another method is to inject the chlorine as a gas under vacuum into a stream of chlorite solution. Chlorine dioxide can also be prepared by adding acid, such as hydrochloric acid, to a chlorite solution. This method can be employed where the use of chlorine is not convenient, but for large-scale applications, the second method is usually preferable.

Regulations

This section discusses regulations associated with the Surface Water Treatment Rule, THM limits, DBPs, and mandatory chlorination.

Surface Water Treatment Rule Requirements

On June 29, 1989, the USEPA enacted the Surface Water Treatment Rule (SWTR). This regulation applies to every public water system in the United States that uses surface water as a source. It also applies to systems that use groundwater that might become contaminated by surface water; these systems are labeled as having "groundwater under the direct influence of surface water," or GWUI.

The purpose of the regulation is to protect the public as much as possible from waterborne diseases, which are most commonly transmitted by contamination of surface water. Because it is difficult to monitor for particular microorganisms, including *Giardia lamblia* and viruses, the SWTR emphasizes treatment techniques as the condition for compliance, rather than establishing maximum contaminant levels (MCLs) for microorganisms.

Because of the variety of water qualities, local conditions, and methods of treatment, the rule does not prescribe a particular method of treatment; instead it offers several alternatives. Any of these can then be used by a water system to meet the overall goal, which is removal or inactivation of essentially all disease-causing organisms. By allowing each water system to choose the best method of treatment for its situation, the rule makes it possible to protect public health and still provide safe water at the least possible cost. However, to ensure that water quality goals are met, the rule contains many operation and monitoring requirements.

Studies indicate that *Giardia* cysts, viruses, and *Cryptosporidium* are among the most resistant waterborne pathogens. Therefore, water systems that attain adequate removal and/or inactivation of these organisms will, to the best of our current knowledge, provide adequate protection from other waterborne disease-causing organisms.

As discussed in previous chapters, most systems with a surface water source must use sedimentation and filtration to ensure adequate removal of microorganisms. Only systems with extremely clean source water may be allowed to operate without filtration, and they may do so only under very stringent operating and monitoring conditions. All surface water and GWUI systems, whether they must provide filtration or not, must practice disinfection under very specific conditions.

The effectiveness of a disinfectant for inactivating *Giardia* cysts and viruses depends on

- the type of disinfectant used
- the disinfectant residual concentration
- the time the water is in contact with the disinfectant
- the water temperature
- the pH of the water (if chlorine is used)

The second and third items in this list are particularly important. The residual concentration C of a disinfectant, in milligrams per liter, multiplied by the contact time T, in minutes, equals the $C \times T$ value. The $C \times T$ values required by various disinfectants to guarantee the necessary reduction in microorganisms are given in publications listed at the end of this chapter. Discussions of determining the contact time T, calculating $C \times T$, and on other details of the SWTR are included in *Water Quality*, also part of this series.

Some of the principal disinfection requirements of the SWTR are as follows:

- Each system's disinfection treatment process must be sufficient to ensure that the total treatment process achieves at least 99.9 percent (3-log) inactivation and/or removal of *Giardia* cysts, and 99.99 percent (4-log) inactivation and/or removal of viruses.

TABLE 7-8 Sampling frequency for disinfectant residual for water entering the distribution system

Population Served	Samples per Day
<500	1
501–1,000	2
1,001–2,500	3
2,501–3,300	4

- The disinfectant residual of water entering the distribution system must be monitored continuously by systems serving a population of more than 3,300. Systems serving less than 3,300 may take grab samples at the frequencies given in Table 7-8.
- The disinfectant residual in water entering the distribution system must not be less than 0.2 mg/L for more than 4 hours during periods when a system is serving water to the public. If it falls below this level, the system must notify the state.
- The disinfectant residual in the distribution system must be measured at the same points and at the same time that total coliforms are sampled. The disinfectant residual in these samples cannot be undetectable in more than 5 percent of the samples each month for any two consecutive months that water is served to the public.

Trihalomethane Limits

All surface waters and groundwaters contain varying levels of organic compounds. These compounds generally come from decaying vegetation, and they are primarily humic and fulvic acids. Typical examples of water with high levels of organic compounds are runoff water from forested land and water from lakes with high algae levels.

When chlorine is added to water containing these organic compounds, a reaction takes place to form complex chloroorganic compounds known as trihalomethanes (THMs). These compounds, the most common of which is chloroform, are considered potential cancer-causing substances. At the time of this publication, the USEPA has set an MCL of 0.1 mg/L for total trihalomethanes (TTHMs) in drinking water. Lower levels of TTHMs have been proposed for the future. Further information on this regulation is included in *Water Quality*, also part of this series.

It is possible to remove THMs from treated water by using activated carbon, but this is an expensive solution. A preferred solution is to prevent the THMs from forming in the first place. Another solution is to use a disinfectant other than chlorine for the initial dose to the untreated water. Other disinfectants will not form THMs; however, they may have other disadvantages and may form other DBPs. Many systems find that excessive THM formation is only a seasonal problem, such as during spring and fall runoff periods. In this case, special operating procedures may be necessary only during those periods.

Although it is important to meet TTHM requirements, it must be kept in mind that microbiological safety of the water is far more important and must not be sacrificed in the process. Major changes in treatment plant disinfection processes should not be made without professional advice and prior approval of the state.

Disinfection By-products

In addition to THMs, it is known that chlorination can produce a variety of other compounds, including haloacetic acids, halonitriles, haloaldehydes, and chlorophenols. Alternative disinfectants including chloramines, chlorine dioxide, and ozone can also react with source water organics to yield new DBPs.

Exactly how much of these compounds is formed, as well as their public health significance, is still not completely known. Analysis of low levels of these compounds is difficult, and health effects studies take time and careful interpretation. It is likely that additional DBPs formed from chlorine and other disinfectants will, in the future, be found to pose a health threat and will have to be eliminated, or at least minimized, during the water treatment process. These would then be regulated by the USEPA under the Safe Drinking Water Act as adequate supporting data become available.

Mandatory Chlorination

Many states currently require all community public water systems to practice chlorination. In addition, USEPA is considering more restrictive mandatory chlorination regulations that will probably affect all public water systems. Under the new requirements, systems having a protected groundwater supply and a history of good compliance with microbiological standards will probably be allowed a variance to avoid chlorination, at the discretion of the state.

Operation of the Chlorination Process

Successful operation of the chlorination process requires an understanding of how each of the system components operates. In addition, the operator must be aware of the safety procedures that must be followed when handling cylinders and when dealing with leaks and equipment breakdowns.

Using Cylinders and Ton Containers

Cylinders

Cylinders should always be stored and used in an upright position. In this position, the cylinders will deliver chlorine gas continuously at a maximum rate of about 42 lb/d (19 kg/d). If the gas is withdrawn at a faster rate, the drop in pressure in the cylinder will cause a drop in cylinder temperature, and frost will form on the outside of the cylinder. This will reduce the withdrawal rate because a cooler temperature retards the vaporization of liquid chlorine. Under extreme conditions of frosting, the valve may freeze and completely stop the flow of gas.

Moderate frosting of a cylinder can be reduced by improving air circulation, simply by placing a fan in the chlorinator room. Heat should never be applied directly to a chlorine cylinder. If the heat is excessive, the cylinder's fusible plug could melt or pressure could increase to the point where the valve fails. If the problem of frosting continues, multiple cylinders must be used to reduce the withdrawal rate from any single cylinder.

Ton Containers

Ton containers are transported, hoisted, stored, and used in the horizontal position. When they are to be used, they must be positioned so that the two valves are oriented with one directly above the other. In this position, the top valve delivers chlorine gas, and the bottom delivers liquid chlorine.

A single ton container can deliver chlorine gas at rates up to 400 lb/d (180 kg/d) against a back pressure of 35 psig (240 kPa) at room temperature without frosting. Liquid chlorine can be delivered at rates up to 9,600 lb/d (4,400 kg/d) if an evaporator is being used. If a plant's requirements exceed 400 lb/d (180 kg/d), the plant can still operate on gas withdrawal by connecting two or more containers to feed at the same time.

The exact maximum withdrawal rate of gas for vacuum systems can be determined from the following formula:

$$\left(\begin{array}{c}\text{chlorine room} \\ \text{temp.} \\ ^\circ\text{F}\end{array} - \begin{array}{c}\text{threshold} \\ \text{temp.} \\ ^\circ\text{F}\end{array}\right) \times \begin{array}{c}\text{withdrawal} \\ \text{factor}\end{array} = \begin{array}{c}\text{maximum} \\ \text{withdrawal} \\ \text{rate}\end{array}$$

The withdrawal factor depends on the size and shape of the cylinder or container and can be obtained from chlorine suppliers or manufacturers.

The threshold temperature is the temperature at which the minimum gas pressure required to operate a gas chlorinator at the point of withdrawal (i.e., the line pressure at the point of application) is reached. The threshold temperature is determined from the vapor pressure curve of liquid chlorine. Table 7-9 is a chart of values from that curve. For example, given a room temperature of 60°F (15°C), a 150-lb (68-kg) cylinder having a withdrawal factor of 1 and a minimum gas pressure of 14 psig (97 kPa), the threshold temperature is 0°F (–17°C).

The maximum withdrawal rate under these conditions is

$$(60 - 0)\,1 = 60 \text{ lb/d } (27 \text{ kg/d})$$

Weighing Procedures

The only reliable method of determining the amount of chlorine remaining in a cylinder or container is to weigh the unit and check its weight against the tare weight (empty weight) stamped on the shoulder of the

TABLE 7-9 Values from the vapor pressure curve for liquid chlorine

Minimum Pressure,		Threshold Temperature,	
psig	*(kPa)*	*°F*	*(°C)*
9	(62)	–10	(–23)
14	(97)	0	(–18)
21	(145)	10	(–12)
28	(193)	20	(–7)
37	(255)	30	(–1)
47	(324)	40	(4)
59	(407)	50	(10)
71	(490)	60	(16)
86	(593)	70	(21)
102	(703)	80	(27)

cylinder or container. The information can be used to determine the feed rate and to decide when to change cylinders or containers. Because the pressure in a cylinder depends on the liquid chlorine temperature, not on the amount of chlorine in the container, the pressure cannot be used to determine when the cylinder is empty and must be changed.

Weighing procedures depend on the scale being used. Simple scales show the combined weight of the container and chlorine. On other scales, the operator enters the tare weight into the scale when the cylinder or container is put in use, and the scale thereafter displays only the weight of the remaining chlorine.

Connecting Cylinders and Ton Containers

When a cylinder is being connected to the chlorine supply line or manifold (using a yoke and adapter), the following procedure should be used:

1. Always wear personal respirator protection when changing cylinders or containers.
2. Never lift a cylinder by its protective hood. The hood is not designed to support the cylinder weight.
3. Secure the cylinder with a safety chain or steel strap in a solid, upright position.
4. Remove the protective hood. If the threads have become corroded, a few sound raps with a wooden or rubber mallet will usually loosen the hood so it can be unscrewed.
5. Remove the brass outlet cap and any foreign matter that may be in the valve outlet recess. Use a wire brush to clear out any pieces of the washer, being careful not to scratch the threads or gasket-bearing surface.
6. Place a new washer in the outlet recess. Do not reuse old washers.
7. Place the yoke over the valve. Insert the adapter in the outlet recess; then, fitting the adapter in the yoke slot, tighten the yoke screw. Make sure the end of the adapter seats firmly against the washer. Use only the cylinder-valve wrench provided by the chlorine supplier for all chlorine cylinder or container valve connections.

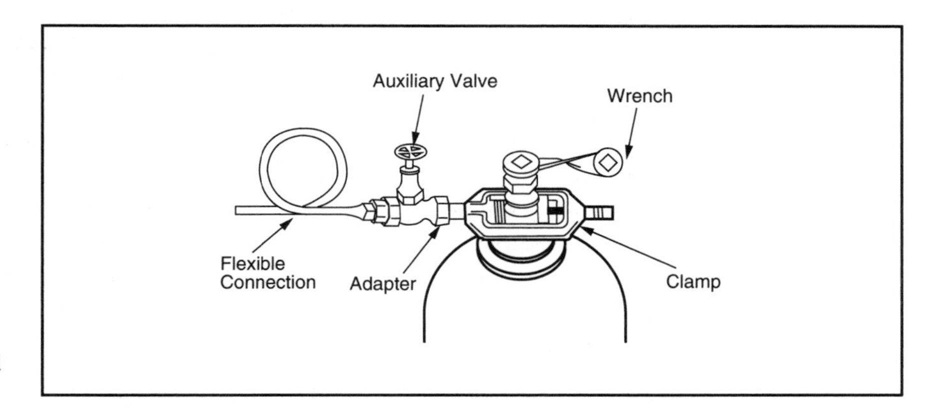

FIGURE 7-42 An installed yoke and auxiliary valve

8. Install the flexible connector, sloping it back toward the chlorine cylinder so that any liquid chlorine droplets will flow back to the cylinder and not to the chlorinator unit (Figure 7-42).

The procedure for connecting ton containers to the piping or manifold is basically the same as that described for cylinders. Note that when the cylinder or ton container is being connected, both the container valve and the auxiliary valve should be closed. Ton containers should have a drip leg (liquid chlorine trap) (Figure 7-18) with a heater installed before the chlorinator. This will vaporize any liquid chlorine initially coming from the eductor at the start of gas operation, and any liquid droplets that may flow out of the container or evaporator during normal operation.

Opening the Valve

Once the cylinder or container has been installed and the flexible connector attached, the valve can be opened and the lines checked for leakage according to the following procedure:

1. Place the valve wrench provided by the chlorine supplier on the cylinder or container valve stem. Stand behind the valve outlet. Grasp the valve firmly with one hand, and give the wrench a sharp blow in a counterclockwise direction with the palm of the other hand. Do not pull or tug at the wrench because this may bend the stem, causing it to stick or fail to close properly.

To open a stubborn valve, follow the normal opening procedure, but use a small block of wood held in the palm of the hand when striking the wrench. If the valve continues to resist opening, return the cylinder to the supplier. Do not — under any circumstances — use a pipe wrench or an ill-fitting wrench because these wrenches will round the corners of the square-end valve stem. Avoid using wrenches longer than 6 in. (150 mm) for opening stubborn valves because they might bend or break the valve stem.

2. Open the valve and close it immediately.
3. The line is now pressurized. All new joints and connections can be checked for leaks using an ammonia solution. Use only commercial 26° Bé ammonia, which can be obtained from a supplier of chemicals or chlorine. Common household ammonia is not strong enough to work properly.

   Hold an opened plastic squeeze bottle of the ammonia beneath the valve and joints, and allow the ammonia fumes to rise up around any suspected leak areas. Ammonia fumes react with chlorine gas to form a white cloud of ammonium chloride, thus making small leaks easy to locate. Do not spray, swab, or otherwise bring ammonia liquid into contact with chlorination equipment. The chlorine will combine with the ammonia and may start corrosion at that point.
4. If no leaks are found, open the cylinder valve. One complete turn will permit the maximum withdrawal rate.
5. Leave the wrench on the valve to allow for easy and rapid shutoff in an emergency. The wrench also indicates to other operators which cylinder is being used. A sign disk should be installed on the stem of the cylinder or container valve (Figure 7-43) to show the closing direction in case of emergency.
6. If the injector and chlorinator are already operating, open the auxiliary valve; the newly connected cylinder or container will start feeding chlorine to the system. Do not open the auxiliary valve until the injector is operating because the necessary vacuum will not be developed and the regulating valves will not function.

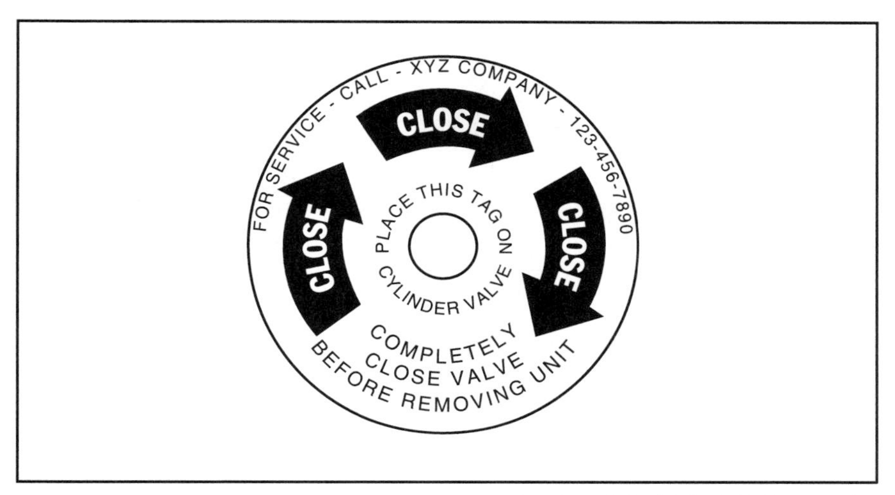

FIGURE 7-43 Sign to show which direction to close the valve in case of an emergency

Courtesy of Capital Controls Company, Inc.

Closing the Valve

When all the chlorine has been released from the cylinder or container, as indicated by the weight reading on the scale, the unit should be disconnected and replaced according to the following procedures:

1. Close the cylinder or container valve. After about 2 minutes, close the auxiliary valve. The delay allows any remaining chlorine gas in the line to be drawn into the injector.

   To close the valve, use the wrench provided, grasping the valve in one hand and tapping the wrench in a clockwise direction with the palm of the other. If the valve does not close tightly on the first try, open and close it lightly several times until the proper seating is obtained. Never use a hammer or any other tool to close the cylinder valve tightly.
2. Disconnect the flexible tubing from the cylinder or container, and replace the brass outlet cap on the valve immediately so that the valve parts will be protected from moisture in the air.
3. Screw the protective hood in place.
4. The outlet cap of each valve is fitted with a gasket that is designed to fit against the valve outlet face. If a valve leaks slightly after closing, the leak can often be stopped by drawing up the valve cap tightly.

5. The flexible copper tubing from the manifold to the container should be supported while the empty container is being replaced. Support it on another container, a wall hook, or a block in order to prevent any kinking or weak spots from developing in the pipe. If the flexible tubing is disconnected for any length of time, there is a danger of moisture forming in the line. Close the open end of the pipe with tape or plastic wrap and a rubber band.

Determining the Chlorine Dosage

The chlorine demand of the water being treated must be determined by performing a chlorine-demand test. By using the test result and knowing the desired residual, the operator can determine the dosage as follows:

$$\text{Cl dosage} = \text{Cl demand} + \text{Cl residual}$$

where all quantities are in milligrams per liter.

Once calculated, the dosage should be converted into pounds (or kilograms) per day, and the chlorinator should be set to deliver that dosage. After the dosage rate is set, the water must be tested regularly to ensure that the proper residual is maintained.

As noted previously, temperature, pH, and contact time are important variables affecting the effectiveness of chlorination. As these factors change, the amount of residual needed will also change. Table 7-10 shows general recommendations for the minimum residuals required for effective disinfection. Notice that recommendations are included for free residual as well as

TABLE 7-10 Recommended minimum concentrations of free chlorine residual versus combined chlorine residual

pH Value	Minimum Concentration of Free Chlorine Residual (Disinfecting Period Is at Least 10 minutes), *ppm*	Minimum Concentration of Combined Chlorine Residual (Disinfecting Period Is at Least 60 minutes), *ppm*
6.0–7.0	0.2	1.0
7.0–8.0	0.2	1.5
8.0–9.0	0.4	1.8
9.0–10.0	0.8	Not recommended
10.0+	0.8+ (with longer contact)	Not recommended

combined residual. Free residual values are based on a contact time of at least 10 minutes. The combined residual concentrations require at least 60 minutes of contact time. However, systems that must comply with the SWTR must meet considerably more complex requirements for determining $C \times T$ values.

Additional information on chlorine dosage, demand, and residual calculations is included in *Basic Science Concepts and Applications*, part of this series.

Chlorination Operating Problems

The following are a number of operating problems related to chlorination:

- chlorine leaks
- stiff container valves
- hypochlorinator problems
- tastes and odors
- sudden change in residual
- THM formation (discussed earlier in this chapter)

Proper maintenance can prevent many problems. Most manufacturers have equipment troubleshooting guides that help locate and correct problems.

Chlorine Leaks

A major concern in the operation and maintenance of the chlorination process is the prevention of chlorine leaks. The most common place for leaks to occur is the pressurized chlorine supply line between the containers and the chlorinator. Every joint, valve, fitting, and gauge in the line is a possible point of leakage.

Some chlorine leaks are readily apparent. Others are very slow, very small, partly hidden, or otherwise difficult to locate. The usual method of detection is to open a bottle of ammonia solution, place the bottle near a suspected leak, and allow the fumes to rise around the suspected area. If there is a sizable leak, the chlorine will combine with the ammonia to form a visible white vapor. Unfortunately, this method will not indicate a very small leak. In addition, very small leaks will often not produce a noticeable odor. Small leaks can go unnoticed for weeks unless the operator periodically looks for two signs: joint discoloration and moisture.

Even the smallest leak will remove cadmium plating from chlorine tubing and fittings. The metal underneath the plating (copper, brass, or bronze) will appear reddish, and a green copper-chloride scum may appear around the edges of the area affected.

Portions of the pressure piping system (for example, the manifold) are often painted. As a result, discoloration of the metal beneath the paint will not be apparent. To locate leaks in painted piping, look for small droplets of water that may form on the underside of joints. Small, almost invisible leaks must be located early; otherwise, the corrosion they cause will often result in a sudden and massive chlorine leak after a period of time.

The best and most reliable way to find chlorine leaks is a chlorine detector, which is sensitive to leaks as small as 1 ppm chlorine in air. Such leaks are not normally detectable by the ammonia technique or by smell.

If a major leak requires shutdown of the system for repair, the tank valve should be closed, the yoke disconnected, and the injector left running with the auxiliary tank valve open, until any remaining chlorine gas is purged from the line. To prevent leaks, the operator should observe the following precautions:

- Install a new gasket every time a cylinder or container is changed.
- Each time a threaded fitting is opened, clean the threads with a wire brush. Then wrap them with polytetrafluoroethylene (PTFE) tape or use one of the following pipe joint compounds: linseed oil and graphite, linseed oil and white lead, or litharge and glycerine. If PTFE tape is used, remove any previous remnants of tape before remaking the joint.
- Replace all chlorine supply line valves annually. Refit and repack the old ones so they are ready for use the following year.

Stiff Container Valves

Container valves are carefully checked before leaving the manufacturer's plant, but occasionally a valve may be stiff to turn or difficult to shut off tightly. This problem is often caused by overly tight packing. Sometimes the valve can be freed by opening and shutting it a few times. If the valve does not operate at all, set the container aside and call the supplier.

Hypochlorinator Problems

Two problems that commonly occur with hypochlorinators are clogged equipment and broken diaphragms.

Clogged Equipment

Clogging, due to calcium carbonate ($CaCO_3$) scaling, occurs primarily in two areas of a hypochlorinator: at the pump head and in the suction and discharge hoses. Scale is most likely to form when the water used for preparing the solution has a high calcium hardness and carbonate alkalinity. Under these conditions, calcium carbonate forms and causes a scale deposit in the pump head, suction hose, and discharge hose. Scale may also form in the solution injector or diffuser.

The scale can readily be removed by pumping a dilute (5 percent) hydrochloric acid solution (also known as muriatic acid) through the pump head, hoses, and diffuser. The hypochlorite solution should be completely flushed out of the system with water before the acid is used.

An associated problem affecting the pump head is the accumulation of dissolved calcium hypochlorite (lime sludge). When the solution tank level is low, or the suction foot valve is too near the bottom of a one-tank installation, the suction hose can draw some of the undissolved chemical up into the pump head and fill the area of the head. This can result in the pump not feeding hypochlorite solution, and it can cause diaphragm rupture. To prevent these problems, it is best to use a two-tank setup when calcium hypochlorite is being used.

Broken Diaphragms

The second most common problem with hypochlorinators is broken diaphragms. It is important that the operator inspect the diaphragm regularly to ensure that it is functioning properly. A visual inspection of the pump head may not reveal a broken diaphragm, but an outflow of solution from the diaphragm hose is a positive indication that the unit is functioning properly. Figures 7-44 and 7-45 indicate the points of diaphragm weakness in the two most common types of hypochlorinators.

Tastes and Odors

The cause of "swimming pool" taste and odor in drinking water is commonly misunderstood to be too much chlorine. Oddly enough, chlorine-like taste and odor are usually caused by too little chlorine. In the discussion of Figure 7-3, it was pointed out that the curve segment between points 2 and 4 is predominantly combined chlorine residual. The combined residual available in this range is the weakest form of chlorine for disinfection, and the accompanying dichloramines, trichloramines, and chloroorganic compounds cause taste and odor.

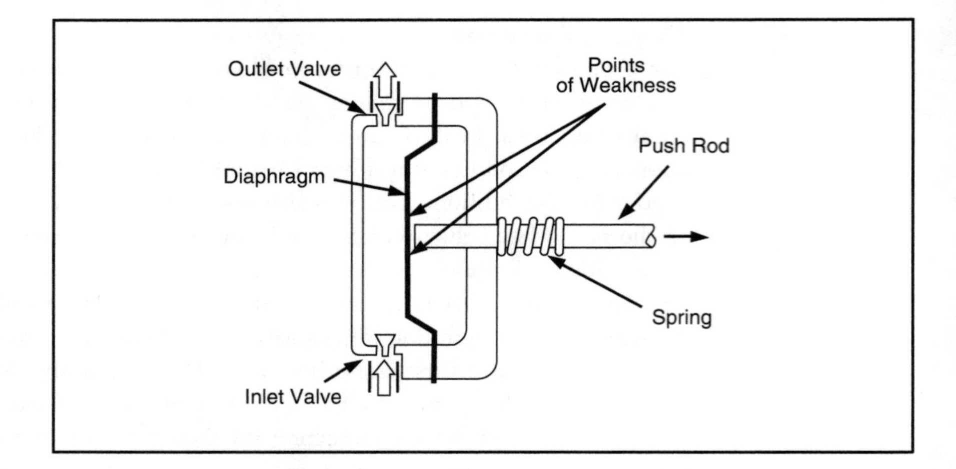

FIGURE 7-44 Points of weakness in a mechanically actuated diaphragm

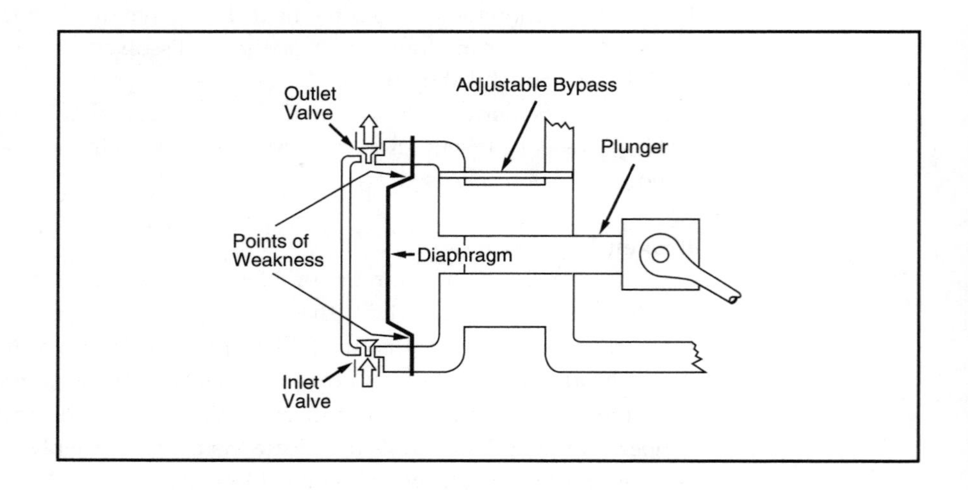

FIGURE 7-45 Points of weakness in a hydraulically actuated diaphragm

By increasing the chlorine dosage beyond the breakpoint (point 4 in Figure 7-3), any taste or odor is usually eliminated or significantly reduced. Free available chlorine residual is free from taste and odor at the concentrations commonly used in disinfection.

Sudden Change in Residual

One important attribute of chlorine is that the residual present in a sample can be measured easily. Based on a chlorine residual test procedure, the strength of the disinfectant remaining in the water at any point in the distribution system can be determined. This test should take less than

5 minutes. Records of the chlorine residual kept over a period of time can help predict the type and amount of residual that will be found at various locations in the distribution system. Any sudden drop in residual is a warning of potential danger, such as a cross-connection that is allowing contaminated water into the drinking water system.

When a sudden drop in residual occurs, chlorination levels should be increased immediately to raise the residual to the desired level. This guards against the possibility of waterborne disease while the problem is being analyzed and corrected. Samples should then be analyzed to identify the contaminants, and the distribution system should be checked to locate the source of contamination or other cause of the drop in residual.

Control Tests

Two types of tests are used to monitor the disinfection process. First, the level of chlorine residual is checked regularly in the treatment plant and at points throughout the distribution system. Second, bacteriological tests are performed on samples from selected points in the distribution system.

Chlorine Residual Test

The chlorine residual test is essential to the successful and efficient operation of the chlorination process. The test results provide the operator with the following three important pieces of information:

- whether or not a residual exists
- the type of residual (free or combined)
- the amount of residual (concentration)

The operator should also periodically monitor temperature and pH. These two factors influence the amount and type of residual formed, which in turn control the effectiveness of disinfection.

The chlorine residual test is one of the quickest and easiest of all operational control tests for a water plant. It can be performed using a field test kit. For laboratory use, the amperometric titrator is the most accurate method for measuring all forms of residual. The DPD method and the operational significance of the chlorine residual are covered in detail in *Water Quality*, also part of this series.

As a general rule, it is best to maintain a free available chlorine residual; it is a far more effective disinfectant than a combined residual.

Chlorine Demand Test

The chlorine demand test should be performed routinely to determine the proper chlorine dosage. The test consists of treating a series of water samples with known but varying chlorine doses. After a specified contact time, the chlorine residual of each sample is determined, which will indicate which dose satisfied the demand and provided the desired residual.

Bacteriological Test

All pathogenic organisms have a common source: the feces of humans and animals. One obvious way to determine the effectiveness of disinfection would be to test the treated water for the presence of pathogens. However, this is not practical because pathogens in water are few in number and difficult to measure, even with sophisticated laboratory equipment.

Fortunately, there is a group of bacteria, known as coliform bacteria, that is relatively easy to measure and whose presence indicates that pathogens might also be present. Because most pathogens are less resistant to chlorine than coliform bacteria, it is assumed that if no coliforms are found in treated water, pathogens are not present. Therefore, routine samples are taken from the raw and treated water at various points and tested for the presence of coliforms. A discussion of the sampling and testing procedures for the total coliform test is included in *Water Quality*, part of this series.

Safety Precautions

Chlorine should never cause an accident or injury if it is used by properly trained operators in adequately equipped plants. The key factors in safely operating a chlorination system are proper safety equipment and procedures.

Proper Safety Equipment

Without adequate equipment and proper training in the use of safety equipment, the operator's life — as well as the lives and well-being of the surrounding community — is needlessly at risk.

It is essential that every chlorination facility be equipped with the following safety devices:

- self-contained breathing equipment
- emergency repair kits
- adequate ventilation equipment

In addition, installation of a chlorine detector is strongly recommended at all chlorination installations, and it should *always* be installed at unattended chlorination stations.

Self-Contained Breathing Apparatus

Self-contained breathing equipment should be available wherever gas or liquid chlorine is in use. Air packs, as shown in Figure 7-46, have a positive-pressure mask with a full, wide-view face piece and a cylinder of air or oxygen carried on the operator's back. The only units that should be purchased are those that have been approved by the National Institute of Occupational Safety and Health (NIOSH). Canister-type gas masks, which were used in the early days of chlorination, are no longer recommended because they are not effective against heavy chlorine concentrations.

Every operator should be familiar with the location and use of the breathing apparatus available at the treatment plant. Operating instructions are provided with each unit. Operators should study and review these instructions periodically and have regular formal training and practice sessions in the use of the equipment.

An air pack is very similar to the equipment used by scuba divers and fire departments. The tank contains a 15- to 30-minute supply of air, depending on tank size. The actual time an air supply will last also depends on an individual's pattern of breathing while under stress. The mask must fit

FIGURE 7-46 Air pack with positive-pressure mask

tightly around the face. Operators with a beard or eyeglasses may find it difficult to fit the mask.

Operators should practice repairing simulated leaks while using breathing equipment, so that they will know how long they can work before the air tank is exhausted. A low–air-pressure alarm on the 30-minute air pack is activated when about 5 minutes of air is left. When the alarm sounds, the wearer should immediately leave the contaminated area to get a fresh cylinder of air.

The air pack should be located at a readily accessible point, away from the area likely to be contaminated with chlorine gas. It should not be located in the chlorine feed or storage rooms. A wall storage cabinet is usually mounted outside these rooms (Figure 7-47). The mask and air supply tanks should be inspected routinely and maintained in good condition. Spare air cylinders should be on-site for use during prolonged emergencies.

The mask and breathing apparatus should be cleaned at regular intervals and after each use. When air tanks are depleted, they should be refilled at stations where proper air compressor equipment is available. Many local fire departments have this equipment.

When stowing the equipment after use, the straps on the masks and backpacks should be extended to their limits. This allows the equipment to be fitted quickly to the next user.

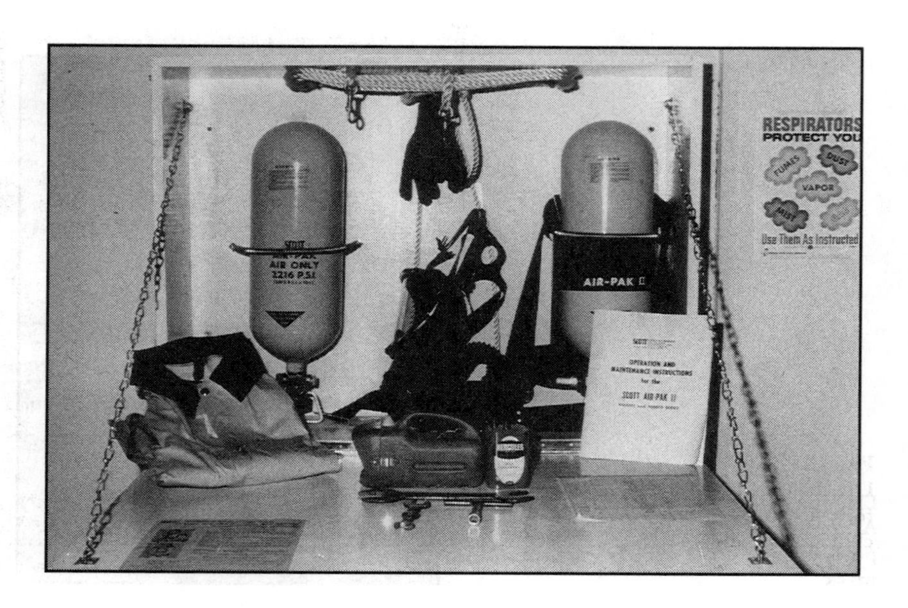

FIGURE 7-47
Storage unit for chlorine air pack

Emergency Repair Kits

Standardized emergency repair kits that meet US and Canadian specifications contain various devices and hardware for stopping leaks from chlorine-shipping containers. Currently, the three standard Chlorine Institute emergency kits are

- emergency kit A for cylinders (Figure 7-48)
- emergency kit B for ton containers (Figure 7-49)
- emergency kit C for tank cars and tank trucks

Each kit is designed to be used in the repair of leaks that can occur in the shipping containers. The kits contain all the equipment needed to cap a leaking valve, seal a sidewall leak, and cap a fusible plug.

Instruction booklets and other materials are available from the Chlorine Institute for training operators in the use of repair kits. It is important that each operator be trained through an established training program.

Adequate Ventilation

Because chlorine gas is 2.5 times as dense as air, it will settle and stay near the floor when leakage occurs. The rooms in which chlorine cylinders are stored and the enclosures that surround chlorinators should have sealed

FIGURE 7-48 Energency kit A for use with chlorine cylinders

Courtesy of the Chlorine Institute, Inc.

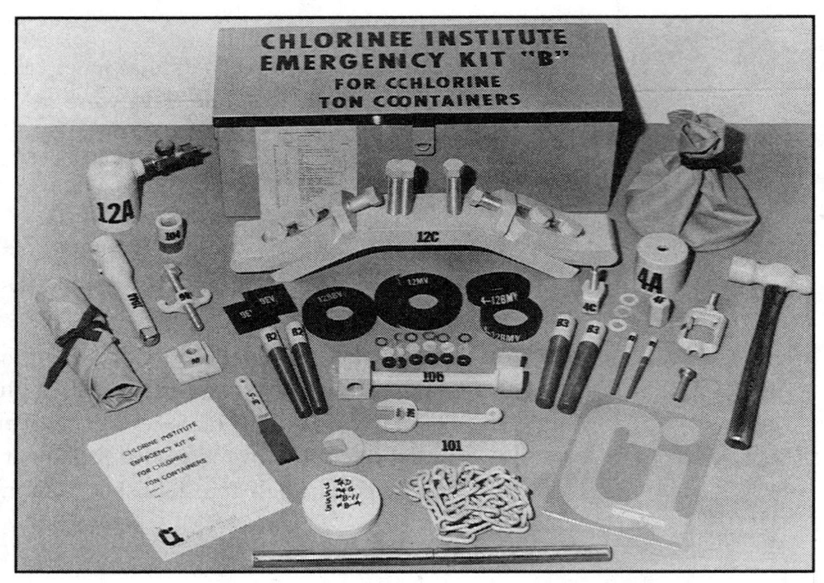

FIGURE 7-49 Emergency kit B for use with chlorine ton containers

Courtesy of the Chlorine Institute, Inc.

walls, and the doors to the rooms should open outward. The rooms should also be fitted with chlorine-resistant power exhaust fans ducted at the floor level and with fresh-air intake vents at the ceiling. In large installations, it is usually desirable to provide a fan near the ceiling to force fresh air into the chlorinator room. The ventilating equipment should be capable of completely changing the volume of air in the room every 1 to 4 minutes. A different required rate may be specified in the regulations of the state drinking water agency.

Exhaust fan switches should be located outside the room and should be wired to room lights so that the lights and fan go on at the same time. It is also good practice to have a window in the door or adjacent to the door, so that the operator can look into the room to detect any abnormal conditions before entering (Figure 7-50).

Chlorine Detector

The installation of a chlorine detector is a wise investment for every treatment plant using chlorine gas. It detects chlorine concentrations so small that they are impossible to smell. It can give early warning so that a small leak can be stopped before becoming larger, and it can provide immediate warning of a large leak.

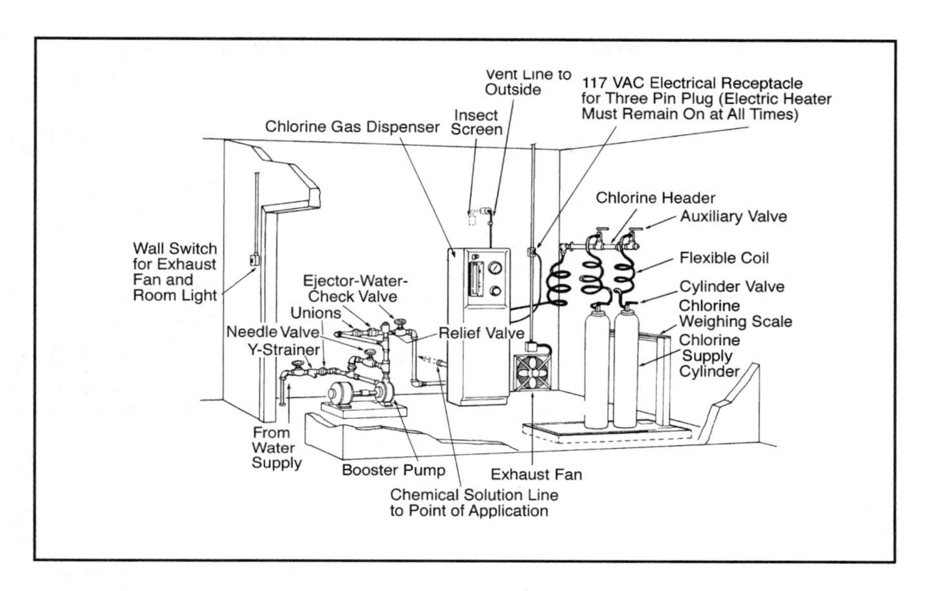

FIGURE 7-50 Schematic of a well-designed chlorine room

Various types of chlorine detectors are available. One type, shown in Figure 7-51, operates by measuring conductivity between two electrodes set in a liquid or solid electrolyte. Samples of air are drawn from the area being monitored, and any chlorine in the sample will cause an increase in the current passing between the electrodes. A sensitive circuit detects this change and activates an alarm.

Proper Safety Procedures

Every facility using chlorine, regardless of size, should have established procedures for handling chlorine and chlorine leaks. These procedures should include

- safety precautions for storing and handling cylinders or containers
- basic steps in connecting and disconnecting cylinders or containers
- procedures to follow in case of a chlorine leak
- emergency procedures to be taken if a chlorine leak threatens nearby residential areas
- first aid procedures for persons exposed to chlorine

Each operator should be thoroughly familiar with all of these procedures through routine, in-plant training programs. These programs should emphasize the use of respiratory protection equipment, leak repair kits, and emergency first aid.

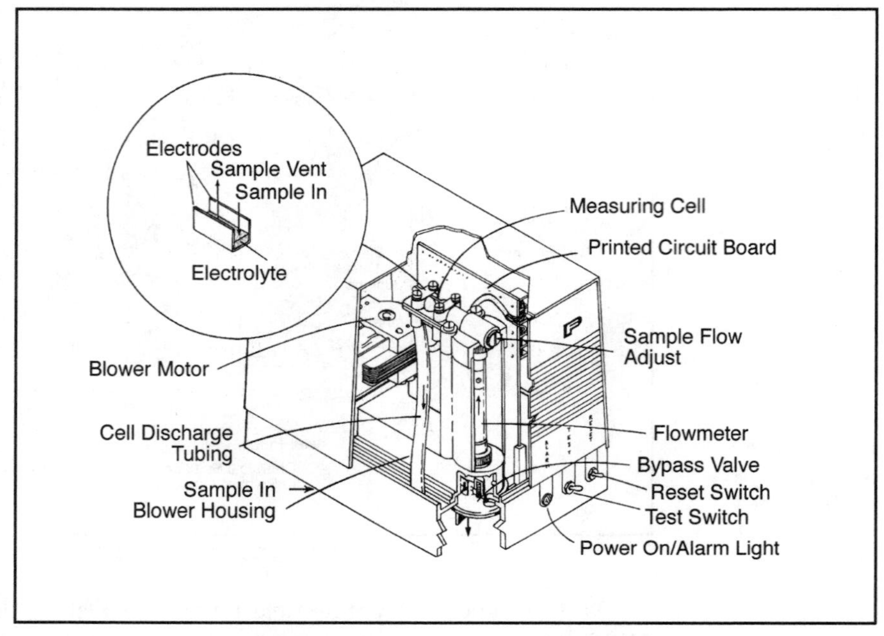

FIGURE 7-51
Chlorine detector

Courtesy of Fischer & Porter Co.

It is helpful to post descriptions of important procedures near the chlorination facilities. Many of the chlorine chemical and equipment manufacturers distribute wall charts on safety and handling procedures that can be used for training and display. Additional information on chlorine handling is also available from the Chlorine Institute (see appendix C).

Emergency Assistance for Chlorine Leaks

The North American Chlorine Emergency Plan (CHLOREP) is a mutual-aid program for chlorine incidents that occur during transportation or at user locations. The program provides immediate emergency response from a local chlorine packager or manufacturer. Through CHLOREP, about 250 emergency response teams are available 24 hours a day to provide emergency telephone instructions to personnel at the scene of an incident or to provide actual assistance if required. Team members are all experts who have been trained through company programs and Chlorine Institute seminars.

The dispatch agencies are as follows:

- the Chemical Transportation Emergency Center (CHEMTREC), operated by the Chemical Manufacturers Association
- the Transportation Emergency Assistance Plan (TEAP), operated by the Canadian Chemical Producers Association
- the Canadian Transport Emergency Centre (CANUTEC) in Ottawa

Dispatchers from these agencies contact the designated CHLOREP representative, who in turn contacts the CHLOREP team leader. The team leader then contacts the emergency caller to determine what expertise and aid are needed. The numbers to call are as follows:

- in the United States (48 contiguous states): 800-424-9300
- in Alaska, the District of Columbia, and Hawaii: 202-483-7616
- in Canada (all provinces): 613-966-6666 (call collect)

Record Keeping

Records for the disinfection process should show the type and amount of disinfectant used, and the bacteriological and other operational control test results. The following is a list of information that should be recorded as part of the disinfection process:

- types of disinfectants in use
- ordering information, including
  - manufacturer's name, address, and phone number
  - shipper's name, address, and phone number
  - type, size, and number of shipping containers
- most recent costs
- current dosage rate, in milligrams per liter or pounds per day
- bacteriological test results
- chlorine and other disinfectant residual test results
- water temperature
- raw-water pH
- daily explanation of unusual conditions, mechanical problems, supply problems, emergencies, or unusual test results

Selected Supplementary Readings

Back to Basics Guide to Disinfection With Chlorine. 1991. Denver, Colo.: American Water Works Association.

Bishop, M.M., et al. 1993. Alternative Disinfection Technologies for Small Systems. ***Opflow,*** 19(2):1.

———. 1993. Improving the Disinfection Detention Time of a Water Plant Clearwell. ***Jour. AWWA,*** 85(3):68.

Bull, R.J., and F.C. Kopfler. 1991. ***Health Effects of Disinfectants and Disinfection By-products.*** Denver, Colo.: American Water Works Association Research Foundation and American Water Works Association.

The Chlorine Manual. Washington, D.C.: The Chlorine Institute, Inc.

Controlling Disinfection By-Products. 1993. Denver, Colo.: American Water Works Association.

DeMers, L., and R.C. Renner. 1992. ***Alternative Disinfection Technologies for Small Drinking Water Systems.*** 1992. Denver, Colo.: American Water Works Association Research Foundation and American Water Works Association.

Dietrich, A.M., R.C. Hoehn, and C.E. Via Jr. 1991. ***Taste and Odor Problems Associated With Chlorine Dioxide.*** Denver, Colo.: American Water Works Association Research Foundation and American Water Works Association.

Dietrich, A.M. 1993. ***Chlorine Dioxide: Drinking Water Issues Comparative Analytical Methods.*** Denver, Colo.: American Water Works Association Research Foundation and American Water Works Association.

Dietrich, A.M., et al. 1992. Tastes and Odors Associated With Chlorine Dioxide. ***Jour. AWWA,*** 84(6):82.

Disinfection By-products: Current Perspectives. 1989. Denver, Colo.: American Water Works Association Research Foundation and American Water Works Association.

Ferguson, D.W., et al. 1990. Comparing PEROXONE and Ozone for the Control of Taste and Odor Compounds, Disinfection By-products, and Microorganisms. ***Jour. AWWA,*** 82(4):181.

Ferguson, D.W., J.T. Gramith, and M.J. McGuire. 1991. Applying Ozone for Organics Control and Disinfection: A Utility Perspective. ***Jour. AWWA,*** 83(5):32.

Finch, G.R., E.K. Black, Lyndon Gyurek, and Miodrag Belosevic. 1994. ***Ozone Disinfection of* Giardia *and* Cryptosporidium.** Denver, Colo.: American Water Works Association Research Foundation and American Water Works Association.

George, D.B., V.D. Adams, S.A. Huddleston, K.L. Roberts, and M.B. Borup. 1990. ***Case Studies of Modified Disinfection Practices for Trihalomethane Control.*** Denver, Colo.: American Water Works Association Research Foundation and American Water Works Association.

Gordon, G., et al. 1990. Minimizing Chlorite Ion and Chlorate Ion in Drinking Water Treated With Chlorine Dioxide. ***Jour. AWWA,*** 82(4):160.

Gordon, G., W.J. Cooper, R.G. Rice, and G.E. Pacey. 2nd ed. 1992. ***Disinfectant Residual Measurement Methods.*** Denver, Colo.: American Water Works Association Research Foundation and American Water Works Association.

Guidance Manual for Compliance With the Filtration and Disinfection Requirements for Public Water Systems Using Surface Water Sources. 1990. Denver, Colo.: American Water Works Association.

Hoehn, R.C., et al. 1990. Household Odors Associated With the Use of Chlorine Dioxide. ***Jour. AWWA,*** 82(4):166.

Jacangelo, J.G., et al. 1989. Ozonation: Assessing Its Role in the Formation and Control of Disinfection By-products. ***Jour. AWWA,*** 81(8):74.

Kirmeyer, G.J., G.W. Foust, G.L. Pierson, J.J. Simmler, and M.W. LeChevallier. 1993. ***Optimizing Chloramine Treatment.*** Denver, Colo.: American Water Works Association Research Foundation and American Water Works Association.

Krasner, S.W., et al. 1989. Free Chlorine Versus Monochloramine for Controlling Off-Tastes and Odors. ***Jour. AWWA,*** 81(2):86.

———. 1989. The Occurrence of Disinfection By-products in US Drinking Water. ***Jour. AWWA,*** 81(8):41.

LeChevallier, M.W., W.D. Norton, R.G. Lee, and J.B. Rose. 1991. **Giardia *and* Cryptosporidium *in Water Supplies.*** Denver, Colo.: American Water Works Association Research Foundation and American Water Works Association.

Manual M3, Safety Practices for Water Utilities. 1990. Denver, Colo.: American Water Works Association.

Manual M20, Water Chlorination Principles and Practices. 1973. Denver, Colo.: American Water Works Association.

Manual of Instruction for Water Treatment Plant Operators. 1975. Albany, N.Y.: New York State Department of Health.

Manual of Water Utility Operations. 8th ed. 1988. Austin, Texas: Texas Water Utilities Association.

The Metropolitan Water District of Southern California and James M. Montgomery, Consulting Engineers, Inc. 1991. ***Pilot Scale Evaluation of Ozone and PEROXONE.*** Denver, Colo.: American Water Works Association Research Foundation and American Water Works Association.

Neden, D.G., et al. 1992. Comparing Chlorination and Chloramination for Controlling Bacterial Growth. ***Jour. AWWA,*** 84(7):80.

The Netherlands Waterworks Testing and Research Institute (Kiwa N.V.) 1986. ***Chlorination By-products: Production and Control.*** Denver, Colo.: American Water Works Association Research Foundation and American Water Works Association.

Reckhow, D., and P.C. Singer. 1990. Chlorination By-products in Drinking Waters: From Formation Potentials to Finished Water Concentrations. ***Jour. AWWA,*** 82(4):173.

Recommended Standards for Water Works. 1992. Albany, N.Y.: Health Education Services.

Tan, L., and G.L. Amy. 1991. Disinfection By-products Control. ***Jour. AWWA,*** 83(5):74.

Teefy, S.M., and P.C. Singer. 1990. Performance and Analysis of Tracer Tests to Determine Compliance With the Surface Water Treatment Rule. ***Jour. AWWA,*** 82(12):88.

Total Coliform Rule. 1990. Denver, Colo.: American Water Works Association.

Treatment Techniques for Controlling Trihalomethanes in Drinking Water. 1982. Denver, Colo.: American Water Works Association.

Von Huben, H. 1991. ***Surface Water Treatment: The New Rules.*** Denver, Colo.: American Water Works Association.

Water Quality and Treatment. 4th ed. 1990. New York: McGraw-Hill and American Water Works Association (available from AWWA).

White, G.C. 1972. ***Handbook of Chlorination.*** New York: Van Nostrand Company.

CHAPTER 8

Fluoridation

Fluoridation is the deliberate adjustment of the fluoride concentration in a public water supply. It is done to maintain the optimal level of fluoride needed by children to develop teeth resistant to decay.

Fluoride is an ion originating from the element fluorine. It is a constituent of the earth's crust and consequently is found naturally, to some degree, in all drinking water sources. A small amount of fluoride in the diet is essential for proper tooth and bone formation.

The benefits of fluoride in reducing tooth decay were discovered through comparisons of the teeth of children from areas that have different concentrations of natural fluoride in their drinking water. Fluoridation is a safe, effective, and economical process endorsed by the American Dental Association, the American Water Works Association, and public health groups worldwide.

Process Description

Reasons for Fluoridation

At optimal levels, fluoride can greatly reduce the incidence of tooth decay among children. The amount of fluoride consumed with drinking water is generally based on the total amount of water consumed each day. Water consumption generally depends on the temperature in a region; people in warm climates tend to drink more water than people in cold climates. Consequently, the optimal fluoride concentration in drinking water varies across the country and is set in relation to the average air temperature. The fluoride levels shown in Table 8-1 are for annual averages of maximum daily air temperatures, which must be determined for a five-year period. Each state

TABLE 8-1 Optimal fluoride concentrations

Annual Average of Maximum Daily Air Temperature,* °F	(°C)	Recommended Control Limits of Fluoride Concentration, *mg/L* Lower	Optimal	Upper
53.7 and below	(12.0 and below)	0.9	1.2	1.7
53.8–58.3	(12.1–14.6)	0.8	1.1	1.5
58.4–63.8	(14.7–17.6)	0.8	1.0	1.3
63.9–70.6	(17.7–21.4)	0.7	0.9	1.2
70.7–79.2	(21.5–26.2)	0.7	0.8	1.0
79.3–90.5	(26.3–32.5)	0.6	0.7	0.8

*Based on temperature data for a minimum of five years.

health department uses this information to establish optimal fluoride levels for the state.

To achieve the maximum benefits of fluoridation, the optimal fluoride concentration in the water supply must be continuously maintained. A drop of only 0.3 mg/L below optimal can reduce fluoride's benefits by as much as two thirds. However, concentrations above 1.5 mg/L over the optimal level do not significantly reduce tooth decay any further and can cause mottling of the teeth.

Effects of Excessive Fluoride in Water

Surface water normally contains only trace amounts of fluoride; groundwater often contains fluoride near the optimal level. Some wells have fluoride levels up to four or five times the optimal level. When children are exposed to excessive levels of fluoride, a condition known as fluorosis occurs. In its mildest form, fluorosis appears as very slight, opaque, whitish areas (called mottling) on the tooth surface. More severe fluorosis causes teeth to darken, turning from shades of gray to black. When the fluoride concentration is over 4 mg/L, teeth are likely to be pitted; they then become more susceptible to cavities and wear. Studies have shown that fluorosis starts to occur when children younger than eight years old regularly drink water containing twice the optimal fluoride level for three months or longer.

When a water source contains fluoride concentrations slightly higher than optimal, the water utility may be required to provide periodic public notification to warn customers that fluorosis may occur in their children's teeth. If levels exceed the maximum contaminant level, the utility must stop supplying the water to the public. Details of regulations concerning excessive fluoride are provided in *Water Quality*, also part of this series.

Fluoridation Facilities

The facilities used for fluoridation are similar to those used for feeding other water treatment chemicals. The type of equipment depends primarily on how much water is to be treated and the type of chemical being used.

Fluoride Chemicals

The three chemical compounds used for fluoridation are

- sodium fluoride
- fluorosilicic acid
- sodium fluorosilicate

The characteristics of these compounds are summarized in Table 8-2. Only chemicals meeting applicable standards should be used for addition to potable water (see appendix A).

Because of the relatively small quantities of chemicals required to maintain the optimal dosage, chemical costs for fluoridation are very small in relation to the overall operation cost of a water system.

Sodium Fluoride

The first compound to be used in practicing controlled fluoridation was sodium fluoride (NaF), which is still widely used today. It is a white, odorless material that comes as free-flowing crystals or in a coarse crystalline form.

The solubility of sodium fluoride is an almost constant 4 g NaF/100 mL water within the common range of water temperatures of 32–77°F (0–25°C). It is available in 100-lb (45-kg) multi-ply paper bags, in drums that hold up to 400 lb (180 kg), and in bulk.

Additional information on sodium fluoride can be found in AWWA Standard B701.

Fluorosilicic Acid

Fluorosilicic acid (H_2SiF_6) was formerly referred to as hydrofluosilicic acid or "silly acid." It is a clear, colorless to straw yellow colored, fuming, and very corrosive liquid. It has a pungent odor, can cause skin irritation, and will even etch glass.

All commercial solutions of fluorosilicic acid have a low pH, ranging from 1.0 to 1.5. In highly alkaline waters, the addition of the acid usually will

TABLE 8-2 Characteristics of fluoride compounds

Item	Sodium Fluoride, NaF	Sodium Fluorosilicate, Na_2SiF_6	Fluorosilicic Acid, H_2SiF_6
Form	Powder or crystal	Powder or very fine crystal	Liquid
Molecular weight	42.00	118.1	144.08
Commercial purity, %	97–98	98–99	20–30
Fluoride ion, % (100% pure material)	42.25	60.7	79.2
Pounds required per mil gal for 1.0 ppm F at indicated purity	18.8, 98%	14.0, 98.5%	35.2, 30%
pH of saturated solution	7.0	3.5–4.0	1.2 (1% solution)
Sodium ion contributed at 1.0 ppm F, ppm	1.17	0.40	0.00
F ion storage space, ft^3/100 lb	22–34	23–30	54–73
Solubility at 77°F, g/100 mL water	4.0	0.762	Infinite
Weight, lb/ft^3	65–90	85–95	10.5 lb/gal, 30%
Shipping containers	100-lb bags, 125–400-lb fiber drums, bulk	100-lb bags, 125–400-lb fiber drums, bulk	13-gal carboys, 55-gal drums, bulk

not appreciably affect the pH of the treated water. However, in low-alkaline (poorly buffered) waters, the addition of fluorosilicic acid can reduce the pH, so its use is not recommended without a study of the possible side effects.

Fluorosilicic acid is available in 13-gal (50-L) or 55-gal (210-L) drums for small users, and in tank cars or trucks for large users. Because it contains a high proportion of water (about 70 percent), fluorosilicic acid is costly to ship compared to the dry chemicals. However, the greater ease of feeding a liquid that is already prepared more than offsets this disadvantage for many users.

If the acid must be diluted, care should be taken to avoid dilutions between the ranges of 10 to 1 and 20 to 1 (parts water to parts acid). Within these ranges, an insoluble silica precipitate often forms, which can clog feeders, orifices, and other equipment.

Additional information on fluorosilicic acid can be found in AWWA Standard B703.

Sodium Fluorosilicate

The most inexpensive chemical available for fluoridation is sodium fluorosilicate, formerly known as sodium silicofluoride (Na_2SiF_6). It is a white or yellowish-white, slightly hygroscopic, crystalline powder with limited solubility in water. Although odorless, it has an acidic taste.

Unlike sodium fluoride, its solubility decreases as water temperature decreases. At 60–70°F (16–21°C), it takes 60 gal (230 L) of water to dissolve 1 lb (0.45 kg) of sodium fluorosilicate. The pH of the saturated solution is quite low, between 3.0 and 4.0, but this is not a problem because the solution is diluted once it is added to drinking water. Sodium fluorosilicate is available in the same size containers as sodium fluoride.

Additional information on sodium fluorosilicate can be found in AWWA Standard B702.

Chemical Feeders

Fluoride is fed into the water system by either a dry feed system or a solution feed system. Either system will feed a specific quantity of chemical into the water at a preset rate. Table 8-3 lists the required equipment and other characteristics for several methods of feeding fluoride.

Dry Feeders

A dry feeder meters a dry powder or crystalline chemical at a given rate. It is generally used to add fluoride to systems that produce 1 mgd (3.8 ML/d) or more. The two basic types are volumetric and gravimetric dry feeders.

Volumetric feeders (Figures 8-1 and 8-2) are simple to operate and less expensive to purchase and maintain than gravimetric feeders. However, they are generally less accurate. The feed mechanism delivers the same volume of dry chemical to the dissolving tank for each complete revolution of the screw or roll. Varying the speed of rotation varies the feed rate.

Gravimetric dry feeders can deliver large quantities and are extremely accurate, but they are relatively expensive. They can readily be adapted to

TABLE 8-3 Fluoridation checklist

Operating Parameters	Sodium Fluoride, Manual Solution Preparation	Sodium Fluoride, Automatic Solution Preparation	Fluorosilicic Acid, Diluted	Fluorosilicic Acid, 23–30%	Sodium Fluorosilicate, Dry Feed	Sodium Fluoride, Dry Feed
Water flow rate	Less than 500 gpm	Less than 2,000 gpm	Less than 500 gpm	More than 500 gpm	More than 100 gpm	More than 2 mgd
Population served by system or each well of multiple-well system	Less than 5,000	Less than 10,000	Less than 10,000	More than 10,000	More than 10,000	More than 50,000
Equipment required	Solution feeder, mixing tank, scales, mixer	Solution feeder, saturator, water meter	Solution feeder, scales, measuring container, mixing tank, mixer	Solution feeder, day tank, scales, transfer pump	Volumetric dry feeder, scales, hopper, dissolving chamber	Gravimetric dry feeder, hopper, dissolving chamber
Feed accuracy	Depends on solution preparation and feeder	Depends on feeder	Depends on solution preparation and feeder	Depends on feeder	Usually within 3%	Usually within 1%
Chemical specifications and availability	Crystalline NaF, dust-free, in bags or drums; generally available	Downflow: coarse crystalline NaF in bags or drums, may be scarce; upflow: fine crystalline NaF, generally available	Low-silica or fortified acid in drums or carboys; generally available	Bulk acid in tank cars or trucks; available on contract	Powder in bags, drums, or bulk; generally available	

Table continued next page

TABLE 8-3 Fluoridation checklist (continued)

Operating Parameters	Sodium Fluoride, Manual Solution Preparation	Sodium Fluoride, Automatic Solution Preparation	Fluorosilicic Acid, Diluted	Fluorosilicic Acid, 23–30%	Sodium Fluorosilicate, Dry Feed	Sodium Fluoride, Dry Feed
Handling requirements	Weighing, mixing, measuring	Dumping whole bags only	Pouring or siphoning, measuring, mixing, weighing	All handling by pump	Bag loaders or bulk-handling equipment required	
Feeding point	Injection into filter effluent line or main	Injection into filter effluent line or main	Injection into filter effluent line or main	Injection into filter effluent line or main	Gravity feed from dissolving chamber into open flume or clearwell, pressure feed into filter effluent line or main	
Other requirements	Solution water may require softening	Solution water may require softening	Dilution water may require softening	Acid-proof storage tank, piping, etc.	Dry storage area, dust collectors, dissolving-chamber mixers, hopper agitators, eductors, etc.	
Hazards	Dust, spillage, solution preparation error	Dust, spillage	Corrosion, fumes, spillage, solution preparation error	Corrosion, fumes, leakage	Dust, spillage, arching, and flooding in feeder and hopper	

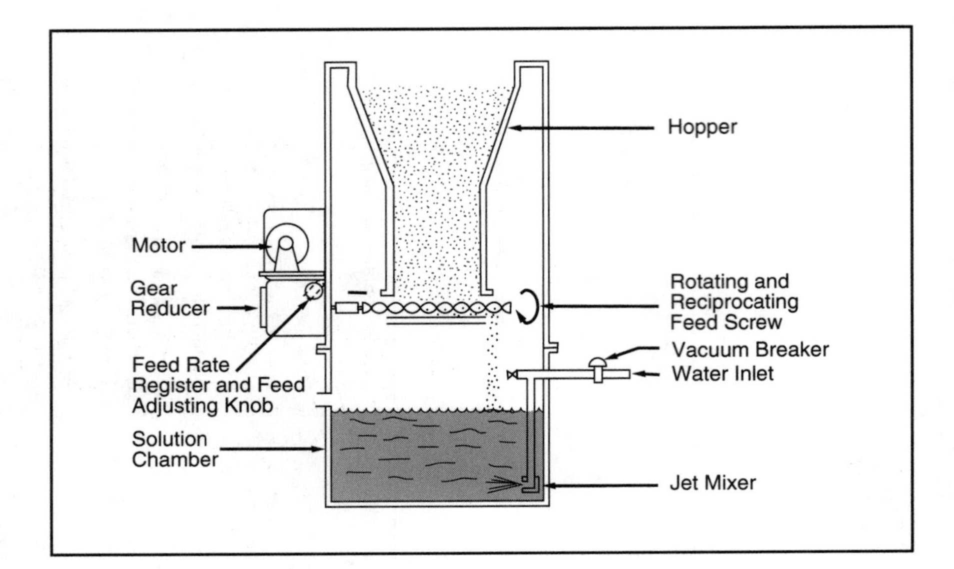

FIGURE 8-1 Screw-type volumetric dry feeder

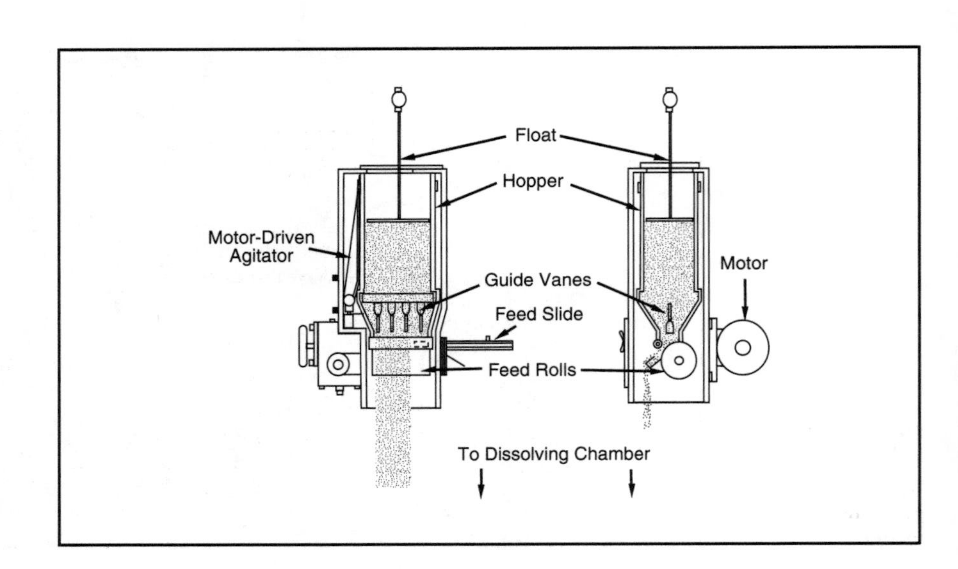

FIGURE 8-2 Roll-type volumetric dry feeder

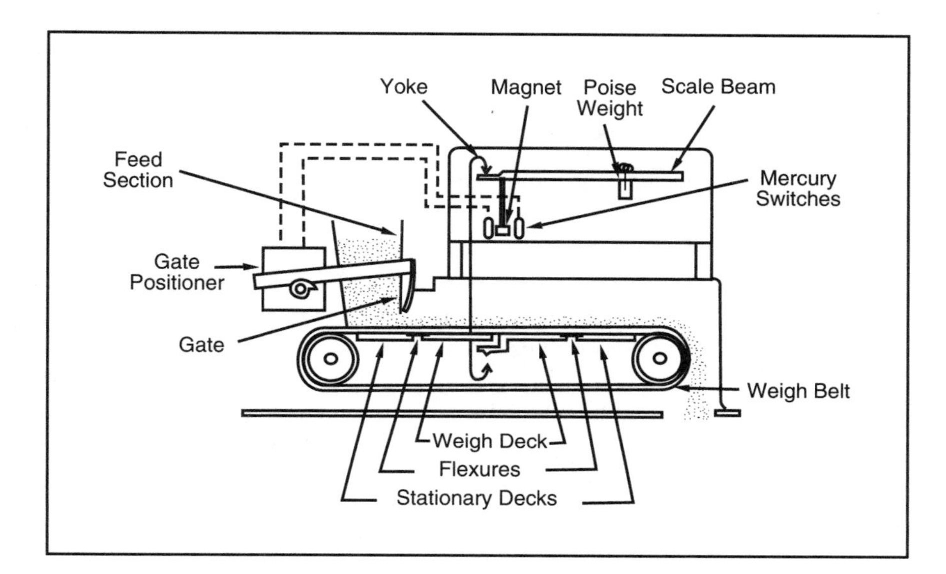

FIGURE 8-3 Belt-type gravimetric dry feeder

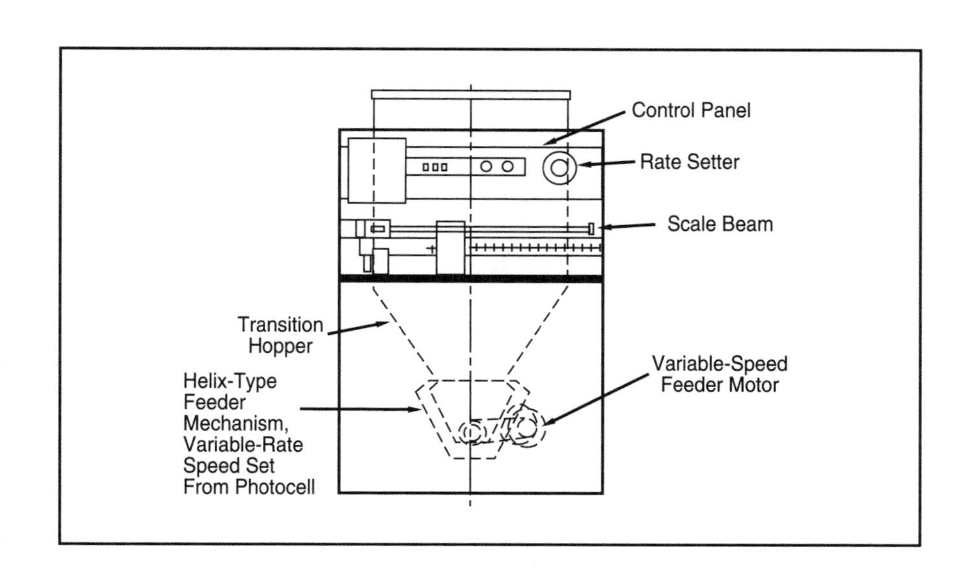

FIGURE 8-4 Loss-of-weight type of gravimetric dry feeder

automatic control and recording. The belt-type feeder (Figure 8-3) delivers a certain weight of material for each revolution of the conveyor belt. The feed rate is adjusted by varying the speed of the belt. The loss-of-weight–type feeder (Figure 8-4) matches the weight lost by the feed hopper to the preset weight of the required dosage. Because gravimetric feeders control the weight of material, not the volume, variations in density have no effect on feed rate. This accounts for the extreme accuracy of this type of feeder.

The steady stream of dry material discharged from a dry feeder falls into a solution chamber or tank, where it is dissolved into water. The tank is usually equipped with a mixer to ensure that the chemical is thoroughly dissolved. The resulting fluoride solution either flows by gravity into the clearwell or is pumped into a pressure pipeline. Either of the dry chemicals can be used in a dry feeder, but sodium fluorosilicate is most often used for dry chemical feed because it is less expensive.

Solution Feeders

Solution feeders are often the most economical way for water systems to fluoridate water (Table 8-3). The feeders are small pumps that feed a sodium fluoride or fluorosilicic acid solution from a tank or saturator into the water system at a preset rate.

Diaphragm pumps (Figure 8-5) and piston pumps are generally used for fluoride feed because they are accurate and can operate against a pressure if the chemical is being fed into a pipeline.

Diaphragm pumps are the same as those used for hypochlorination systems. They consist of a flexible pumping diaphragm made of rubber, plastic, or thin metal, which is actuated by a reciprocating shaft driven by a cam. To vary the feed rate, either the cam speed or the pump shaft stroke is adjusted. The pumps can usually be electronically controlled based on flow rate.

Piston pumps are similar, but instead have a rigid piston that moves back and forth within a cylinder that serves as the metering chamber. By varying the stroke length, the pump can vary the feed rate. Some piston pumps are capable of delivering chemicals at very low rates and against high pressure.

Manual Solution Feed

The manual solution feed method, as shown in Figure 8-6, is sometimes used by small systems feeding sodium fluoride. The chemical is added to the mixing tank for dissolving. Then the concentrated solution is transferred to the day tank and fed by a metering pump. The day tank is placed on a

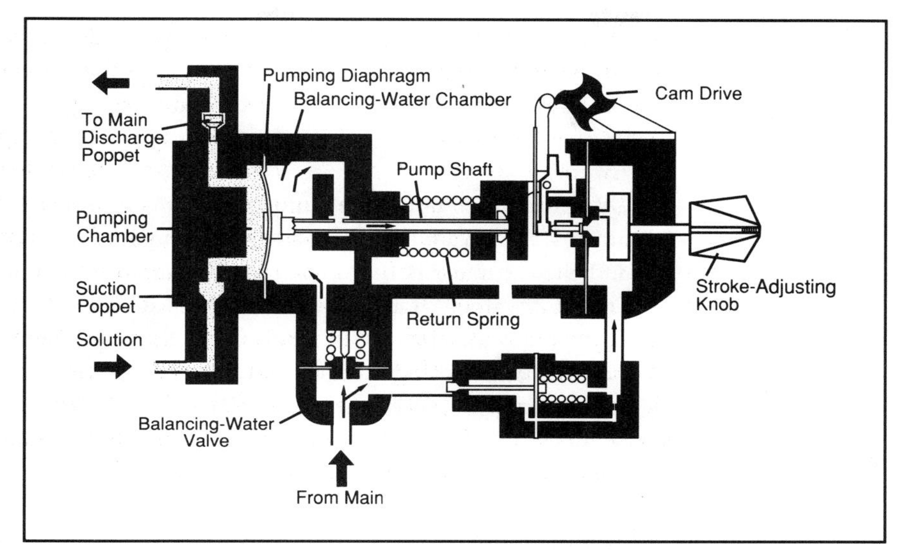

FIGURE 8-5 Diaphragm chemical feed pump

Courtesy of Wallace & Tiernan, Inc.

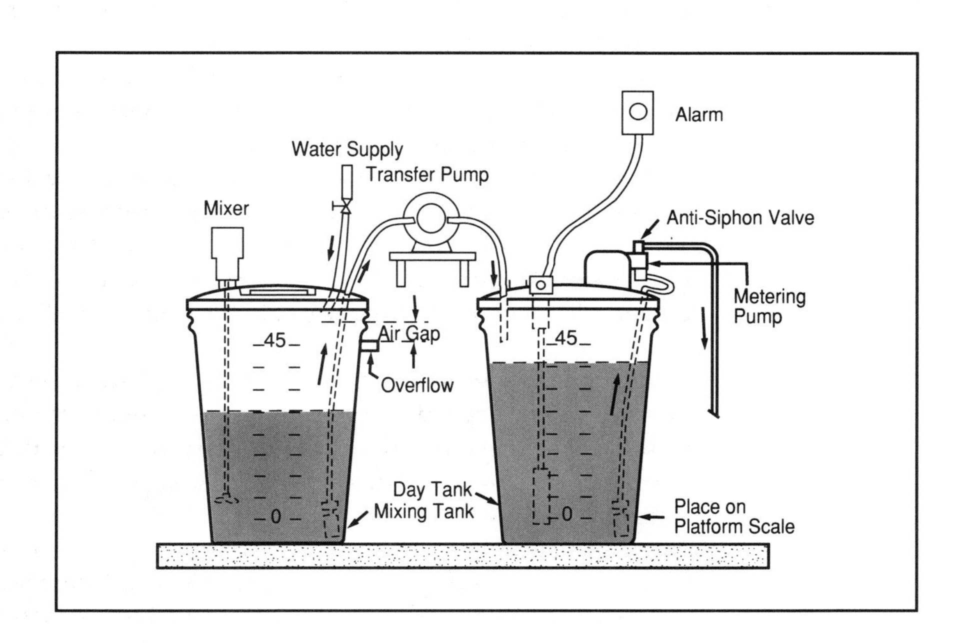

FIGURE 8-6 Manual solution feed installation

platform scale, which allows one to determine how much solution is fed. The major disadvantage of this system is that the sodium fluoride must be manually weighed and added to the mixing tank.

Saturators

A saturator retains the simplicity of the manual feed system but eliminates the disadvantage of constant chemical handling. The principle behind the saturator is that a saturated solution of 40 g/L sodium fluoride will result if water at normal temperatures is allowed to trickle through a bed of sodium fluoride crystals. In other words, no matter how much chemical is in the saturator tank, the fluoride solution will stabilize at 4 percent. This solution of known concentration can then be pumped to the water supply at a controlled rate by a metering pump. A saturator is ideal for small systems because of its low initial cost, low maintenance requirements, and ease of operation.

There are two types of saturators. Upflow saturators (Figure 8-7A) have water introduced at the bottom of a 50-gal (190-L) polyethylene tank beneath crystals of sodium fluoride. The water moves slowly upward, forming the saturated solution.

Downflow saturators (Figure 8-7B) are constructed so that the sodium fluoride crystals rest on a bed of sand and gravel, which in turn rests on a collection system such as a perforated pipe manifold. Water enters the top of the tank and moves slowly through the chemical layer; it is a saturated solution by the time it reaches the collection manifold. The layer of sodium fluoride crystals should be maintained at least 6 in. (150 mm) thick.

Acid Feed Systems

An acid feed system is generally the simplest installation. Fluorosilicic acid can be fed directly from the shipping container into the water supply, as shown in Figure 8-8. The shipping container rests on a scale, which allows one to determine the amount of acid used. However, for water systems treating less than 500 gpm (32 L/s), it is usually impractical to feed the acid directly from the container because the metering pump cannot be set at a low enough rate to provide a constant feed. In those cases, the dilution system shown in Figure 8-9 can be used. The acid is then diluted to one-half strength or some other proportion, so the metering pump can be operated at a moderate rate.

Larger water systems purchase fluorosilicic acid in tank truck loads and store it in large tanks. In this case, the acid is normally pumped periodically

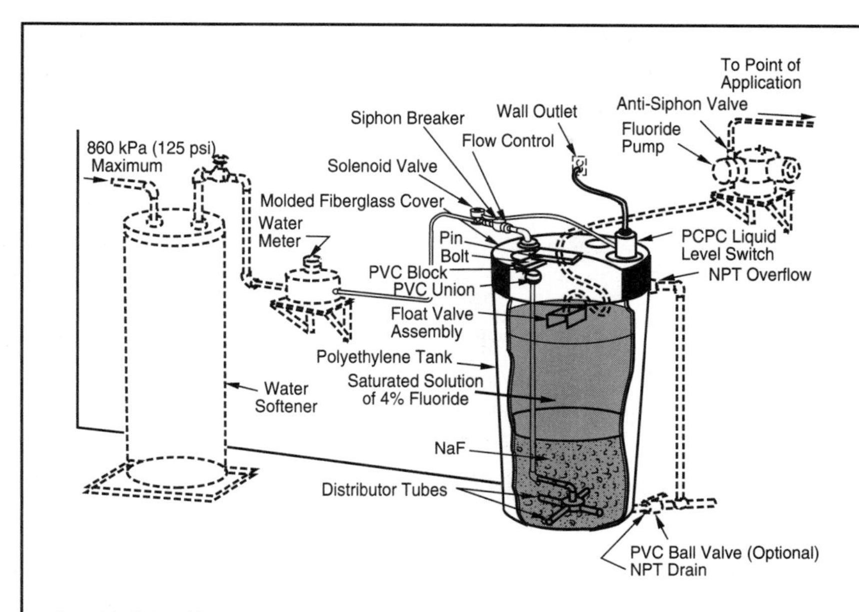

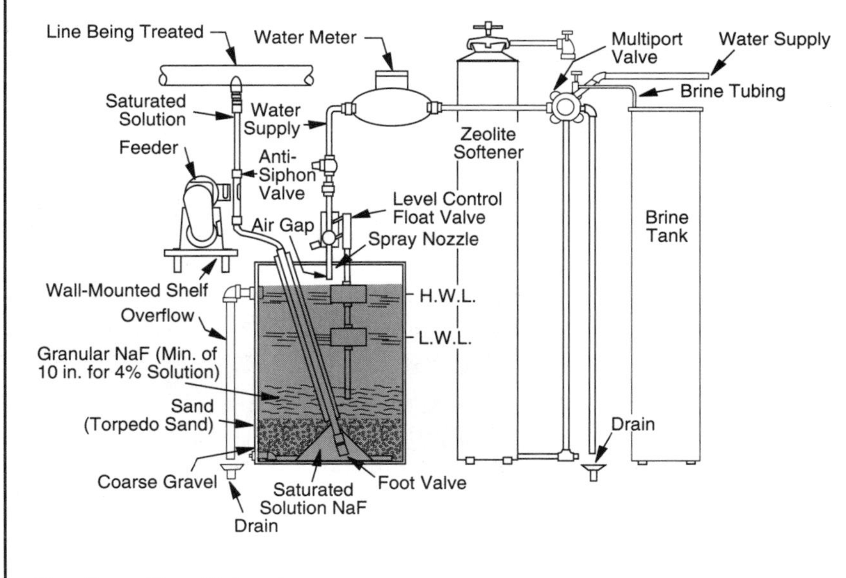

FIGURE 8-7 Two types of saturators

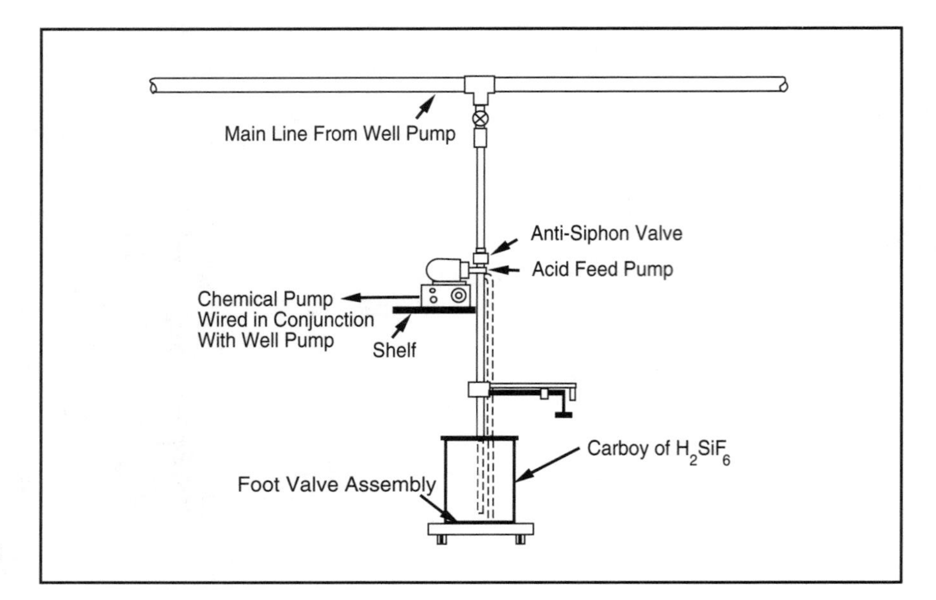

FIGURE 8-8 Acid feed installation

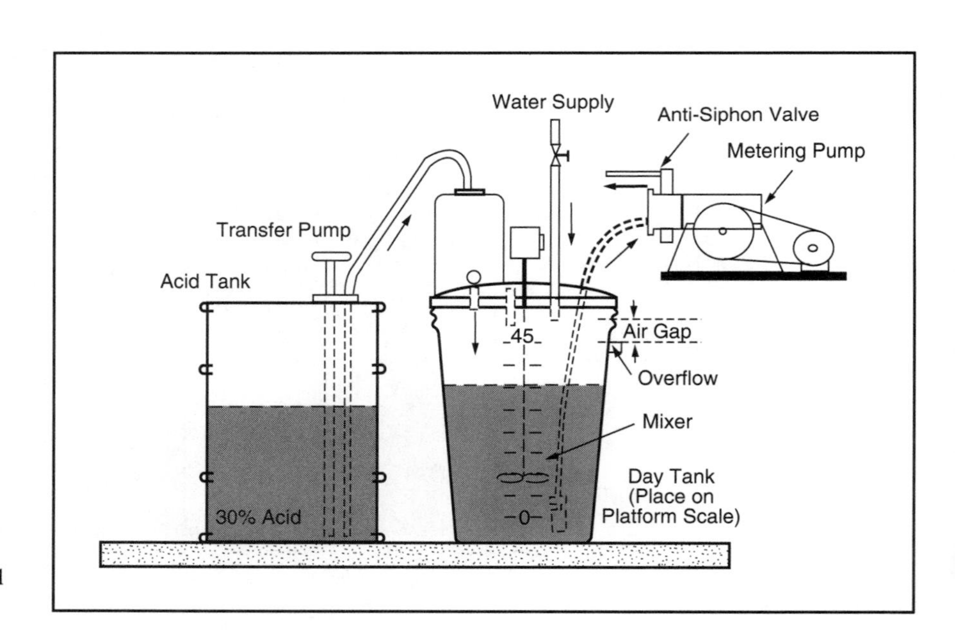

FIGURE 8-9 Diluted-acid feed system

from storage to a day tank on a scale, so that the amount used can be monitored. From there it is metered to the water supply.

Auxiliary Equipment

The following are descriptions of the more common types of auxiliary equipment used with the feed systems described in this chapter.

Scales

Scales are needed for determining the quantity of solution being fed and the quantity of dry fluoride compound or acid delivered by the feeders. They are necessary in all systems except those using saturators. The most common type is the platform scale, on which a solution tank, a carboy of acid, or an entire volumetric dry feeder is placed.

The scale must have sufficient capacity to weigh a full tank and give measurements to the nearest pound (half kilogram) or better. An example of a platform scale installation is shown in Figure 8-10.

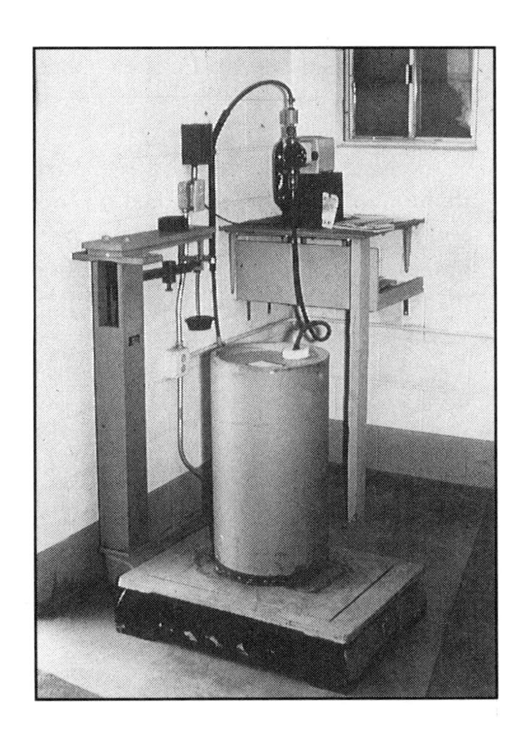

FIGURE 8-10 Platform scale installation

Softeners

When sodium fluoride solutions are used with hard water, insoluble compounds of calcium and magnesium fluoride can form. These compounds can clog the feeder, the feeder suction line, the gravel bed in a downflow saturator, and other equipment.

For this reason, water being fed a sodium fluoride solution should be softened if the hardness exceeds 75 mg/L as $CaCO_3$. Only the water used in preparing the solution (the makeup water) need be softened. Because this amount of water is quite small, a household-type ion exchange water softener is usually adequate. It can be installed directly in the pipeline used for solution makeup (Figure 8-7A).

Dissolving Tanks

The dry chemical discharged from a feeder must be continuously dissolved in a chamber located beneath the feeder. This chamber, referred to as the solution pot, dissolving tank, solution tank, or dissolving chamber, may be a part of the feeder or may be a separate unit. The chemical can be pumped to the water system only when it is completely dissolved. Slurry feed cannot be used because the buildup of undissolved chemical can cause inaccurate feed rates and clogging and deposits in the tanks or water system.

Mixers

Whenever solutions are prepared, it is important that the solution be thoroughly mixed. A fractional horsepower mixer with a stainless-steel shaft and propeller is satisfactory for preparing sodium fluoride solution. All immersed parts of a mixer used for fluorosilicic acid must be plastic coated or made of a corrosion-resistant alloy.

A jet mixer, like the one illustrated in Figure 8-1, is sometimes used for dissolving sodium fluorosilicate, but a mechanical mixer is usually preferred. Because of sodium fluorosilicate's low solubility, particularly in cold water, and the limited detention time in the dissolving tank, violent agitation is usually necessary to prevent discharge of a slurry. Mechanical mixers should never be used in conjunction with saturators.

Water Meters

The solution makeup water for a saturator must be metered in order for the fluoride feed rates to be determined accurately (Figure 8-7). Because the flow is very small, the smallest available meter is usually sufficient.

Flowmeters

Measuring the total flow through a treatment plant serves two important functions for effective fluoridation. First, the meter indicates the flow rates at which the feeder must be set to provide the correct concentration. Second, it can be designed to provide a signal that will allow automatic adjustment (known as pacing) of the feed rates so that those rates will be proportional to flow.

Day Storage Tanks

A day storage tank is a plastic tank that holds enough fluoride solution for one day, or in the case of a large plant, enough for one shift. A tank is essential for large systems that feed acid from bulk storage. Acid is siphoned or pumped into the small tank located on a platform scale, and from there it is metered to the system. As shown in Figures 8-6 and 8-9, day tanks are also used with manual solution and diluted-acid feed systems.

Hoppers

Most dry feeders come equipped with a small hopper. In large installations, an additional or extension hopper is provided over the main hopper to provide more storage capacity. As shown in Figure 8-11, this extension hopper is located one floor over the feeder. The chemicals are then stored on the upper floor and can be conveniently loaded into the hopper.

In small plants, the hopper should be large enough to hold slightly more than one bag or drum of chemical. A hopper that holds less than one bag will cause additional dust and spillage when partially filled bags are handled. Hoppers for large plants are designed to hold several bags at one time. An electric vibrator should be installed on the hopper to keep the chemicals flowing and to prevent "bridging."

Bag Loaders

When the hopper of a dry feeder is directly above the feeder and the operator must lift the bag of chemical a considerable height to fill the hopper, a bag loader is usually necessary. A bag loader is a hopper extension large enough to hold a single bag of chemical (Figure 8-12). The front of the loader is hinged to swing down to a more accessible height. The operator fastens the bag by running an attached rod through the bottom of the bag. The bag is then opened, and the loader is swung back into position. Although this device minimizes dust, the operator should still wear a dust mask during the operation.

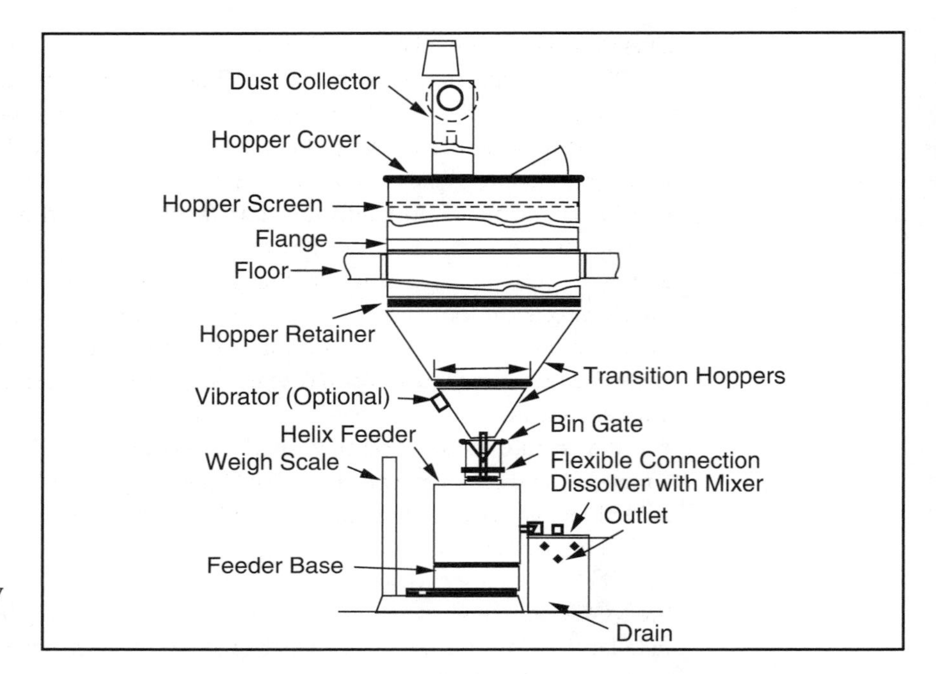

FIGURE 8-11 Dry feed hoppers and dust collectors

FIGURE 8-12 Bag loader

Dust Collectors and Wet Scrubbers

The handling of powdered dry chemicals always generates dust. When small quantities of fluoride are being handled, ordinary care will minimize dust, and good housekeeping plus an exhaust fan will keep the storage and loading areas relatively dust-free. However, when larger quantities are handled, dust prevention and collection facilities should be provided.

A dust canopy, completely enclosing the hopper-filling area and equipped with an exhaust fan, will prevent dust from spreading throughout the loading area. To prevent dust from escaping into the air outside the plant, dust filters are incorporated into the exhaust system. The hoppers on some large feeders have dust collectors and exhaust fans incorporated into the hopper unit.

Wet scrubbers can also be used to remove dust from exhausted air. The air flows through a chamber in which there is a continuous water spray. The air is thus "scrubbed" clean, and the dust particles are carried down the drain with the wastewater.

Weight Recorders

If a platform scale is used to weigh the dry chemicals or solutions, a recorder can be attached to keep a record of the weight of chemical fed. Many volumetric dry feeders have recorders available as an accessory along with the scale.

Alarms

Alarms can be included with either solution or dry feed systems to alert the operator of problems, such as a low level in the solution day tank or a low dry chemical level in the feeder hopper. The alarms are triggered by level switches, flow switches, or pressure switches.

Vacuum Breakers

Whenever there is a water connection to a chemical solution, a cross connection is possible. That is, under certain conditions, concentrated chemicals could flow back into the potable water system. Particularly vulnerable locations are the supply line to a dissolving tank or the discharge of a solution feeder.

The simplest way to prevent backflow is to provide an air gap in the water fill connection. An air gap is a vertical separation between the water line and the tank or device receiving the water (Figures 8-6 and 8-9).

When pressure must be maintained in the system, a vacuum breaker (or anti-siphon device) must be used. This is a valve that is kept closed by water pressure. However, if the water pressure fails, the valve opens to the atmosphere so that the chemical will not be siphoned into the water system. The vacuum breaker should be installed as close to the chemical feeder as possible. In some states, mechanical vacuum breakers are not permitted and air gaps must be used. Additional information on cross-connection control is provided in *Water Distribution and Transmission*, also part of this series.

Regulations

There are no federal regulations requiring water systems to add fluoride to drinking water. The reason is that under the Safe Drinking Water Act, the US Environmental Protection Agency (USEPA) is empowered only to restrict contaminants in drinking water that might be adverse to public health. Fluoridation is, of course, intended to improve public health. However, the practice is strongly endorsed by the USEPA.

Most states require water systems serving municipalities to provide fluoridation, and the other states strongly promote voluntary compliance, usually through the state health department. Some states allow individual systems to avoid fluoridation if there is strong feeling against the practice, as demonstrated by a referendum vote. Details on state requirements should be obtained from the state public water supply supervision program office.

Operation of the Fluoridation Process

Proper operation and maintenance of equipment are required to ensure uninterrupted and unvarying feed of fluoride chemicals. The following paragraphs give general guidelines for operation and maintenance.

Dry Feeders

Dry chemical feeders should be inspected and cleaned routinely to prevent breakdowns. The belts, rolls, and disks or screws must be regularly inspected for signs of wear. Worn parts should be replaced before failure occurs. A lubrication schedule should be established based on the manufacturer's recommendations. The dissolving tank should periodically be inspected for precipitate buildup and cleaned if required. If the deposits are a result of hardness, the makeup water may have to be softened.

The calibration of dry feeders should be checked occasionally, particularly if the plant is having a problem maintaining a constant fluoride concentration. To do this, a small set of scales and a watch that indicates seconds are needed. A shallow pan or sheet of cardboard is inserted between the feeder's measuring mechanism and dissolving chamber to collect a sample over a short period of time. Several samples are collected in intervals, such as 5 minutes each; each sample is weighed and all of them are totaled. The weights of the individual samples will indicate the uniformity of feed, which should show less than 10 percent variation. The total should check against the rate setting on the feeder.

Saturators

The operator must periodically check to ensure that at least 6 in. (150 mm) of chemical is maintained in the saturator at all times. Saturators that treat over 100 gpm (6.3 L/s) require a chemical depth of at least 10 in. (0.25 m). Lines drawn on the outside of the translucent containers will help in determining when to add the chemical.

Only crystalline sodium fluoride should be used in downflow saturators. Powdered sodium fluoride will quickly clog the gravel bed. Any form of sodium fluoride can be used with upflow saturators, but crystalline grade is most often used because less dust is produced. Sodium fluorosilicate should never be used with a saturator because it will not dissolve to produce a 4 percent solution. The saturator tank and gravel bed in a downflow saturator should be cleaned from one to three times a year depending on how rapidly the hardness scale builds up.

Metering pumps are quite reliable, but they should receive routine maintenance according to the manufacturer's recommendations. The pump head should periodically be dismantled for cleaning, and the check valves and diaphragm should be inspected and replaced if worn or cracked.

Fluoride Injection Point

The fluoride injection point should normally be located so that the chemical is applied after water has received complete treatment. In particular, it should not be fed before alum coagulation, lime softening, or ion exchange processes. If water treatment chemicals containing calcium (lime or calcium hypochlorite) are used in the treatment process, the fluoride injection point should be as far away as possible to prevent precipitation.

The feed point used in most treatment plants is just ahead of the clearwell. If fluorosilicic acid is applied into a horizontal pipeline, the point of application should be located in the lower half of the pipe. Injectors should be cleaned one to three times a year to prevent blockage caused by scale formation.

FIGURE 8-13
Poor storage of chemicals

Chemical Storage

Chemical storage areas must be kept clean and orderly. Poor storage conditions, as shown in Figure 8-13, can cause safety hazards and loss of chemicals. The storage area for fluoride chemicals should be isolated from areas used to store other chemicals to avoid any possible mixup when feeders are being charged. Bags of dry chemicals should be piled neatly on pallets as close to the feeding equipment as possible (Figure 8-14). Whenever possible, whole bags should be emptied into hoppers because partially emptied bags are difficult to store without spilling and generating dust.

Fluoridation Operating Problems

Two problems commonly encountered in fluoridation are (1) varying fluoride concentration, and (2) measured concentrations that differ from those computed based on the feed rate.

FIGURE 8-14
Proper chemical storage

Variable Fluoride Concentration

Fluoride levels detected in the distribution system will vary considerably for a period of time after fluoride is first fed at the treatment plant. This is partly because the fluoridated water is diluted by being mixed with the unfluoridated water held in storage.

Varying concentrations detected in the system may also indicate that a dry feeder needs recalibration. A variation of 0.2–0.3 mg/L over a two- to three-day period is not a serious problem, but the operator should investigate the variation because it may indicate a more serious problem.

Low Fluoride Concentration

When the fluoride concentrations measured by laboratory tests are consistently lower than those calculated based on the feed rate, a number of problems may be present.

Assuming that the calculations, weight, and flow data are correct, the problem may be caused by something interfering with the laboratory test procedure. For example, aluminum introduced by alum in the coagulation process sometimes interferes noticeably with the test method and causes erroneously low readings. The fluoride concentration in samples should be rechecked by a comparison of results with those of the state or local health department laboratory to determine if this is the cause of the problem.

One of the most common causes of low readings is a fluoride underdose due to inadequate chemical depth in a saturator or incomplete mixing in a dissolving tank. Deposits of undissolved chemical are an indication of incomplete mixing in a dissolving tank.

If the fluoride level is low in a sample collected from the distribution system, unfluoridated water could be mixing with the treated water. In some systems, groundwater sources are used periodically to supplant normal flows from the treatment plant. If the groundwater is low in natural fluoride, it will lower the concentration detected in parts of the system.

If fluoride is added before filtration, a significant fluoride loss will occur, resulting in low readings. Either the injection point will have to be moved downstream of the filters or a higher concentration of fluoride will have to be used to compensate for the loss.

High Fluoride Concentration

If testing indicates a fluoride concentration consistently higher than the calculated concentration, several problems could be indicated.

Polyphosphates used for water stabilization can cause high readings when the SPADNS method is used (see *Water Quality*, part of this series, for more information on this and other fluoride test methods). This can be checked by the electrode method or by a comparison of results with those of the state or local health department.

Sometimes high readings occur because the natural fluoride in the water has not been measured or considered in the dosing calculations. The natural fluoride level in some surface water sources can vary greatly. If this is the case, the level may have to be measured daily so the correct dosage can be calculated. The natural fluoride level in most groundwater varies only slightly from month to month.

Danger of Fluoride Poisoning

During the past 25 years, there have been several reported episodes of acute fluoride poisoning in the United States, caused by fluoride equipment improperly designed or operated, or malfunctioning. These cases were all caused by a fluoride feeder accidentally or purposely not being turned off when the supply pump stopped. As a result, high concentrations of fluoride built up in the supply line, and later produced concentrations as high as 375 mg/L in water delivered to the system when the pump was turned back on. Although there were no fatalities, several hundred people became ill during these episodes.

In addition, at least one death has been blamed on excessive levels of fluoride in a public water system. In this case, several patients became ill while they were connected to kidney dialysis machines, and one later died. The official cause of death was ruled "acute fluoride intoxication," caused by a water system fluoride level about 15 times above normal as a result of a malfunction of fluoridation equipment. Although there were no other reports of sickness in this case, the occurrence does point out the extreme consequences that can result from not carefully controlling and monitoring fluoride feed.

Control Tests

The types of testing for fluoride concentration that are necessary on a water system include

- measurement of the fluoride level before fluoride is added (to determine the natural fluoride level)
- measurement of the level after fluoridation to confirm that the correct amount of chemical is being added

Daily Control Testing

The fluoride concentration of treated water should be measured daily to ensure that the level being delivered to customers conforms to the limits specified by the state.

If experience has shown that a surface water source has a variable natural fluoride content, daily testing of the raw water may be necessary in order to calculate the dosage needed for the fluoridation equipment. If experience has shown this dosage to be relatively stable, the raw water will need to be checked only periodically to determine whether there has been any change.

If a water system is drawing from several wells, the natural fluoride may vary from well to well, so the utility must compensate based on which wells are being used.

Continuous Control Testing

Continuous testing can be performed by automatic monitors. An automatic monitor is usually connected to a recording chart, and it also has connections for an alarm that will alert the operator if the level varies greatly from preset limits.

Plants using a water source that has variable natural fluoride levels can improve their operation by using a continuous monitor on the raw water. In this way, they can anticipate changes and make adjustments to their fluoride solution feed rate.

Monitors equipped with control equipment can also be used to automatically control the fluoride feeder. Even though a plant has an automatic monitor, manual sampling should still be performed periodically as a check on the accuracy of the monitor.

Safety Precautions

Fluoridation chemicals present a potential health hazard to water plant operators through overexposure from ingestion, inhalation, or bodily contact.

Ingestion

Dry fluoride chemicals can be ingested through contaminated food or drink. Fluoride chemicals could be mistaken for sugar or salt if meals are eaten in areas where fluorides are stored or applied. No one should be allowed to eat, drink, or smoke in areas where fluoride chemicals are stored, handled, or applied.

Personnel who handle fluoride should be instructed not to touch their faces until they have washed thoroughly.

Inhalation

Accidental inhalation of dust is very possible in a water plant using dry chemicals. The causes of dust should be minimized, and operators should wear dust masks when filling feeder hoppers and disposing of bags.

Bags should not be dropped. An even slit should be made at the top of the bag to avoid its tearing down the side, and the contents should be poured gently into the hopper. Crystalline chemicals should be used when possible, because they produce much less dust than the powdered forms. Even if masks are worn and there is no visible dust, the area should be well ventilated.

Bodily Contact

To avoid bodily contact, personnel should wear the following when handling fluoride chemicals:

- chemical goggles
- respirator or mask approved by the National Institute of Occupational Safety and Health (NIOSH)

- rubber gloves with long gauntlets, a rubber apron, and rubber boots
- clothing that covers the skin as completely as possible
- tight covers over open cuts or sores

Acid Handling

Fluorosilicic acid requires very special handling. Acid spilled on the skin or splashed into the eyes is a serious hazard, as is inhaling the vapors.

Fluoride is not absorbed through the skin, but the acid is very corrosive and can burn the skin. If acid is splashed on the body, it should immediately be rinsed with water. An emergency shower and eye wash should be provided for this purpose in areas where acid must be handled.

Record Keeping

Typical records that should be maintained to monitor the fluoridation process include the following:

- daily analyses of raw-water fluoride concentration unless it is known to be stable. If there is more than one source and the levels are known to differ, the concentration must be measured for each source.
- daily analysis of finished water fluoride concentration
- daily records of the amount of chemical fed (in pounds or kilograms)
- records by the operating shift of the chemical feed rate setting
- daily computation of the theoretical concentration, based on the weight of chemical fed and the volume of water produced, as well as a comparison with analyzed values

Selected Supplementary Readings

AWWA Standard for Sodium Fluoride, ANSI/AWWA B701 (latest edition). Denver, Colo.: American Water Works Association.

AWWA Standard for Sodium Fluorosilicate, ANSI/AWWA B702 (latest edition). Denver, Colo.: American Water Works Association.

AWWA Standard for Fluorosilicic Acid, ANSI/AWWA B703 (latest edition). Denver, Colo.: American Water Works Association.

Fluoridation Facts. 1993. Chicago, Ill.: American Dental Association.

How to Maintain Fluoride Feeding Equipment. ***Opflow,*** 19(11):6.

Manual of Instruction for Water Treatment Plant Operators. 1975. Albany, N.Y.: New York State Department of Health.

Manual M3, Safety Practices for Water Utilities. 1990. Denver, Colo.: American Water Works Association.

Manual M4, Water Fluoridation Principles and Practices. 4th ed., 1995. Denver, Colo.: American Water Works Association.

Recommended Standards for Water Works. Albany, N.Y.: Health Education Services.

Should Your Community Fluoridate Its Drinking Water? ***Opflow,*** 19(11):1.

Water Quality and Treatment. 4th ed. 1990. New York: McGraw-Hill and American Water Works Association (available from AWWA).

CHAPTER 9

Control of Corrosion and Scaling

Many water systems must apply special chemical treatment because their source water either causes damaging corrosion or deposits scale on pipelines and plumbing fixtures. The treatment process for controlling these problems is known as *stabilization*. Many more systems must provide corrosion control treatment under new federal and state regulations enacted to protect the public from the health dangers of lead and copper in drinking water.

Water system operators are cautioned that complicated interactions often occur in the control of water corrosion and scaling. A seemingly simple change to improve one characteristic may have an adverse effect on some other water characteristic or treatment process. This chapter offers an overview of the need for corrosion and scaling control and their related processes. However, because of all the complex variables involved, it is not intended to be a guide for the best treatment method for any particular system. It is best to get professional guidance and state approval before beginning any new stabilization treatment.

Purposes of Corrosion and Scaling Control

Corrosion and scaling are controlled for the following reasons:

- to protect public health
- to improve water quality
- to extend the life of plumbing equipment
- to meet federal and state regulations

Protecting Public Health

Corrosive water can leach toxic metals from distribution and household plumbing systems. Lead and copper are the metals most likely to be a problem because they are commonly used in plumbing systems.

In addition, corrosion of cast-iron mains can cause the formation of iron deposits, called tubercules, in the mains. These deposits can protect bacteria and other microorganisms from chlorine, allowing them to grow and thrive. Changes in water velocity or pressure can then cause the microorganisms to be released, creating a potential for disease outbreaks. Some bacteria shielded by the tubercules can also accelerate the corrosion process.

Improving Water Quality

Corrosive water attacking metal pipes can cause taste, odor, and color problems in a water system. Red-water problems occur when iron is dissolved from cast-iron mains by corrosive water. The iron will stain a customer's plumbing fixtures and laundry and make the water's appearance unappealing for drinking and bathing. The dissolved iron also acts as a food source for a group of microorganisms called iron bacteria, which can cause serious taste-and-odor problems. Corrosion of copper pipes can cause a metallic taste, as well as blue-green stains on plumbing fixtures and laundry.

Extending the Life of Plumbing Equipment

Unstable water can also result in significant costs to water systems and customers. Aggressive water can significantly reduce the life of valves, unprotected metal, and asbestos–cement (A–C) pipe. It can also shorten the service life and performance of plumbing fixtures and hot water heaters.

Buildup of corrosion products (a process known as tuberculation) or uncontrolled scale deposits can seriously reduce pipeline capacity and increase resistance to flow. This in turn reduces distribution system efficiency and increases pumping costs. If scale deposits or tuberculation go unchecked, pipes can become completely plugged, requiring expensive repair or replacement. Scaling can also increase the cost of operation of hot water heaters by increasing their fuel consumption, as shown in Table 9-1.

Meeting Federal and State Regulations

As detailed later in this chapter, the Lead and Copper Rule enacted by the US Environmental Protection Agency (USEPA) in 1991 requires water systems to check if their water is corrosive enough to cause lead and copper corrosion products to appear in customers' water at levels exceeding the new

TABLE 9-1 Estimated effect of scale on boiler fuel consumption

Scale Thickness, *in.*	*(mm)*	Fuel Consumption, *% increase*
1/50	(0.5)	7
1/16	(1.6)	18
1/8	(3.2)	39

action level. If the level is exceeded, the system is required to take action to reduce the corrosivity of the water.

Water System Corrosion

Corrosion can be broadly defined as the wearing away or deterioration of a material due to chemical reactions with its environment. The most familiar example is the formation of rust (oxidized iron) when an iron or steel surface is exposed to moisture. Corrosion is usually distinguished from erosion, which is the wearing away of material due to physical causes such as abrasion. Water that promotes corrosion is known as corrosive or aggressive water.

In water treatment operations, corrosion can occur to some extent with almost any metal that is exposed to water. Whether corrosion of a material will be extensive enough to cause problems depends on several related factors, such as the type of material involved, the chemical and biological characteristics of the water, and the electrical characteristics of the material and its environment.

The relationships among these factors, as well as the process of corrosion itself, are quite complex. As a result, it is difficult to make general statements about what combinations of water and equipment will or will not have corrosion problems. The discussions in this chapter cover only basic principles; the operator faced with persistent corrosion in a given installation may require the assistance of corrosion-control specialists.

Chemistry of Corrosion

The chemical reactions that occur in the corrosion of metals are similar to those that occur in an automobile battery. In fact, corrosion generates an electrical current that flows through the metal being corroded. The chemical and electrical reactions that occur during concentration cell corrosion of iron pipes are illustrated in Figure 9-1.

As shown in Figure 9-1A, minor impurities and variations (present in all metal pipes) have caused one spot on the pipe to act as an electrical anode in relation to another spot that is acting as an electrical cathode. At the anode, atoms of iron (Fe^{+2}) are breaking away from the pipe and going into solution in the water. As each atom breaks away, it ionizes by losing two electrons, which travel through the pipe to the cathode.

In Figure 9-1B, it is shown that chemical reactions within the water balance the electrical and chemical reactions at the anode and cathode. Many of the water molecules (H_2O) have dissociated into H^+ ions and OH^- radicals. This is a normal condition, even with totally pure water. The Fe^{+2} released at the anode combines with two OH^- radicals from dissociated water molecules to form $Fe(OH)_2$, ferrous hydroxide. Similarly, two H^+ ions from the dissociation of the water molecules near the cathode pick up the two electrons originally lost by the iron atom, then bond together as H_2, hydrogen gas.

The formation of $Fe(OH)_2$ leaves an excess of H^+ near the anode, and the formation of the H_2 leaves an excess of OH^- near the cathode. This change in the normal distribution of H^+ and OH^- accelerates the rate of corrosion and causes increased pitting in the anode area (the concentration cell), as shown in Figure 9-1C.

If the water contains dissolved oxygen (O_2) — most surface water does — then $Fe(OH)_3$, ferric hydroxide, will form (Figure 9-1D). Ferric hydroxide is common iron rust. The rust precipitates, forming deposits called tubercules (Figure 9-1E). The existence of tubercules further concentrates the corrosion, increasing both pitting at the anode and growth of the tubercule. Tubercules can grow into large nodules (Figure 9-2), significantly reducing the carrying capacity of a pipe. During rapid pressure or velocity changes, some of the $Fe(OH)_3$ can be carried away, causing "red water."

Factors Affecting Corrosion

The rate of corrosion depends on many site-specific conditions, such as the characteristics of the water and pipe material. Therefore, there are no established guidelines that determine the rate at which a pipe will be corroded.

Chemical reactions play a critical role in determining the rate of corrosion at both the cathode and the anode. Any factor that influences these reactions will also influence the corrosion rate.

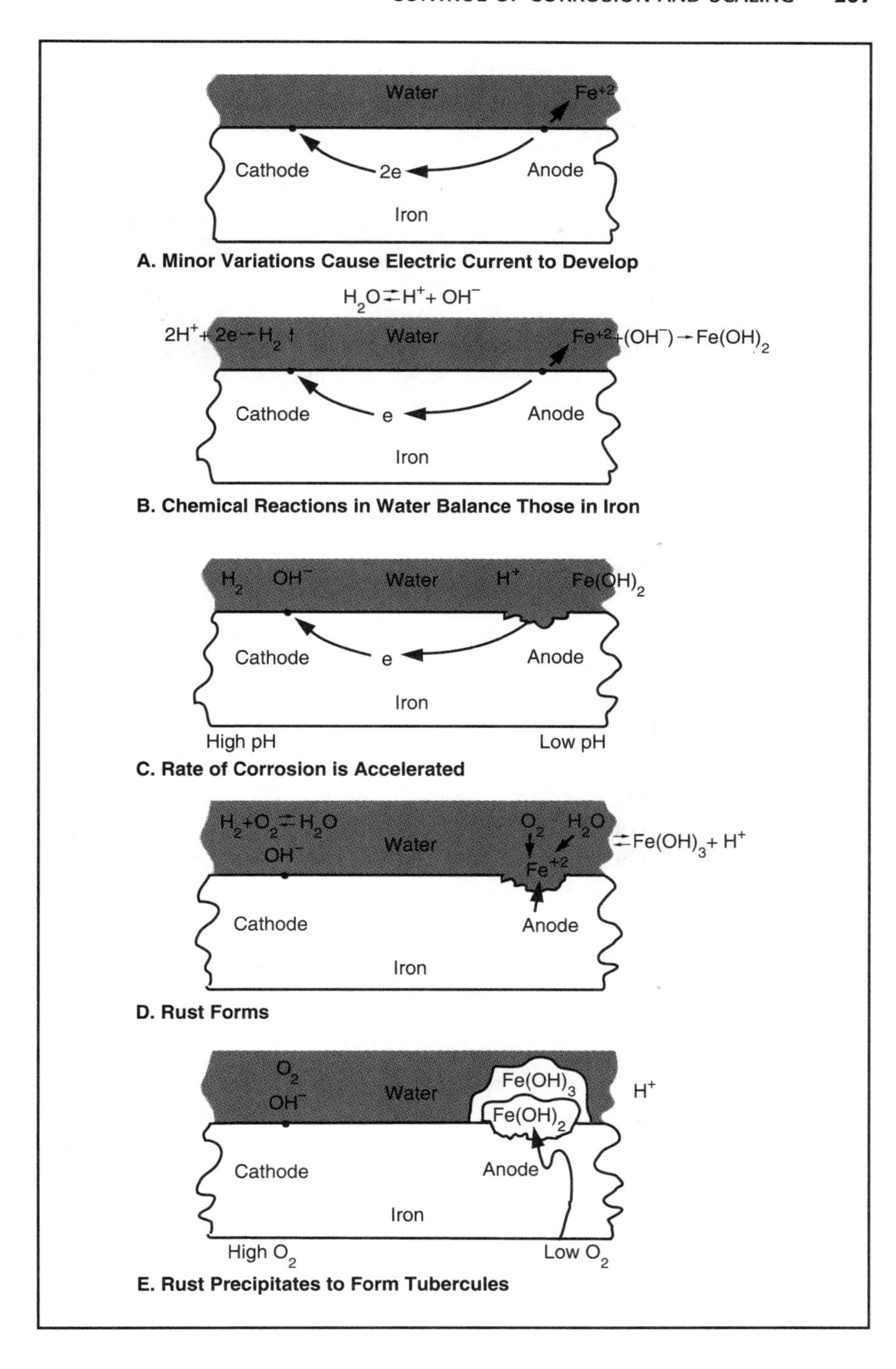

FIGURE 9-1 Chemical and electrical reactions that occur during corrosion of iron pipe

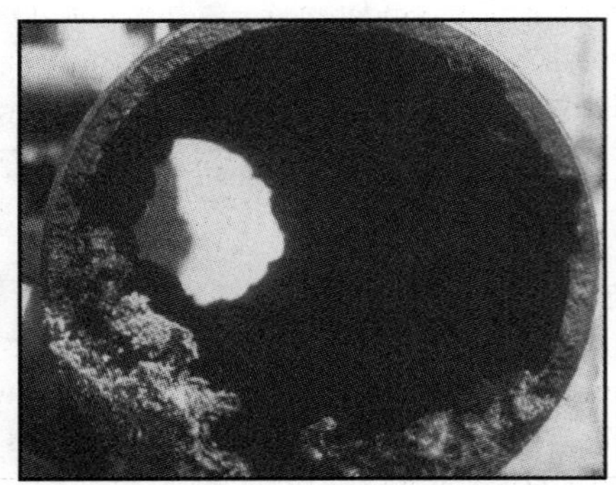

FIGURE 9-2
Tuberculated pipe

Courtesy of Girard Industries

Dissolved Oxygen

The concentration of dissolved oxygen (DO) in water is a key part of the corrosion process. As the concentration of DO increases, the corrosion rate will also increase.

Total Dissolved Solids

The total dissolved solids (TDS) concentration is important because electrical flow is necessary for the corrosion of metal to occur. Pure water is a poor conductor of electricity because it contains very few ions. But as the TDS is increased, water becomes a better conductor, which in turn increases the corrosion rate.

Alkalinity and pH

Both the alkalinity and the pH of the water affect the rate of chemical reactions. Generally, as pH and alkalinity increase, the corrosion rate decreases.

Temperature

Because chemical reactions occur more quickly at higher temperatures, an increase in water temperature generally increases the corrosion rate.

Flow Velocity

The velocity of water flowing past a piece of metal can also affect the corrosion rate, depending on the nature of the water. If the water is corrosive, higher flow velocities cause turbulent conditions that bring dissolved oxygen to the corroding surface more rapidly, which increases the corrosion rate. However, if chemicals are being added to stabilize the water, the higher velocities will decrease the corrosion rate, allowing the chemicals (such as $CaCO_3$) to deposit on the pipe walls more quickly.

Type of Metal

Metals that easily give up electrons will corrode easily. Table 9-2 lists metals commonly used in water systems, with those at the top being most likely to corrode. This listing is called the galvanic series of metals. Where dissimilar metals are electrically connected and immersed in a common flow of water, the metal highest in the galvanic series will immediately become the anode, the other metal will become the cathode, and corrosion will occur. This is termed *galvanic corrosion* (as opposed to concentration cell corrosion).

The rate of galvanic corrosion will depend largely on how widely separated the metals are in the galvanic series. Widely separated metals will exhibit extremely rapid corrosion of the anode metal (the highest in the series), and the cathode metal will be protected from corrosion. A common example of galvanic corrosion occurs when a brass corporation cock is

TABLE 9-2 Galvanic series for metals used in water systems

Corroded End (Anode)	MOST ACTIVE
Magnesium	+
Magnesium alloys	
Zinc	
Aluminum	Corrosion Potential
Cadmium	
Mild steel	
Wrought (black) iron	
Cast iron	
Lead–tin solders	
Lead	
Tin	
Brass	
Copper	
Stainless steel	—
Protected End (Cathode)	LEAST ACTIVE

tapped into a cast-iron main and attached to a copper service line. The copper will be protected at the expense of the brass and cast iron.

Electrical Current

If electrical current is passed through any corrodible metal, corrosion will be accelerated. The two causes of electrical current in water mains are improperly grounded household electrical systems and electric railway systems.

Bacteria

Certain types of bacteria can accelerate the corrosion process because they produce carbon dioxide (CO_2) and hydrogen sulfide (H_2S) during their life cycles, which can increase the corrosion rate. They can also produce slime, which will entrap precipitating iron compounds and increase red-water problems and the amount of tuberculation.

Two groups of bacteria cause the most problems. Iron bacteria such as *Gallionella* and *Crenothrix* can form considerable amounts of slime on pipe walls, particularly if the water contains enough dissolved iron to allow them to survive. The iron present in the water can be naturally occurring in the source water or can be due to corrosion of the pipe material. Beneath this slime layer, CO_2 production by the bacteria can significantly lower the pH, which will speed up the corrosion rate.

The periodic sloughing of these slime accumulations can cause other major problems such as tastes and odors. The slimes can also prevent the effective deposition of a protective calcium carbonate ($CaCO_3$) layer by enmeshing it within the slime layer. As this layer sloughs away, it carries away the $CaCO_3$ and leaves the pipe surface bare.

Sulfate-reducing bacteria such as *Desulfovibrio desulfuricans* can accelerate corrosion when sulfate (SO_4) is present in the water. They reduce the SO_4 under anaerobic conditions, which occur under the slime layer where oxygen is depleted by other bacteria. The products formed are iron sulfide (Fe_2S_2) and hydrogen sulfide (H_2S), which causes obnoxious odors and black-colored water. The carbon dioxide formed can also lower the pH of the water.

All of these factors interact with each other and with metal pipes, tanks, and various equipment in water plants, the distribution system, and the customer's plumbing. As the factors change, the corrosion rate will also change. The only factors over which the operator has significant control are pH, alkalinity, and the bacteriological content of the water. The techniques used to control these parameters are discussed later in this chapter.

Types of Corrosion

Corrosion in water systems can be divided into two broad classes: localized and uniform. Localized corrosion, the most common type in water systems, attacks metal surfaces unevenly. It is usually a more serious problem than uniform corrosion because it leads to a more rapid failure of the metal. Two types of corrosion that produce pitting are galvanic corrosion and concentration cell corrosion (discussed previously).

Uniform corrosion takes place at an equal rate over the entire surface. It usually occurs where waters having very low pH and low alkalinity act on unprotected surfaces.

Scale Formation

The formation of mild scale on the interior of pipes can protect the pipe from corrosion by separating the corrodible pipe material from the water. However, uncontrolled scale deposits can significantly reduce the carrying capacity of a distribution system. Figure 9-3 shows a pipe that is almost completely blocked by scale.

Chemistry of Scale Formation

Scale is formed when the divalent metallic cations associated with hardness, primarily magnesium and calcium, combine with other minerals dissolved in the water and precipitate to coat pipe walls. (See chapter 11 for further discussion of hardness.) The most common form of scale is calcium carbonate ($CaCO_3$). Other scale-forming compounds include magnesium carbonate ($MgCO_3$), calcium sulfate ($CaSO_4$), and magnesium chloride ($MgCl_2$).

FIGURE 9-4
Scaling of pipe

Courtesy of Johnson Controls, Inc.

Factors Affecting Scale Formation

Water can hold only so much of any given chemical in solution. If more is added, it will precipitate instead of dissolve. The point at which no more of the chemical can be dissolved is called the saturation point. The saturation point varies with other characteristics of the water, including pH, temperature, and total dissolved solids (TDS).

The saturation point of calcium carbonate depends primarily on the pH of the water. For example, if water with a certain temperature and TDS concentration can maintain 500 mg/L of $CaCO_3$ in solution at pH 7, then the same water will hold only 14 mg/L of $CaCO_3$ in solution if the pH is raised to 9.4.

Temperature also affects the saturation point, although not as dramatically as pH. The solubility of $CaCO_3$ in water decreases as temperature increases. The most common example is when the higher temperature in hot water heaters and boilers causes scale to precipitate out of the water and build up on pipe and tank walls. Because the presence of other minerals in the water affects the solubility of $CaCO_3$, the TDS concentration must be known in order to determine the $CaCO_3$ saturation point. As the TDS concentration increases, the solubility of $CaCO_3$ increases.

Corrosion and Scaling Control Methods

The basic methods used for stabilizing water to protect against the problems of corrosion or scaling are

- adjustment of pH and alkalinity
- formation of a calcium carbonate coating
- use of corrosion inhibitors and sequestering agents

The selection of which method or methods are finally used on any water system depends on both the chemical characteristics of the raw water and the effects of other treatment processes being used.

The type of source water, number of sources, and the hydraulics and flow patterns of the system can also have a bearing on the choice of corrosion control measures that should be taken. Systems that have multiple sources with different chemistry may have particularly complex problems. The pH of the water is sometimes critical in the use of chemical corrosion control measures; some measures work properly only within a narrow pH range. The principal treatment techniques for corrosion and scaling control are summarized in Table 9-3.

TABLE 9-3 Summary of treatment techniques for controlling corrosion and scaling

Treatment	Application	Effectiveness	Comments or Problems
To Prevent Corrosion			
Lime alone or lime with sodium carbonate or sodium bicarbonate	Increase pH Increase hardness Increase alkalinity	Most effective in water with low pH and hardness Excellent protection for copper, lead, and asbestos–cement pipe in stabilized waters Good protection for galvanized and steel pipe	May be best overall treatment approach Oversaturation may cause calcium deposits
Sodium hydroxide	Increase pH	Most effective in waters with sufficient hardness and alkalinity to stabilize water May provide adequate protection against lead corrosion in low-alkalinity, soft waters	Should not be used to stabilize waters without the presence of adequate alkalinity and hardness May cause tuberculation in iron pipes at pH 7.5–9.0
Sodium hydroxide and sodium carbonate or sodium bicarbonate	Increase pH Adjust alkalinity	Most effective in water with low pH and sufficient hardness Excellent protection for lead corrosion in soft waters at pH 8.3	Combination of high alkalinity and hardness with low pH is more effective than combination of high pH with low hardness and alkalinity
Inhibition with phosphates (primarily sodium zinc phosphate and zinc ortho-phosphate)	Formation of protective film on pipe surfaces	Effective at pH levels above 7.0 Good protection for asbestos–cement pipe Addition of lime may increase effectiveness of treatment for copper, steel, lead, and asbestos	May cause leaching of lead in stagnant waters May encourage the growth of algae and microorganisms May cause red water if extensive tuberculation is present May not be effective at low pH levels

Table continued next page

TABLE 9-3 Summary of treatment techniques for controlling corrosion and scaling (continued)

Treatment	Application	Effectiveness	Comments or Problems
Inhibition with silicates	Formation of protective film on pipe surfaces	Most effective in waters having low hardness and pH below 8.4 Good protection for copper, galvanized, and steel pipe	May increase the potential of pitting in copper and steel pipes May not be compatible with some industrial processes
To Prevent Scale Formation			
Carbon dioxide or sulfuric acid	Decrease pH Decrease alkalinity	Effective with high-pH, high-alkalinity water such as lime-softened water	Overfeeding can cause low pH and corrosion
Sequestering with phosphates (primarily sodium hexametaphosphate and tetrasodium polyphosphate)	Sequester scale-forming ions	Effective in controlling scale formation from lime-softened waters and iron in the source water	Can loosen existing deposits and cause red-water complaints Compounds lose sequestering ability in hot water heaters, causing precipitation of $CaCO_3$ or iron

Adjustment of pH and Alkalinity

In general, soft waters that have a pH of less than 7 and are slightly buffered will be corrosive to lead and copper. Water that has too much alkalinity can also be quite corrosive. Water that is nominally corrosive naturally can also have the corrosivity increased by the addition of other water treatment chemicals. For instance, gaseous chlorine will reduce pH levels.

In general, a moderate increase in pH and alkalinity levels can reduce corrosion, and a decrease can prevent scale formation. The formation of a protective film on the interior of lead and copper pipes is also usually aided by increasing pH.

Lime is generally used to increase both pH and alkalinity because it is less expensive than other chemicals having the same effect. Soda ash (sodium carbonate) can also be added along with the lime to further increase the alkalinity. Sodium bicarbonate is sometimes used instead of sodium carbonate because it will also increase alkalinity without as much of an increase in pH. The increased alkalinity buffers the water against pH changes in the distribution system. This has proven particularly effective in controlling corrosion of lead and copper service pipes. Instead of lime, caustic soda with soda ash or sodium bicarbonate can also be used to increase pH and alkalinity.

As discussed in chapter 11, lime-softened water can cause severe scale problems if it is not stabilized. Stabilization after softening is accomplished by the addition of carbon dioxide (a process called *recarbonation*) or sulfuric acid. Both chemicals lower the pH so that calcium carbonate will not precipitate in the distribution system.

Formation of a Calcium Carbonate Coating

Because corrosion attacks the surface of a pipe, a protective coating on the pipe surface can inhibit corrosion. A coating of cement, plastic, or asphaltic material is commonly applied to the interior of pipes and tanks and on metal equipment to protect them from corrosion. Although these coatings form an effective barrier against corrosion, very aggressive water can sometimes attack the coatings. In addition, if there are any breaks in the coating, corrosion will be particularly severe at these points.

As a result, many systems apply an additional protective coating by controlling the chemistry of the water. A common protective-coating technique is to adjust the pH of the water to a level just above the saturation point of calcium carbonate. When this level is maintained, calcium carbonate will precipitate and form a protective layer on the pipe walls. This process must be closely controlled. A pH that is too low may result in corrosion, and a pH that is too high may result in excessive precipitation, which will cause a clogging of service pipes and a restriction of flow in the distribution system. Lime, soda ash, sodium bicarbonate, or sodium hydroxide can be used to raise the pH level. Lime is often used because it also adds needed calcium (hardness) and alkalinity.

Use of Corrosion Inhibitors and Sequestering Agents

Some waters do not contain enough calcium or alkalinity to make the formation of calcium carbonate coatings economical. Water obtained primarily from snowmelt is an example, having alkalinity and calcium concentrations as low as 2 mg/L as $CaCO_3$. In this event, other chemical compounds can be used to form protective coatings.

The most common compounds are polyphosphates and silicates. The chemical reactions by which these compounds combine with corrosion products to form a protective layer are not completely understood; however, the chemicals have proven successful in many water systems.

Some polyphosphates can also be used as sequestering agents to prevent scale formation. These compounds sequester, or chemically tie up, the scale-forming ions of calcium and magnesium so that they cannot react to form scale. Because these compounds remain in solution, they are eventually ingested by consumers. Therefore, any sequestering agent selected must be suitable for use in drinking water.

Polyphosphates also sequester iron, whether it is dissolved in water from the source or from corrosion of the system. This prevents the precipitation of the iron compounds, so red water will not result. However, this effect does not prevent corrosion — it merely prevents the corrosion by-products from being noticed.

Corrosion and Scaling Control Facilities

Whatever method or methods are used to control corrosion and scaling, the facilities required for the process consist primarily of selected chemicals and the equipment required to store, handle, and feed these chemicals. A wide range of chemicals is available for stabilization. Some of these are primarily intended for industrial or other nonpotable applications. Therefore, all chemicals used to stabilize drinking water must be approved for such use. Chemicals must also comply with NSF International and American Water Works Association (AWWA) standards to ensure quality. See appendix A for further details. Table 9-4 lists the chemicals commonly used for stabilizing potable water.

Chemical Storage and Handling

Chemicals should be purchased in quantities that will maintain a 30-day minimum supply at all times. This practice will guard against interruptions of service due to temporary shortages and other unforeseen events. In determining the quantity of chemical to be stored at any one time, the operator should consider storage space available, discounts for purchasing large quantities, and length of time the chemical can be held without losing potency or caking so that it becomes unusable.

TABLE 9-4 Chemicals commonly used for stabilization of potable water

Treatment Method	Chemical Name	Chemical Formula
Increase pH and alkalinity	Unslaked lime (quicklime)	CaO
	Slaked lime (hydrated lime)	$Ca(OH)_2$
	Sodium bicarbonate	$NaHCO_3$
	Sodium carbonate (soda ash)	Na_2CO_3
	Sodium hydroxide (caustic soda)	$NaOH$
Decrease pH and alkalinity	Carbon dioxide	CO_2
	Sulfuric acid	H_2SO_4
Formation of protective coatings	Sodium silicate (water glass)	$Na_2O(SiO_2)_n$*
	Sodium hexametaphosphate (sodium polyphosphate, glassy)	$(NaPO_3)_n \cdot Na_2O$†
	Sodium zinc phosphate	$(MPO_3)_n \cdot M_2O$‡
	Zinc orthophosphate	$Zn_3(PO_4)_2$
Sequestering agents	Sodium hexametaphosphate (sodium polyphosphate, glassy)	$(NaPO_3)_n \cdot Na_2O$†
	Tetrasodium pyrophosphate	$Na_4P_2O_7 \cdot 10H_2O$

*Typically $n = 3$.
†Typically $n = 14$.
‡M = Na and/or ½ Zn, typically $n = 5$.

Lime

Lime is available in either unslaked or slaked form. Unslaked lime, also called quicklime or calcium oxide (CaO), is available in a variety of sizes from a powder to pebble form. Although available in bags, it is considerably more economical and easier to handle in bulk. As a result, quicklime is generally used only by treatment plants that use large quantities. It should not be stored for more than three months because it can deteriorate. Quicklime is noncorrosive in the dry form and can be stored in steel or concrete bins. A minimum of two bins should be available to provide for maintenance and ease in unloading. The bulk deliveries are transferred from the rail cars or trucks to the bins by mechanical or pneumatic conveyors.

For treatment plants that do not use large quantities of lime, hydrated (or slaked) lime — calcium hydroxide, $Ca(OH)_2$ — is more cost effective. Hydrated lime is a finely divided powder available in 50- or 100-lb (23- or 45-kg) bags and in bulk shipments. The bulk form is unloaded and stored like quicklime and should be used within three months. The bagged hydrated lime can be stored for up to one year without serious deterioration. Both types should be stored in a dry, well-ventilated area. This is particularly important for quicklime because moisture will start the slaking process, generating a tremendous amount of heat, which could start a fire if combustible materials are nearby.

Soda Ash

Soda ash (sodium carbonate, Na_2CO_3) is a white, alkaline chemical. It is available as granules and powder in bulk or in 100-lb (45-kg) bags. Because soda ash absorbs moisture and will readily cake, it should be stored in a dry, well-ventilated area. Dry soda ash is not corrosive, so it can be stored in steel or concrete bins. Sodium carbonate should not be stored near products containing acid. If acid contacts the chemical, large quantities of carbon dioxide can be released, creating a safety problem.

Sodium Bicarbonate

Sodium bicarbonate ($NaHCO_3$) is also commonly known as baking powder. It is a white, alkaline chemical that is available in either powdered or granular form. It can be purchased in 100-lb (45-kg) bags or in barrels up to 400 lb (180 kg). It must be stored in a cool, dry, well-ventilated area because it decomposes rapidly as the temperature nears 100°F (38°C).

As with sodium carbonate, sodium bicarbonate should not be stored near products containing acid. Corrosion-resistant materials must be used for storing and transporting sodium bicarbonate solution.

Sodium Hydroxide

Sodium hydroxide (NaOH) is also commonly called caustic soda. It is available as a liquid or in dry form in flakes or lumps. Regardless of the form, it must receive careful handling because of the many hazards involved. As a liquid, caustic soda is available in solution concentrations of 50 percent or 73 percent NaOH. The 50 percent solution will begin to crystallize if its temperature drops below 54°F (12°C). The 73 percent solution will crystallize as its temperature drops below 145°F (63°C). Special storage facilities are therefore required to maintain the temperature of the solution high enough to prevent crystallization. This can be accomplished by further dilution or by installing heaters in the storage tanks.

The manufacturer's recommendations should be followed closely to prevent problems. The storage tanks should be constructed of nickel–cadmium or nickel–alloy steel, lined with caustic-resistant material such as rubber or polyvinyl chloride (PVC). Tanks made of PVC or fiberglass can also be used.

If the dry form of caustic soda is used, it is normally dissolved immediately upon delivery in a storage tank. The dilution concentration is at a point where crystallization will not be a problem at local temperatures.

With either the dry or the liquid form, the addition of water for dilution generates a considerable amount of heat. Therefore, the rate of dilution must be carefully controlled so that boiling or splattering of the solution does not occur. A source of flushing water should be readily available in case of spills.

Sulfuric Acid

Sulfuric acid (H_2SO_4) is a corrosive, dense, oily liquid available in strengths containing 62 percent, 78 percent, or 93 percent H_2SO_4. It can be purchased in 55- or 110-gal (210- or 420-L) barrels or in bulk in tank cars or trucks. The acid must be stored in corrosion-resistant tanks.

If diluted, the acid must always be added to the water (not the water to the acid). If the water is added to the acid, there will be splattering and rapid generation of heat.

Sodium Silicate

Sodium silicate is also frequently called water glass. It is an opaque, syrupy, alkaline liquid, available in barrels and in bulk. The barrels can be stored, or the liquid can be added immediately to storage tanks. Bulk deliveries are unloaded directly into storage tanks. The tanks can be constructed of steel or plastic because sodium silicate is not corrosive.

Carbon Dioxide

Carbon dioxide (CO_2) is a colorless, odorless gas that can be generated on-site or purchased in bulk in a liquified form. On-site generation involves producing the CO_2 in a furnace or in an underwater burner.

Liquified CO_2 is stored in insulated, refrigerated pressure tanks ranging in capacity from 6,000 to 100,000 lb (2,700 to 45,000 kg). The liquified CO_2 must be kept at about 0°F (–18°C) and 300 psig (2,070 kPa [gauge]) so that it will remain in liquid form. Carbon dioxide vapor forms above the liquid surface and is withdrawn and piped to the application point. Each tank has a built-in vaporizer, which helps maintain a constant vapor supply. The storage tank should be located as close as possible to the application point.

Phosphate Compounds

A number of phosphate compounds can be used for stabilization. Three of the most common are sodium hexametaphosphate (sodium polyphosphate, glassy), sodium zinc phosphate, and zinc orthophosphate. Sodium hexametaphosphate is available in bags or drums in a form that looks like broken glass. The other two compounds are generally available in dry form in 50-lb (23-kg) bags and in liquid form in drums or tank trucks. The concentrated liquid solutions are slightly acidic, so they should be stored in corrosion-resistant tanks constructed of fiberglass, PVC, or stainless steel.

Chemical Feed Equipment

Lime

The type of equipment needed for lime feeding depends on the type of lime being used. If slaked lime is used, a gravimetric or volumetric dry feeder adds the lime to a solution chamber, where water is added to form a slurry (Figure 9-4). When unslaked lime is used, the dry feeder adds lime to a slaker, where the lime and water are mixed together to slake the lime and form a slurry ($CaO + H_2O \rightarrow Ca(OH)_2$). This slurry is often called "milk of lime." Slakers are described in more detail in chapter 11. The hoppers or bins supplying the dry feeders must be equipped with agitators or vibrators, because lime does not flow freely and will pack and "bridge" during storage.

Regardless of the type of lime used, the slurry must be added to the water. It is best to minimize the distance from the feeders to the application point because lime slurry will cake on any surface. Open troughs or flexible feed lines should be used to carry the slurry to the application point because

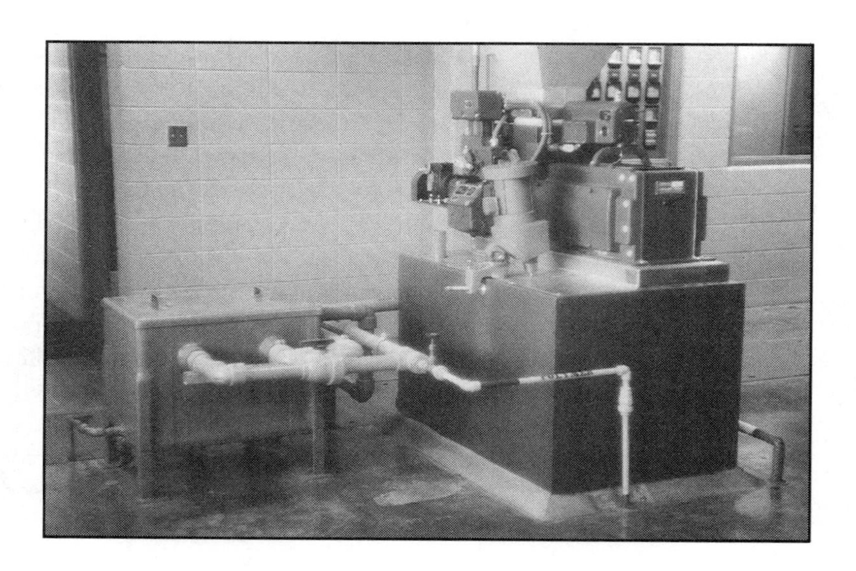

FIGURE 9-4
Lime-feeding equipment

they must frequently be cleaned and flushed. Solution feeders used with the slurry should be designed for easy cleaning.

Soda Ash and Sodium Bicarbonate

Soda ash and sodium bicarbonate can be fed with gravimetric or volumetric dry feeders. Agitators should be provided on the hoppers to prevent the chemical from bridging, which will block the flow.

Soda ash solutions can be fed using conventional solution feeders and lines. Because soda ash goes into solution slowly, it is necessary to provide larger than usual dissolving chambers to provide proper detention time, as well as good mixers to achieve thorough mixing. Sodium bicarbonate solutions are caustic. Therefore, metering pumps, dissolving chambers, and pipes should be constructed or lined with caustic-resistant materials, such as stainless steel or PVC.

Sodium Hydroxide

Sodium hydroxide solutions can be fed by metering pumps designed for handling corrosive and caustic solutions. All piping should be of caustic-resistant material such as PVC. Valves and fittings should not contain any copper, brass, bronze, or aluminum because these materials will deteriorate rapidly.

Sulfuric Acid

Sulfuric acid can be fed directly from the shipping container by a corrosion-resistant metering pump and piping. The diluted acid solution is even more corrosive than the pure acid and must be handled carefully.

Sodium Silicate

Sodium silicate can be fed directly from the shipping container or from a day tank, by using a conventional metering pump and piping.

Carbon Dioxide

If carbon dioxide is generated on-site, the feeding system consists of the generation equipment and related piping. A typical generation system consists of a combustion unit located beneath the surface of the water in a reaction basin. The CO_2 is produced and mixed by the same unit. The submerged combustion unit is also discussed in chapter 11.

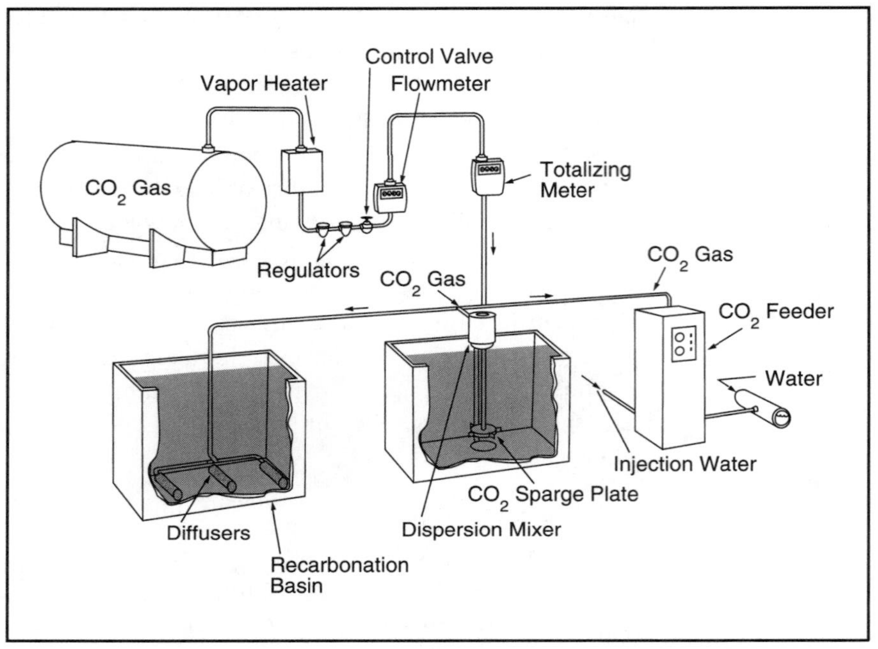

FIGURE 9-5 Liquid CO_2 recarbonation system

Courtesy of TOMCO Equipment Co.

Another common system uses gas drawn from a CO_2 storage tank. The equipment necessary is shown in Figure 9-5. The gas can be fed to a recarbonation basin or directly to a pipeline for pH control.

Phosphate Compounds

Phosphate compounds purchased in liquid form are often fed directly from the shipping container; they can also be diluted in a day tank. If dry compounds are used, they are normally dissolved in the storage tanks by placing them in stainless-steel baskets suspended in the tanks. Metering pumps must be designed to handle corrosive materials, and piping, valves, and fittings should be made of PVC or stainless steel.

Limestone Contactors

A limestone bed contactor is a treatment device in which water flows through a container packed with crushed limestone and thereby dissolves calcium carbonate. The dissolving of the calcium carbonate increases the pH and alkalinity of the water. This treatment has been shown to reduce the corrosion of lead and copper in water systems having nominally corrosive water.

The contactors are simple, low cost, and require minimal maintenance, so they are especially suitable for corrosion control in small water systems.

Regulations

On June 7, 1991, the USEPA enacted a regulation called the Lead and Copper Rule, designed to reduce exposure of the public to excessive lead and copper in drinking water. The rule requires all community and nontransient, noncommunity water systems to set up special programs to monitor for lead and copper in drinking water at customer taps. Where excessive amounts are found, the system must provide special treatment to reduce the corrosivity of the water; in addition, it may have to provide a special customer education program.

It is relatively rare for there to be an appreciable amount of either lead or copper in raw groundwater or surface water. The principal source of excessive lead and copper in customers' water is from corrosion of materials in household plumbing systems. When water is in contact with lead or copper for a period of time, some of the metal is dissolved. The more aggressive the water and the longer the contact time, the higher the lead and copper level will usually be.

Details of the required monitoring, reporting, action levels, and other provisions of the rule are included in *Water Quality*, also part of this series.

Operation of the Control Processes

Some water systems already practicing corrosion control treatment may find that the procedures are not sufficient to meet the requirements of the Lead and Copper Rule. Many additional systems will find that stabilization is necessary to meet regulations or for other reasons. The first step for the operators of these systems is to review the processes most appropriate for their system.

Selecting a Corrosion or Scaling Control Process

The information in the following paragraphs should be used in deciding on a process.

Distribution System Conditions

A prime objective of a control program is to reduce corrosion and scaling in the distribution system. The effects of corrosion or scaling can be evaluated to some extent by examining (1) records of main breaks and leaks, (2) information on how well older valves operate, and (3) information on any known problems of reduced flow rates in mains.

Whenever possible, pieces of pipe, valves, and pipe wall sections removed when pressure taps are made should be tagged with their removal location and date and stored for future reference. Inspection of these pieces can often help when the condition of the distribution system piping is being evaluated. Likewise, pieces of water service pipe can be inspected for signs of corrosion or scaling.

Home Plumbing and Customer Complaints

Customer complaints should be recorded based on type of complaint, location, and date. These records can often identify parts of the system or times of the year when particularly serious problems occur. Complaints of red water, dirty water, and loss of pressure are prime examples. Local plumbers are also usually a good source of information on conditions and problems with home plumbing that may be related to excessive corrosion or scale formation. Examples are premature replacement of hot water heaters and replacement of plumbing clogged with scale and precipitates.

Meeting Regulation Requirements

Systems that fail to meet the lead and copper action levels required by the Lead and Copper Rule *must* take corrective action as directed by the state. Most systems will probably need only slight adjustment in treatment, but some systems will have to provide extensive treatment changes in order to comply with the requirements. The steps taken to meet these requirements will probably also reduce general corrosion in the distribution system, as well as customer complaints related to water corrosivity.

Water Quality Data

Water quality data for the source water and for several distribution system samples should be examined. These data should cover the following:

- pH
- alkalinity
- conductivity
- temperature

- iron
- hardness
- chloride
- sulfate
- total dissolved solids
- fluoride
- color
- zinc
- copper
- lead
- cadmium
- dissolved oxygen
- carbon dioxide

Data for the raw water should be compared with data for samples taken from points within the system. If water from the system shows a significant increase in constituents such as iron, copper, zinc, lead, or cadmium, then the water is probably causing corrosion.

Water quality data can be used to calculate stability indexes or indicators. The most common of these is the Langelier saturation index (LI). The LI provides an indication of whether the water is likely to form or dissolve calcium carbonate. It is not a direct measure of the potential of the water to be corrosive or scaling. The LI is calculated by the following equation:

$$LI = pH_{actual} - pH_s$$

Where:

pH_{actual} = measured pH of the water
pH_s = theoretical pH at which the water will be saturated with calcium carbonate

The value of pH_s is calculated using a formula that takes into account calcium ion concentration, alkalinity, pH, temperature, and total dissolved solids.

If the LI equals zero, the water is considered stable. If the LI is a positive value, calcium carbonate will precipitate and the water has scale-forming tendencies. If the LI is a negative value, calcium carbonate will be dissolved and the water has corrosive tendencies. Additional information on LI calculation is given in *Water Quality*, also part of this series.

Determining the Best Corrosion Control Treatment for a System

System operators can follow these general suggestions when determining the best type of corrosion control treatment to use on a water system:

- Do not initiate treatment changes without first checking with the office of the state drinking water program. The officials have experience with many different systems and can provide valuable advice on treatment methods and possibly prevent a costly mistake.
- If a corrosion inhibitor is to be used, obtain the recommendations of the proposed supplier to ensure there is no conflict in water chemistries that might cause turbidity or reduce effectiveness of the chemical.
- Do not make treatment changes without first checking with industries, hospitals, other customers, and the local waste treatment authority to ensure the change in water chemistry will not adversely affect them.
- Check with other water systems that use the same water source or have similar water quality to see if they have initiated corrosion control measures. If so, obtain information on their effectiveness. In particular, small systems may be able to use the experience of larger systems that must comply with the new requirement sooner.
- Unless the effects of a treatment change are known for sure, do not experiment on a whole-system basis. If the results of a proposed treatment change are unknown, run pipe loop experiments under controlled conditions for a time to evaluate the effectiveness and possible side effects. Although pipe loop tests can at times be misleading, they can be of help in determining the best treatment method. If a portion of the distribution system can be isolated and treated separately, it may be possible to run a controlled evaluation on that section.
- Consider the advantages and disadvantages of handling and feeding various chemicals, and the advantages and disadvantages of liquid and dry chemicals. Important considerations are
  - — chemical storage space and facilities required
  - — chemical shelf life
  - — ease or problems in handling chemicals
  - — chemical feed equipment required

— hazards or special handling requirements
— any extra advantages of a specific chemical, such as improvements to water quality in other ways
— the cost of chemical treatment per unit of water furnished to the system

Adjusting pH and Alkalinity

Table 9-5 shows the alkalinity adjustments possible with chemicals commonly used to adjust pH or alkalinity. The dosage in a given system depends on the stabilization goals identified as appropriate for that system's water.

Lime

The most common chemical used for pH and alkalinity adjustment is lime because it is readily available, inexpensive compared with other chemicals, and relatively easy to feed. However, lime systems can create many maintenance problems if not operated properly. These problems generally relate to scaling caused by the lime slurry. Because of the high pH of the slurry and the low solubility of lime in water, $CaCO_3$ will precipitate on anything the slurry touches. Solution chambers, piping, and pumps must be cleaned frequently to prevent clogging and equipment damage.

The lime slurry is usually added to the filtered-water conduits or into the clearwell. The length of lime slurry-feeding lines should be minimized by locating the feeder as close as possible to the application point. Pumps can be eliminated if the slurry can flow by gravity into a reaction chamber. However, for low rates of lime application, a suitable metering pump should be used. The parts of the pump that contact the slurry must be cleaned routinely to maintain accurate feed rates.

TABLE 9-5 Alkalinity adjustment by chemical treatment

Chemical, *1 mg/L*	Alkalinity Change, *as* $CaCO_3$
Hydrated lime	Increase 1.35 mg/L
Soda ash	Increase 0.94 mg/L
Caustic soda	Increase 1.23 mg/L

Caustic Soda

Caustic soda can be used to increase both pH and alkalinity. The NaOH solution is fed by chemical metering pumps into a filtered-water conduit or the clearwell. Because NaOH is a strong caustic chemical, the pumps, tanks, and feed lines should be routinely inspected for leaks to prevent safety hazards and damage to equipment. Any leaks should be repaired immediately and spills cleaned up promptly.

Other Chemicals

Soda ash or sodium bicarbonate is often fed in combination with lime for additional alkalinity. Neither chemical poses difficult operation or maintenance problems. Carbon dioxide is used primarily for recarbonation of lime-softened water. Sulfuric acid is used to lower pH and requires the same operating precautions as caustic soda.

Creating Protective Coatings

Lime Coating

Lime, alone or in combination with soda ash or sodium bicarbonate, can be added to precipitate a $CaCO_3$ scale on the pipe walls in the distribution system. This coating will protect pipes from corrosion. If the process is not properly controlled, excessive scale can result, which will increase head loss in the distribution system and clog home plumbing systems.

For best results, the protective coating should be dense and provide uniform coverage. An acceptable coating can be produced by treated water that meets all of the following characteristics:

- Alkalinity and calcium concentrations are both maintained at a minimum of 40 mg/L as $CaCO_3$.
- The water is slightly oversaturated with $CaCO_3$ (4–10 mg/L over the saturation concentration).
- The pH is held within the range of 6.8 to 7.3.

In water that has a widely fluctuating pH, stabilization should never be used to try to dissolve previously deposited $CaCO_3$ or to maintain a certain thickness. Such attempts will result only in increased corrosion or scaling problems.

Polyphosphate Coating

A protective coating can be formed on pipe walls by polyphosphates. Sodium zinc phosphate and zinc orthophosphate are typically used for this purpose. The dosage will depend on local conditions, but will generally range between 0.5 and 3 mg/L. If sodium hexametaphosphate is used, the dosage can be as high as 10 mg/L. For all of the compounds, a somewhat higher dose is applied during the first few weeks to help spread the chemical throughout the system.

Polyphosphate solution is usually transferred from the bulk storage or mixing tank to a day tank sized to store about one day's feeding requirement. Phosphate is a bacterial nutrient, and bacterial growth in the storage tanks can be a problem. A small amount of chlorine added to the storage tanks will inhibit bacterial growth. The solution is pumped from the day tank and added to the water either just after filtration or in the clearwell through a diffuser.

Sodium silicate can also be used to form a protective film; it is fed in much the same way as the polyphosphate compounds. The initial dosage is generally 15–30 mg/L, which is then reduced to 5–10 mg/L as a continuous dosage.

Operational Control

Once a stabilization program is started, monitoring and record keeping are essential to control the process and determine if it is effective.

Water Quality Analyses

The water quality analyses must include those parameters needed to control the stabilization process. In most cases, these will be at least the pH and total alkalinity. However, the data needed to calculate the LI (or one of the other indexes) should also be collected. These data include the concentrations of the calcium ions and the total dissolved solids (total filterable residue), as well as the water temperature.

In-plant monitoring. The pH of the treated water entering the distribution system should be monitored by a continuously recording pH meter. The pH must be kept within the proper range at all times to prevent excessive corrosion or scaling. The total alkalinity concentration should be determined at least every eight hours, or more frequently if the quality of the source water is rapidly changing. The LI (or other index) of the raw and treated water should be calculated daily.

To calculate the chemical dosage necessary for proper pH and alkalinity adjustment, the pH and total alkalinity of the water entering the stabilization process must also be monitored. Samples should be tested at least every 8 hours. It is important that the samples be taken just before the application point of the stabilization chemicals. The raw water entering the plant cannot

TABLE 9-6 Important water characteristics based on water-main materials

Water-Main Material	Water Characteristic
Ductile and cast iron	Color, conductivity, dissolved oxygen, iron, manganese, pH (alkalinity and calcium if main is cement-lined)
Steel	Color, conductivity, dissolved oxygen, iron, manganese, pH
Concrete cylinder	Alkalinity, calcium, conductivity, pH
Asbestos–cement or cement-lined	Alkalinity, asbestos fibers, calcium, conductivity, pH
Lead and galvanized steel	Alkalinity, cadmium, color, conductivity, dissolved oxygen, iron, lead, pH, zinc

provide meaningful data because other water treatment chemicals (such as alum, chlorine, and fluoride adjustment chemicals) will change the pH.

Distribution system monitoring. Water quality analyses should periodically be conducted on samples from points in the distribution system to help determine if the stabilization process is effective. These analyses should take into consideration the materials in the distribution and plumbing systems, particularly if corrosion is a problem. Table 9-6 lists some of the analysis parameters that should be included for various types of pipe materials. For example, the pH and alkalinity are important characteristics to measure if A–C pipe or cement-lined pipe is in use. If leaching of the cement is occurring because of aggressive water, the pH and alkalinity will increase through the distribution system (because cement contains lime). For lead and galvanized pipe, lead and cadmium levels should also be tested because these metals are toxic to humans and may be present as a result of corrosion. Samples from household taps should also be taken as required by state and federal regulations.

Pipe and Coupon Testing

A useful method for determining what is happening in the distribution system is to inspect pipe specimens taken from the system. This inspection allows the operator to determine the extent of corrosion or scaling. A number of procedures can be used to determine the corrosion or scaling rate based on pipe specimens; one procedure involves calculating the depth of pits in the pipe or the pipe section's loss of weight. In addition, the scale on the pipe can

FIGURE 9-6 Coupons after they have been cleaned

Courtesy of Technical Products Corporation

be analyzed to help determine why a pipe is being protected or corroded or why excessive scale is being deposited.

Another valuable technique is the coupon test, which measures the effects of the water on a small section of metal (the coupon) inserted in a water line. After a minimum of 120 days, the inserts are removed, cleaned, weighed, and examined. The weight loss or gain of the coupon can provide an indication of the corrosion or scaling rate. Figure 9-6 shows some coupons after they have been cleaned. Coupons cannot provide the day-to-day information needed to adjust the chemical feed rates. However, they can be used along with water quality data to provide valuable information for long-term stability control.

Common Operating Problems

Scaling or corrosion of plant equipment is the most common in-plant operating problem. Problems within the distribution system include excessive scaling, persistence of red-water problems, and failure to meet Lead and Copper Rule requirements.

Excessive Scaling

The chief cause of excessive scaling is poor control of the stabilization process. This is certainly the case when lime (or lime with soda ash or sodium bicarbonate) is being added to form a $CaCO_3$ coating. If the pH is not kept within a narrow range near the saturation pH, excessive calcium carbonate can precipitate and cause clogging of household plumbing.

The same problem can occur if the pH of lime-softened water is not adjusted down to near the saturation pH, either by recarbonation or by addition of sulfuric acid. Properly monitoring the treated water and making necessary adjustments to the chemical feeders can prevent the scaling from becoming a problem.

A similar scaling problem can result when lime is added in conjunction with alum. This is often done to ensure that the pH is in the proper range (6.0–7.8) for effective alum coagulation. However, if too much lime is added and the pH goes above 7.8, the alum will remain in solution and pass into the distribution system, where it will then precipitate and form a scale. A water system can prevent this problem by monitoring the pH and closely controlling the amount of lime fed.

Persistence of Red-Water Problems

A frustrating problem for the water operator is the persistence of red water and other corrosion-related problems, even when corrosion control treatment is being practiced. This can occur for a number of reasons, but in most cases the problems are related to poor flow velocity, tuberculation on the pipe surface, or the presence of iron bacteria.

Flow velocity. The velocity of water flowing in the distribution system plays an important role in corrosion control. There must be enough velocity to carry the chemicals throughout the system and bring them into contact with the pipe surfaces. Where the velocity is low, in dead ends and in areas where water use is low, the stabilization chemicals will not form a protective coating or react with the corrosion by-products. A regular main-flushing program should concentrate on such problem areas to prevent the buildup of corrosion by-products until the dead ends and other problem areas can be eliminated.

Tuberculation. If tuberculation is already present in mains, the rough, uneven surface will prevent a uniform protective coating from forming. As velocity and pressure changes occur, the tubercules can break away, which will also destroy the protective layer. The addition of polyphosphates can actually increase the problems because these chemicals can loosen the deposits and cause them to break away from the pipe. A better approach is to clean the mains and then start the stabilization process.

Iron bacteria. Iron bacteria can interfere with stabilization. The slimy covering these bacteria produce can prevent the stabilization chemicals from reaching the pipe surface. If an attempt is made to lay down a protective layer, that layer sticks to the slime and is destroyed as the slime sloughs off during pressure and velocity changes. The pipes must be cleaned to remove these slime deposits. In addition, proper disinfection to prevent the bacteria from thriving in the pipe must be performed. Once these measures are accomplished, corrosion control can be effective.

Interference With Other Treatment Processes

There are many complex reactions in water chemistry, and the total effect of making a seemingly simple adjustment is not always predictable. A treatment change concerning one parameter so as to meet Lead and Copper Rule requirements may have side effects that adversely affect other water treatment processes or the finished water quality. The following are some examples:

- Some corrosion control methods could raise total trihalomethane levels or promote the formation of other undesirable disinfection by-products.
- An increase in pH may make it more difficult to meet disinfection $C \times T$ values required by the Surface Water Treatment Rule.
- A corrosion control chemical containing sodium may cause a system to exceed the sodium standard in those states that have established a primary standard for sodium.
- An inhibitor chemical may require simultaneous changes in other water parameters, such as the pH, for it to perform properly.
- The addition of a chemical containing phosphate to the water could promote biological growth in an open reservoir.
- A reaction may occur between inhibitors and other minerals in the water to cause a suspended precipitate or undesirable deposits on pipe walls.
- Increasing pH before coagulation and filtration can, under some conditions, adversely affect these processes.
- Some chemicals applied for scaling, iron, or manganese control may result in *higher* lead or copper levels.
- Under certain conditions, changing the water chemistry to reduce corrosion of lead and copper piping could create a problem with other piping. For example, it might cause red water.

Effects on Customer Uses

The effect of a change in water chemistry on how a customer uses the water must also be considered. Industrial users should be consulted before any treatment changes are made to see if the proposed change will affect their processes. In particular, a change in pH or the addition of zinc, phosphate, or silicate could have a major impact on industries that incorporate water into their product — for example, pharmaceutical, beverage, food-packaging, and food-processing firms. Hospitals should also be notified in case a change will affect dialysis equipment or their other water uses. In addition, compounds that precipitate in the distribution system can clog point-of-use treatment devices in homes and in industries.

Effect on Waste Treatment Facilities

Most of the water from a water system eventually ends up at a wastewater treatment plant. Consideration must therefore be given to the effects of any changes in water treatment on wastewater treatment facilities. For instance, the addition of a product containing zinc to drinking water could cause a waste treatment plant to exceed its effluent standard for zinc.

Safety Precautions

In operating corrosion or scaling control processes, the operator is exposed to caustic or acidic chemicals and solutions. When dry chemicals such as lime, soda ash, and sodium bicarbonate are being handled, care must be taken to minimize dust; chemical dusts are irritating to the respiratory system, skin, and eyes. Dust collectors should be installed on storage hoppers and cleaned regularly. A vacuum cleaner should be used routinely to clean dust around the feeders and hoppers. When handling dry chemicals, operators should wear proper protective clothing, including a close-fitting respirator and tight-fitting safety glasses with side shields.

In hot weather, when workers are perspiring, chemical burns become more of a problem, especially where quicklime dust is present. A long-sleeve shirt with buttoned collar and pants with the pant legs covering the top of the shoes or boots are recommended. Gloves and headgear should also be worn. A protective cream can also be placed on exposed skin near the face, neck, and wrists. The operator should shower immediately after handling dusty, dry chemicals. Dusty clothing should be laundered before it is worn again.

Proper storage of chemicals can prevent many safety problems by minimizing dust and reducing opportunities for dangerous chemical reactions. The chemicals should be stored in dry areas that have adequate ventilation. Each chemical should be stored in its own area, not together with other chemicals and equipment. If bags are used, they should be stored on pallets and as close to the feeder hopper as possible.

The chemical solutions made from the dry chemicals (such as lime slurry) or liquid chemicals (acid or caustic soda) must be handled with extreme care. Most of these solutions can cause burns and serious damage to the eyes. Face shields, rubber aprons, rubber boots, and rubber gauntlets should always be worn when working with these solutions. Spills should be cleaned up immediately to prevent accidental contact with other chemicals and damage to equipment. Storage tanks, dilution tanks, pumps, and piping should be inspected routinely to check for leaks.

Operators need to be familiar with the proper first aid procedures to follow for chemical burns from dust or solutions. These procedures should be posted, and first aid training in these procedures should be conducted. The plant should have safety showers and eye-washing facilities available for emergencies.

Record Keeping

Record keeping is especially important for controlling corrosion or scaling because a wide range of information from different sources should be used to monitor the process. For example, records at the treatment plant should be used in conjunction with records from the distribution system to provide an understanding of how the stabilization program is working.

Treatment plant information to be recorded includes

- current inventory of stabilization chemicals
- amount of chemical being fed (in milligrams per liter and pounds or kilograms per day)
- quantity of water being treated
- pH, alkalinity, and other control test results
- calculation of the LI (or other index)
- maintenance on feeders, solution lines, and diffusers

Distribution system records should include

- water quality analyses
- results of coupon tests

- results of examining pipe specimens removed from the system
- customer complaints relating to corrosion or scaling
- results of examining pipe specimens taken from household plumbing

Selected Supplementary Readings

Back to Basics Guide to Corrosion Control For Lead and Copper. 1993. Denver, Colo.: American Water Works Association.

Benjamin, M.M., S.H. Reiber, J.F Ferguson, E.A. Vanderwerff, and M.W. Miller. 1989. ***Chemistry of Corrosion Inhibitors in Potable Water.*** Denver, Colo.: American Water Works Association Research Foundation and American Water Works Association.

Corrosion Control for Operators. 1986. Denver, Colo.: American Water Works Association.

Dollar, Floyd. 1992. Polyphosphates Eliminate Rusty Water Complaints. ***Opflow,*** 18(6):1.

Economic and Engineering Services, Inc. and Kennedy/Jenks/Chilton Inc. ***Economics of Internal Corrosion Control.*** 1989. Denver, Colo.: American Water Works Association Research Foundation and American Water Works Association.

Lead and Copper. 1991. Denver, Colo.: American Water Works Association.

Lead and Copper: How to Comply for Small and Medium Systems. 1993. Denver, Colo.: American Water Works Association.

Lead and Copper Rule Guidance Manual. Vols. 1 and 2. 1992. Washington, D.C.: US Environmental Protection Agency.

Lead Control Strategies. 1989. Denver, Colo.: American Water Works Association.

Lime Handling, Application and Storage. Arlington, Va.: National Lime Association.

Orlando Utilities Commission and CH2M Hill, Inc. 1995. ***Evaluation of the Effects of Electrical Grounding on Water Quality.*** Denver, Colo.: American Water Works Association Research Foundation and American Water Works Association.

Water Quality and Treatment. 4th ed. 1990. New York: McGraw-Hill and American Water Works Association (available from AWWA).

CHAPTER 10

Iron and Manganese Control

Iron and manganese, which are natural constituents of soil and rock, are normally present in insoluble forms. When rain percolates into the soil, dissolved oxygen is removed by the decomposing organic matter. The water is then able to dissolve some of the iron and manganese. As a result, dissolved iron is present in many groundwater supplies. Manganese is present only occasionally, and then usually along with iron. At times, iron and manganese in the oxidized form are present in surface water, especially during times of lake turnover.

Excessive Iron and Manganese

Excessive iron and manganese can result in aesthetic and operational problems.

Aesthetic Problems

Iron and manganese in the concentrations that occur naturally in groundwater and surface water have no known adverse health effects. However, the aesthetic problems they can cause may be quite serious from a consumer's standpoint. Iron and manganese in raw water are generally in the dissolved state; the water is clear, and the substances are not noticeable except that they can cause a taste at high concentrations. When they are oxidized, iron and manganese change and discolor the water from turbid yellow to black, depending on their concentration and the presence of other contaminants.

When a groundwater system pumps water directly from wells to the distribution system and uses no disinfection or other treatment, dissolved iron in the water usually first becomes oxidized when it is exposed to air. After a customer fills a glass, bathtub, or washing machine with water, the iron gradually oxidizes and changes color. This not only makes the water unpalatable for consumption, but it also stains porcelain fixtures and discolors laundry (Figure 10-1).

The reaction between the high levels of iron and the tannic acid in tea and coffee can also cause customer complaints. In some cases the beverage will darken so that it looks like ink.

Operational Problems

If a disinfectant is added to the water, or if iron and manganese are fully or partially oxidized by any means before entering the distribution system (Figure 10-2), the oxidized iron and manganese will precipitate in the distribution system. The following problems may occur:

- Much of the precipitate could settle out in the mains. The worst problems will usually be in dead-end mains, where velocity is the lowest. If the iron and manganese problem in the system is not very serious, sometimes only the customers on dead ends will continually have a problem with rusty water.
- Sudden demands for extra water, such as the opening of a fire hydrant, may disrupt the normal flow in the system. The sediment that has accumulated on the bottom of mains will then be put back into suspension. Parts of the system, or even the whole system, will then have rusty water for a few hours or even a day or two. Customers will register complaints during and after the event — particularly those who were doing laundry at the time.

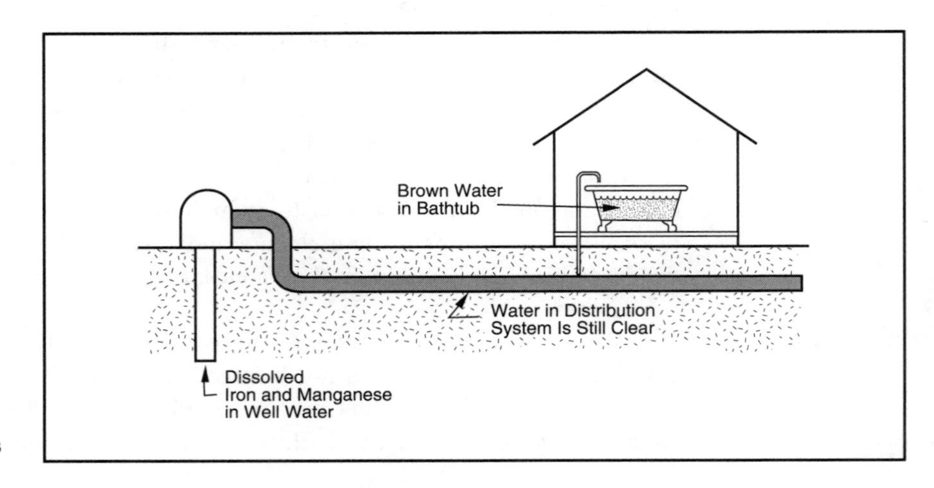

FIGURE 10-1 Iron and manganese oxidized after being exposed to air in customer's plumbing fixtures

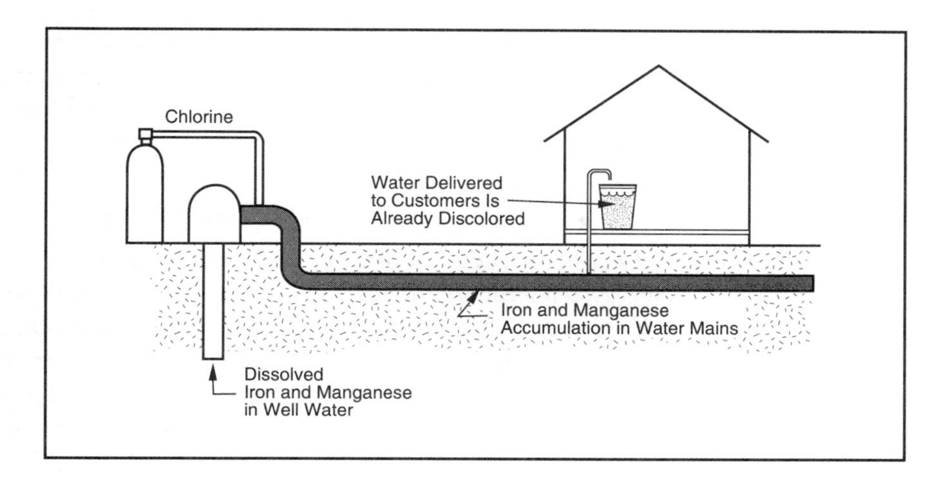

FIGURE 10-2 Iron and manganese oxidized by addition of chlorine

The presence of iron and manganese in the distribution system, in either the dissolved or oxidized state, can also provide a food source for bacterial growth in the system. The bacterial slimes that form can have the following detrimental effects:

- reduction in pipeline flow capacity
- clogging of meters and valves
- further discoloration of the water as a result of the bacterial growth
- the generation of objectionable tastes and odors
- increased chlorine demand

Maximum Desirable Levels

The secondary maximum contaminant level (SMCL) for iron established by the US Environmental Protection Agency (USEPA) is 0.3 mg/L. This is the level at which most customers find its presence undesirable. It is generally better if the level can be maintained below 0.2 mg/L. At concentrations above 0.5 mg/L, the taste is noticeable and disagreeable to most people, and staining of fixtures is quite serious. The presence of manganese is a problem because it creates brown spots on laundry. It is also a major problem for industries that incorporate water into their product because it will react with other chemicals to form undesirable tastes, odors, or colors. The maximum desirable level of magnesium is 0.05 mg/L, and the point at which it creates an undesirable taste is about 5 mg/L.

It should be remembered that the presence of iron in a water distribution system may also be caused by corrosion of metal pipes in the system. In this case, the problem must be corrected by corrosion control, as discussed in chapter 9.

Control Processes

Occasionally, the problem of iron in a water system is relatively mild. Some systems report that customer complaints are not excessive, and the only control method necessary is to set up a regular schedule of flushing accumulated sediment from dead-end mains — sometimes as often as weekly. These systems usually also experience occasional problems when the accumulated sediment is disrupted between flushings by a change in system flow.

Water systems requiring increased control of iron and manganese generally use one of the following methods:

- precipitation and filtration
- ion exchange processes
- sequestration

Both iron and manganese are also effectively removed by lime softening, but the process is generally too costly to use unless water softening is also desired. This process is discussed in detail in chapter 11.

Precipitation

In the precipitation process, the soluble forms of iron and manganese are oxidized to convert them to insoluble ferric and manganic compounds. This process is similar to the rusting of iron, which creates an iron oxide. Oxidized iron and manganese are almost completely insoluble and will precipitate as ferric oxide or oxyhydroxides and manganese hydroxide. After the oxidation reaction is complete, the precipitates must then be removed from the water by a combination of settling and filtration.

Oxidation

Oxygen in some form must be added to water to oxidize the insoluble iron and manganese. This can be accomplished either by aeration or by the addition of an oxidizing chemical to the water. The relative effectiveness of various water treatment chemicals in oxidizing iron and manganese is shown in Table 10-1.

TABLE 10-1 Relative effectiveness of various water treatment chemicals in oxidizing iron and manganese

Chemical	Iron Removal	Manganese Removal
Chlorine	Effective	Somewhat effective
Chloramine	Not effective	Not effective
Ozone	Effective	Effective
Chlorine dioxide	Effective	Effective
Potassium permanganate	Effective	Effective
Oxygen (from aeration)	Effective	Not effective

Aeration. Aeration of raw water (discussed in detail in chapter 14) is very successful in oxidizing iron if the pH of the water is greater than 6.5, but the process proceeds rather slowly at a lower pH. Aeration of manganese in water is generally not very effective if the pH is below 9.5. In addition, a contact time of up to 60 minutes may be necessary after aeration to form a filterable floc.

Chemical oxidation. Iron and manganese can be oxidized by the addition of chlorine (either as chlorine gas or calcium hypochlorite), ozone, potassium permanganate, or chlorine dioxide.

To achieve a manganese removal to less than a level of 0.05 mg/L by using chlorine, the pH must be at least 8.0. The oxidation of manganese by chlorine also essentially stops at a temperature below 40°F (4°C). Although chlorine works well for oxidizing iron at a normal pH, it is important to consider the possibility of excessive levels of trihalomethanes (THMs) also being formed in the process.

Very effective oxidation of iron and manganese can be achieved by using ozone. However, ozone alone is rarely used for this purpose because of the high cost of equipment, operation, and maintenance. Ozone can also be used to control trace organics, if this happens to be a problem in the water. Another factor that must be considered is that excessive amounts of ozone can oxidize manganese to permanganate, which will cause pink water.

Potassium permanganate is very effective in oxidizing both iron and manganese, and the reaction is relatively rapid. Another benefit of using permanganate is that it also reacts with hydrogen sulfide, cyanides, phenols, and other taste-and-odor compounds if they are present. In addition, no THMs will be formed. Care must be taken not to overfeed permanganate, however, or purple water will be discharged to the distribution system.

Detention

After the iron and manganese have been exposed to oxygen by either aeration or a chemical oxidant, it is usually necessary to hold the water for a period of time in order to allow the oxidation process to be completed before the water is released to the precipitate removal stage. A number of relationships involving the concentrations of iron or manganese and the pH, temperature, and oxidant used affect the detention time.

In cases where there is a low concentration of iron only, the action may be almost instantaneous. In other cases, particularly where manganese is present, it may require up to an hour for complete oxidation to take place. If the reaction has not been completed when the water is filtered, the soluble forms will pass right through the filter and will later precipitate in the distribution system.

Removal

After the precipitates of iron and manganese are formed by the oxidation process, they must be removed by filtration.

Granular media filters. Granular media filters are generally used for removing iron and manganese precipitates. If the solids concentration is relatively low (under approximately 5 mg/L), the water can usually be processed directly by filtration, without sedimentation (Figure 10-3). If the solids concentration is higher, the water must be passed through a

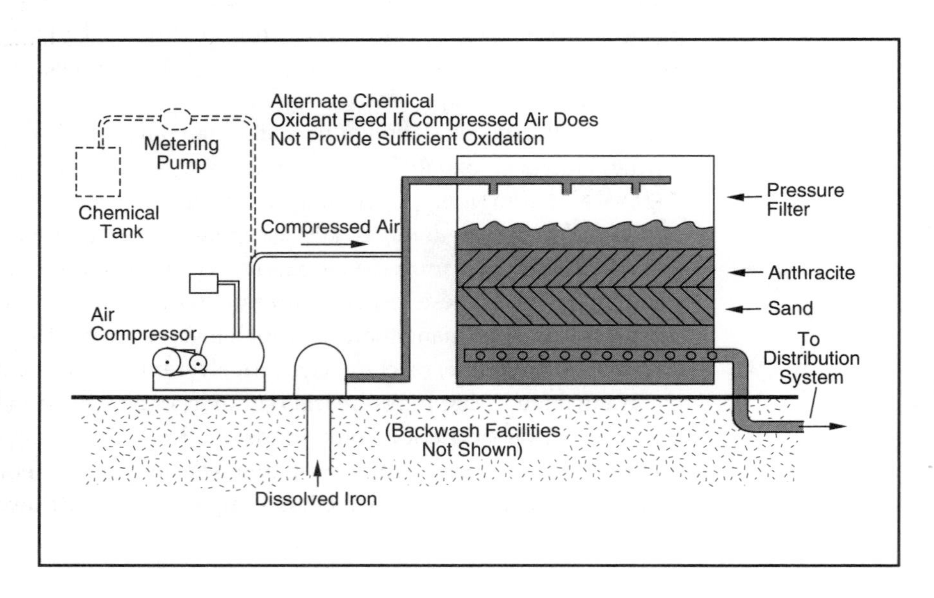

FIGURE 10-3 Low concentration of iron oxidized with compressed air or oxidant chemical and removed with a pressure filter

sedimentation basin to remove as much precipitate as possible before the water is filtered. Lime is sometimes added to raise the pH and facilitate iron precipitation. If the loading is not properly reduced through sedimentation, filter backwashing will be excessive.

Manganese greensand filters. Manganese greensand filters use a special type of medium that removes iron and manganese by a combination of both adsorption and oxidation. In the process, permanganate is added ahead of the greensand filter so that the grains of the medium become coated with oxidation products. The oxidized greensand then adsorbs the dissolved iron and manganese from the water, after which the substances are oxidized with permanganate and removed by the filtering action of the filter bed. A potassium permanganate backwash is then used to regenerate the bed, or permanganate can be fed continuously in a small dose.

Greensand grains are somewhat smaller than silica sand, so the head loss can quickly become excessive under a heavy loading. The length of filter runs can be increased by adding a layer of anthracite above the greensand.

Removal by Ion Exchange

Depending on the type of exchange material used, iron will be removed along with calcium, magnesium, and sodium in an ion exchange unit. Ion exchange is economical to use only on water that is low in hardness and dissolved solids. Only the soluble forms of iron and manganese are removed by the ion exchange medium. Therefore, there must be little or no previously oxidized material entering the exchanger. If there are precipitates already in the water, they will clog the bed and foul the ion exchange medium. The process is not recommended for use with water containing a very high level of iron and manganese. Manufacturers should be contacted for their recommendations.

Sequestration

Sequestration is effective only for groundwater that has a relatively low level of dissolved iron and manganese and no dissolved oxygen. It is not usually recommended if the concentration of iron, manganese, or a combination of the two exceeds 1.0 mg/L. In the process, polyphosphates or sodium silicates are added before the water is exposed to air or disinfectants.

If effective, sequestration tends to keep iron and manganese soluble in the finished water. It does not *remove* the iron and manganese; therefore, bacterial slimes may still form in the distribution system as a result of bacterial growth. The total phosphates applied should not exceed the amount specified by the chemical supplier. In addition, a chlorine residual of at least 0.2 mg/L should be maintained in the system at all times.

Chemical Reactions

An explanation of the chemical reactions that occur when iron and manganese are oxidized with chemicals is included in *Basic Science Concepts and Applications*, part of this series.

All chemicals used in iron and manganese control should meet American Water Works Association (AWWA) and NSF International standards, as discussed in more detail in appendix A.

Control Facilities

This section discusses the types of equipment used in controlling iron and manganese.

Aeration Equipment

Aeration equipment that can be used for the oxidation of iron and manganese is detailed in chapter 14. The medium used in an aerator, such as coke or broken stone in a cascade aerator, becomes coated with iron or manganese hydroxide over a period of time. It has been found that this promotes catalytic precipitation of iron and manganese from the raw water. The process of aeration not only exposes the dissolved iron and manganese to oxygen, but also removes carbon dioxide from the water, which enhances the oxidizing action because it raises the pH of the water. It takes only about 0.14 mg/L oxygen per milligram per liter of iron to oxidize iron so long as the pH is above about 6.5. Oxidation of manganese by aeration is not effective at a pH below about 9.5.

It is not usually advisable to aerate water any more than necessary to achieve oxidation. Although it is not corrosive when it is oxygen-free, low-alkalinity water becomes quite corrosive when it is heavily saturated with oxygen.

Most aeration schemes require that the water be pumped twice. It must first be pumped from the source and through the aeration device, and then pumped again to produce the required distribution system pressure. Line diagrams of typical installations are shown in Figures 10-4 and 10-5.

If only a low level of iron is present and it has oxidized quickly, it may be possible to aerate and filter the iron-bearing water under pressure. As illustrated in Figure 10-3, compressed air can be pumped into the water entering the filter or into the space above the filter. It can produce sufficient

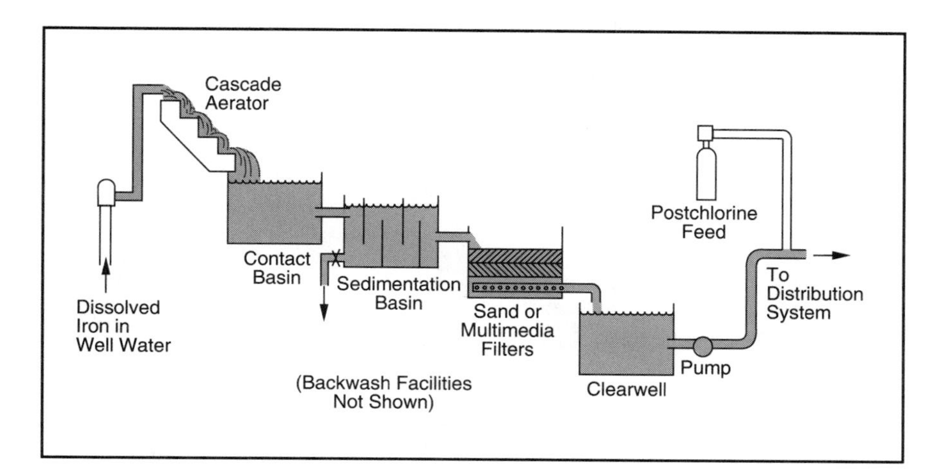

FIGURE 10-4 Iron removal using aeration, sedimentation, and filtration

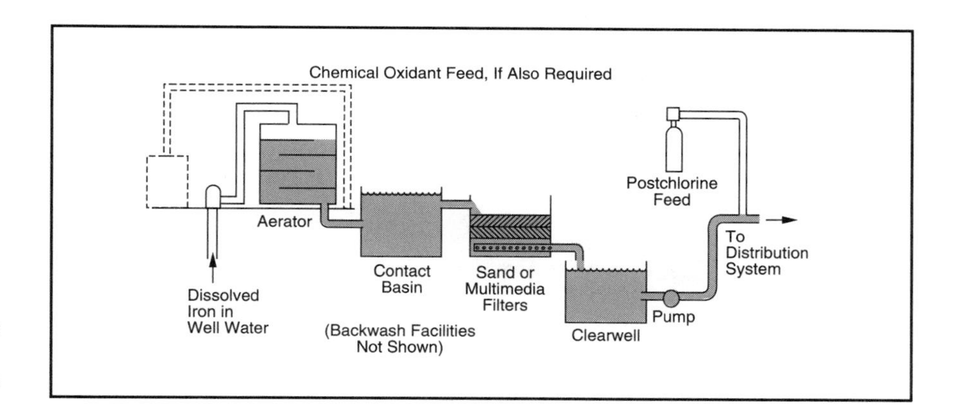

FIGURE 10-5 Iron removal by oxidation using a contact basin but no sedimentation

oxidation to precipitate the iron in the short time before the water passes through the filter media. The filter in this type of installation uses relatively coarse media.

Chemical Oxidation Equipment

The equipment used for feeding chemical oxidants is generally the same as described in chapter 7. Chlorine gas from a chlorinator or chlorine dioxide can be applied directly to the raw-water pipeline or contact basin. Liquid chlorine can be fed using a metering pump.

Potassium permanganate can also be injected into a pipeline or gravity-fed into a detention basin (Figure 10-6). The permanganate can be fed as

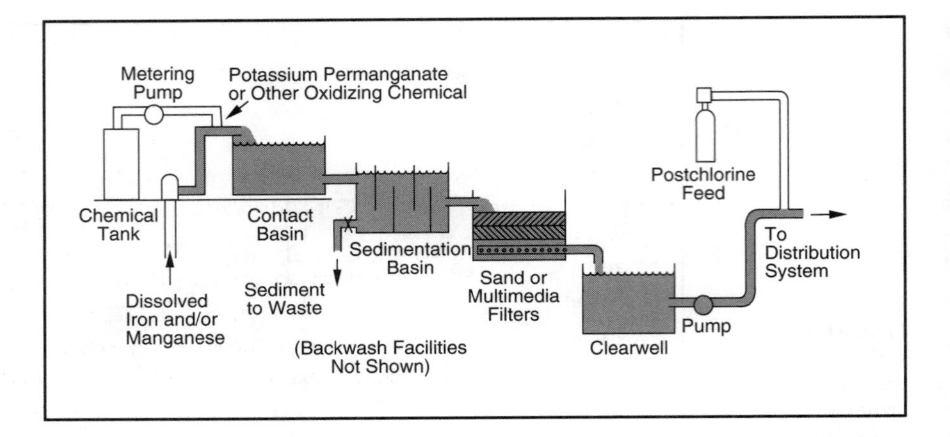

FIGURE 10-6 Iron and manganese removed by chemical oxidation, sedimentation, and filtration

needed by a dry chemical feeder, or batches of concentrated solution may be prepared and fed as a liquid. If ozone is used, transfer of oxygen to the water must take place in a reaction basin.

If an oxidant other than chlorine is used for the iron and manganese removal process, a system could, by choice or by state mandate, be required to add postchlorination before the water leaves the treatment plant to maintain a residual in the distribution system.

Detention Equipment

Unless pilot tests have shown it to be unnecessary, a detention or contact chamber is needed to hold the water temporarily after aeration or the addition of an oxidant. A contact time of 20 minutes is about average, but as long as 1 hour may be required under some circumstances to allow the oxidation reaction to be completed.

The chamber is usually a concrete, steel, or fiberglass tank that has sufficient capacity to hold the water for the required time at the maximum flow rate. If an aerator is used to provide oxidation, the detention tank can be placed directly below the aerator. If the concentration of iron and manganese is relatively low, flow from the contact chamber can be applied directly to the filter. If the level is higher, the waste then flows to a sedimentation basin.

Sedimentation Basins

Unless shown to be unnecessary by a pilot-plant study or past experience, a sedimentation basin is required after the oxidation process to allow as much precipitate as possible to be removed before filtration.

Sedimentation basins are generally constructed similarly to basins for conventional filtration systems, as detailed in chapter 5. Plain sedimentation is sufficient on some systems, but the required settling period may be as long as 12 to 24 hours.

If iron or manganese removal is particularly difficult, additional treatment may be required to hasten sedimentation. Some common methods are as follows:

- Lime, or lime plus alum, can be added and mixed for 20–30 minutes before sedimentation. This raises the pH and facilitates precipitation. If the pH has been sufficiently increased to cause a chemical balance of the water, recarbonation may be required for stabilization before the water is released to the distribution system.
- Alum can be mixed into the water before sedimentation.
- Alum and bentonite can be mixed into the water before sedimentation.
- Alum and specially treated silica can be mixed into the water before sedimentation.

Sand Filtration Equipment

Filtration using either sand alone or sand capped with anthracite can be used for open or tank-type filters, as discussed in detail in chapter 6. The media that work most efficiently may not be the same gradation as used in the conventional filtration of surface water.

Manganese Greensand Equipment

Granular filtration using manganese greensand includes a rather standard-looking filter unit, except that the unit uses special filtering media. The greensand is usually capped with about 6 in. (50 mm) of anthracite to act as a roughing filter medium. No preliminary contact or sedimentation chambers are necessary. Potassium permanganate is continuously fed as far in advance of the filter as possible. Other oxidizing agents or processes such as chlorination or aeration can be used prior to the permanganate feed in order to reduce the cost of materials.

The filter must be furnished with provisions for air-backwash, and also with sample taps at various points opposite the media bed for monitoring the progression of permanganate through the filter. The filters are backwashed with a permanganate solution to regenerate the bed.

Ion Exchange Equipment

Ion exchange processes are described in detail in chapter 12. The equipment used for iron and manganese removal is the same as that for softening, but the media are specifically selected for iron and manganese removal. If there is previously oxidized iron or manganese in the water entering an ion exchange unit, it will be deposited on the resin bed and will be mostly removed by backwashing. However, there will be a buildup of precipitates that will foul the media over a period of time, which will create a definite loss in ion exchange capacity. Ion exchange should be used for iron and manganese removal only when the water has low hardness and dissolved solids and only when the iron and manganese are in the dissolved state.

Sequestering Chemical Feed Equipment

Polyphosphates and sodium silicates used for sequestering iron and manganese are normally purchased as dry chemicals and prepared as a solution. The solution is then fed into the system by a metering pump (Figure 10-7). The feed point should be located immediately after the water leaves the well to minimize oxidation of the iron and manganese before the sequestering chemical is added.

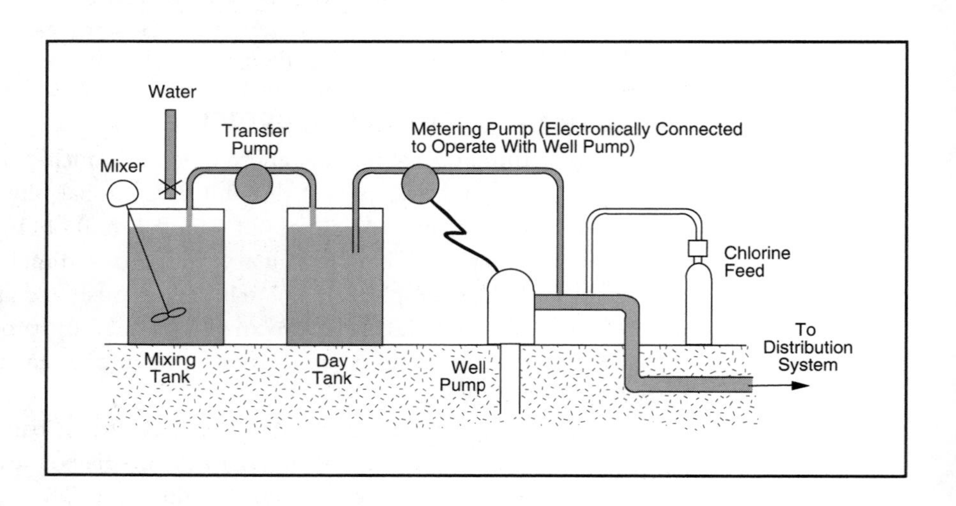

FIGURE 10-7 Sequestering chemical feed equipment

Regulations

There are no federal regulations setting a mandatory limit on the concentration of iron or manganese in drinking water. However, the following SMCLs have been established by the USEPA:

- iron: 0.3 mg/L
- manganese: 0.05 mg/L

Secondary regulations are not mandatory, but they are strongly recommended to avoid tastes, odors, color, staining of fixtures, or other objectionable qualities of the water. There is also fear that customers who are bothered by aesthetically displeasing water from a public water system will use water from another source that is aesthetically pleasing but may not be chemically or microbiologically safe.

Many states have established mandatory limits on the amount of iron and manganese in drinking water. In some cases the limit is the same as the federal secondary limit, and in other cases it is somewhat more stringent. The state public water supply program should be consulted for information on current state requirements.

All chemicals used in iron and manganese control must meet AWWA and NSF International standards, as discussed in more detail in appendix A.

Process Operation

The operation of the equipment used in iron and manganese control processes is covered in other chapters of this volume as follows:

- Liquid chemical feed equipment: chapter 4
- Sedimentation basins: chapter 5
- Filtration equipment: chapter 6
- Oxidant feed equipment: chapter 7
- Ion exchange equipment: chapter 12
- Aeration equipment: chapter 14

Process Monitoring

Equipment should always be available to test for the iron content of water to a minimum of 0.1 mg/L and the manganese content to a minimum of 0.05 mg/L. If the iron and manganese concentration in the raw water is variable, the water may have to be tested rather frequently to adjust the process for thorough yet economical treatment.

When polyphosphate or sodium silicate is used in the appropriate analysis, equipment must be available for periodic testing of process control and to ensure that the maximum limit is not exceeded.

Operating Problems

Although the iron and manganese levels in groundwater are generally rather stable, in a few cases they have drastically increased over a period of years. The design of a control method should therefore be flexible enough that the treatment can be changed to meet the demands of worsening water quality.

Record Keeping

If iron and/or manganese are being removed, the following important records should be maintained:

- results of periodic analysis of raw water to monitor for changes in the iron and manganese concentration
- concentration of the oxidizing chemical being fed
- detention time being provided between oxidation of the water and sedimentation
- concentration of any supplemental chemicals added to improve sedimentation
- sedimentation time being provided
- length of filter runs
- quantity of water treated
- results of periodic analysis of distribution system samples for iron and manganese concentration
- details of all distribution system flushing done to control discolored water
- all customer complaints of discolored water

If iron and manganese are being controlled by the addition of a sequestring chemical, records should include

- results of periodic analysis of raw water to monitor for changes in the iron and/or manganese concentration
- brand of chemical being used and the concentration being fed (noted daily)

- quantity of water treated
- details of all distribution system flushing done to control discolored water
- all customer complaints of discolored water

Selected Supplementary Readings

Knocke, W.R., Suzanne Occiano, and Robert Hungate. 1989. ***Removal of Water Soluble Manganese by Oxide-Coated Filter Media.*** Denver, Colo.: American Water Works Association Research Foundation and American Water Works Association.

Knocke, W.R., J.E. Van Benschoten, M. Kearney, A. Soborski, and D.A. Reckhow. 1990. ***Alternative Oxidants for the Removal of Soluble Iron and Manganese.*** Denver, Colo.: American Water Works Association Research Foundation and American Water Works Association.

Knocke, W.R., et al. 1991. Kinetics of Manganese and Iron Oxidation by Potassium Permanganate and Chlorine Dioxide. ***Jour. AWWA,*** 83(6):80.

Manual of Water Utility Operations. 8th ed. 1988. Austin, Texas: Texas Water Utilities Association.

Recommended Standards for Water Works. 1992. Albany, N.Y.: Health Education Services.

Robinson, R.B., G.D. Reed, and B. Frazier. 1992. Iron and Manganese Sequestration Facilities Using Sodium Silicate. ***Jour. AWWA,*** 84(2):77.

Robinson, R.B., G.D. Reed, D. Christodos, B. Frazier, and V. Chidoombariah. 1990. ***Sequestering Methods of Iron and Manganese Treatment.*** Denver, Colo.: American Water Works Association Research Foundation and American Water Works Association.

Smith, S.A. 1992. ***Methods for Monitoring Iron and Manganese Biofouling in Water Wells.*** Denver, Colo.: American Water Works Association Research Foundation and American Water Works Association.

Water Quality and Treatment. 4th ed. 1990. New York: McGraw-Hill and American Water Works Association (available from AWWA).

CHAPTER 11

Lime Softening

Water contains various amounts of dissolved minerals, some of which impart a quality known as hardness. Consumers frequently complain about problems attributed to hard water, such as the formation of scale in cooking utensils and hot water heaters.

The first section of this chapter discusses the occurrence, chemistry, and effects of hard water. The remainder of the chapter describes the softening processes that remove the hardness-causing minerals by precipitation.

The precipitation process most frequently used is generally known as the lime process, or lime–soda ash process. Because of the special facilities required and the complexity of the process, it is generally applicable only to medium- or large-sized water systems where all treatment can be accomplished at a central location. This process will provide softened water at the lowest cost. Lime softening can be used for treatment of either groundwater or surface water sources.

The other commonly used method of softening involves the ion exchange process, as detailed in chapter 12. This process has the advantages of a considerably lower initial cost and ease of use by small systems or by larger systems at multiple locations. The principal disadvantage is that operating costs are considerably higher. Ion exchange processes can typically be used for direct treatment of groundwater, so long as turbidity and iron levels are not excessive. For treatment of surface water, the process normally must be preceded by conventional treatment.

Softening can also be accomplished using membrane technology, electrodialysis, distillation, and freezing. Of these, membrane methods (as described in chapter 15) seem to have the greatest potential. The process is still somewhat experimental, but it might find wide use in the future as the technology improves.

Effects of Hard and Soft Water

This section discusses the problems associated with water that is too hard or too soft.

Occurrence of Hard Water

Hard water is caused by soluble, divalent, metallic cations (positive ions having a valence of 2). The principal chemcials that cause water hardness are calcium (Ca) and magnesium (Mg). Strontium, aluminum, barium, iron, manganese, and zinc can also cause hardness in water, but they are not usually present in large enough concentrations to contribute significantly to the total hardness.

As shown in Figure 11-1, water hardness varies considerably in different geographic areas of the contiguous 48 states. This is due to different geologic formations, and is also a function of the contact time between water and these formations. Calcium is dissolved as water passes over and through limestone deposits. Magnesium is dissolved as water passes over and through dolomite and other magnesium-bearing minerals. Because groundwater is in contact with these formations for a longer period of time than surface water, groundwater is normally harder than surface water.

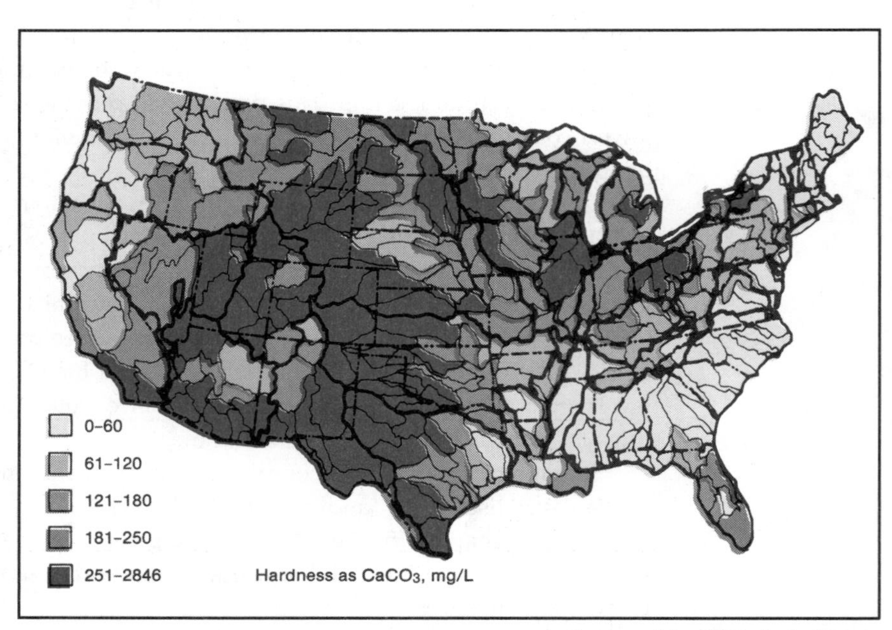

FIGURE 11-1 Average water hardness in the continental United States

Courtesy of the US Geological Survey

Expressing Hardness Concentration

Water hardness is generally expressed as a concentration of calcium carbonate, in terms of milligrams per liter as $CaCO_3$. The degree of hardness that consumers consider objectionable will vary, depending on other qualities of the water and on the hardness to which they have become accustomed. Table 11-1 shows two different classifications of the relative hardness of water.

Types of Hardness

Hardness can be categorized by either of two methods: calcium versus magnesium hardness and carbonate versus noncarbonate hardness.

The calcium–magnesium distinction is based on the minerals involved. Hardness caused by calcium is called calcium hardness, regardless of the salts associated with it, which include calcium sulfate ($CaSO_4$), calcium chloride ($CaCl_2$), and others. Likewise, hardness caused by magnesium is called magnesium hardness. Calcium and magnesium are normally the only significant minerals that cause hardness, so it is generally assumed that

$$\text{total hardness} = \text{calcium hardness} + \text{magnesium hardness}$$

The carbonate–noncarbonate distinction, however, is based on hardness from either the bicarbonate salts of calcium or the normal salts of calcium and magnesium involved in causing water hardness. Carbonate hardness is caused primarily by the bicarbonate salts of calcium and magnesium, which are calcium bicarbonate, $Ca(HCO_3)_2$, and magnesium bicarbonate, $Mg(HCO_3)_2$. Calcium and magnesium combined with carbonate (CO_3) also contribute to carbonate hardness. Noncarbonate hardness is a measure of calcium and magnesium salts other than carbonate and bicarbonate salts. These salts are calcium sulfate, calcium chloride, magnesium sulfate

TABLE 11-1 Comparative classifications of water for softness and hardness

Classification	mg/L as $CaCO_3$*	mg/L as $CaCO_3$†
Soft	0–75	0–60
Moderately hard	75–150	61–120
Hard	150–300	121–180
Very hard	Over 300	Over 180

Source: Adapted from Sawyer 1960 and Briggs and Ficke 1977.

*Per Sawyer (1960).

†Per Briggs and Ficke (1977).

($MgSO_4$), and magnesium chloride ($MgCl_2$). Calcium and magnesium combined with nitrate (NO_3) may also contribute to noncarbonate hardness, although it is a very rare condition. For carbonate and noncarbonate hardness,

$$\text{total hardness} = \text{carbonate hardness} + \text{noncarbonate hardness}$$

When hard water is boiled, carbon dioxide (CO_2) is driven off. Bicarbonate salts of calcium and magnesium then settle out of the water to form calcium and magnesium carbonate precipitates. These precipitates form the familiar chalky deposits on teapots. Because it can be removed by heating, carbonate hardness is sometimes called "temporary hardness." Because noncarbonate hardness cannot be removed or precipitated by prolonged boiling, it is sometimes called "permanent hardness."

Objections to Hard Water

Scale Formation

Hard water forms scale, usually calcium carbonate, which causes a variety of problems. Left to dry on the surface of glassware and plumbing fixtures, including shower doors, faucets, and sink tops, hard water leaves unsightly white scale known as water spots. Scale that forms on the inside of water pipes will eventually reduce the flow capacity or possibly block it entirely. Scale that forms within appliances and water meters causes wear on moving parts.

When hard water is heated, scale forms much faster. In particular, when the magnesium hardness is more than about 40 mg/L (as $CaCO_3$), magnesium hydroxide scale will deposit in hot water heaters that are operated at normal temperatures of 140–150°F (60–66°C). A coating of only 0.04 in. (1 mm) of scale on the heating surfaces of a hot water heater creates an insulation effect that will increase heating costs by about 10 percent.

Effect on Soap

The historical objection to hardness has been its effect on soap. Hardness ions form precipitates with soap, causing unsightly "curd," such as the familiar bathtub ring, as well as reduced efficiency in washing and laundering. To counteract these problems, synthetic detergents have been developed and are now used almost exclusively for washing clothes and dishes. These detergents have additives known as sequestering agents that "tie up" the hardness ions so that they cannot form the troublesome precipitates. Although modern detergents counteract many of the problems of hard water, many customers prefer softer water. These customers can install individual softening units or use water from another source, such as a cistern, for washing.

Aesthetic Concerns About Soft Water

Although various problems are caused by hard water, very soft water (near zero hardness) is also not desirable. It can leave a soap-scum feeling on the skin, and it is very corrosive to the water system and plumbing fixtures. For this reason, water systems that provide softening also limit the softening action or blend hard and softened water to provide moderately soft water to the system.

Although moderately soft water is usually preferable for consumer use, it still has a number of disadvantages. There may still be substantial corrosion of pipes and household appliances, which will shorten their service life. In addition, it may be determined that corrosion of lead and copper pipes and fittings will keep the system from meeting the requirements of the Lead and Copper Rule. The problems and corrective steps for reducing corrosion are discussed in chapter 9.

Removal of Other Contaminants

Although chemical precipitation is normally selected as a treatment method because of its efficiency in softening, it is also useful for the removal of other contaminants. It is particularly effective at removing iron and manganese, heavy metals, radionuclides, dissolved organics, and viruses.

The Decision to Install Softening

The decision to add softening treatment to an existing or new water system depends on an analysis of the various considerations discussed above, including any other water problems that should or must be corrected. An example would be a water system that has a raw-water source that is relatively hard and also exceeds the maximum contaminant level (MCL) for radium. In this case, there may not be enough demand for softening to justify installing the necessary equipment, but softening can be provided at little additional cost in conjunction with the mandated radium removal.

If treatment for softening alone is the only consideration, the public must demand and support it because of the resulting increase in water rates. The cost of a new installation and the continuing added operating costs must be compared with the cost savings and advantages to customers. The savings to customers include prolonged life of plumbing equipment, reduced costs for washing soap and detergent, and improved aesthetic qualities of the water, which are often hard to quantify.

When water from a public system has been relatively hard over a period of years and central softening treatment is not provided, many homeowners and businesses install their own softeners. Some systems estimate that as many as 80 or 90 percent of their customers have installed softeners. Those systems would obviously find it difficult to justify installing central treatment for the sole purpose of softening the water. A special dilemma arises when one of these systems is required to install treatment (for instance, for radium removal) that will incidentally also accomplish softening. This creates the difficult public relations problem of informing customers that their softeners will no longer be necessary.

Softening Processes

This section discusses the processes that use precipitation, the use of coagulants in softening, and the chemical reactions involved in the lime–soda ash process.

Processes Using Precipitation

Several different softening processes involving precipitation can be used for potable water treatment.

Lime Softening

When the raw water contains little or no noncarbonate hardness, it can normally be softened through the use of lime alone.

Lime–Soda Ash Softening

Both lime and soda ash are necessary when there is a nominal amount of magnesium hardness in the water. When the lime and soda ash are added, the minerals form nearly insoluble precipitates, which are then removed by the conventional processes of flocculation, sedimentation, and filtration. Because the precipitates are very slightly soluble, some hardness remains in the water, amounting to about 50–85 mg/L as $CaCO_3$. This works out to be an advantage because completely soft water is undesirable. Treatment is typically performed through the single-stage process shown in Figure 11-2A. This is the process predominantly used in the United States and will be described in detail in this chapter.

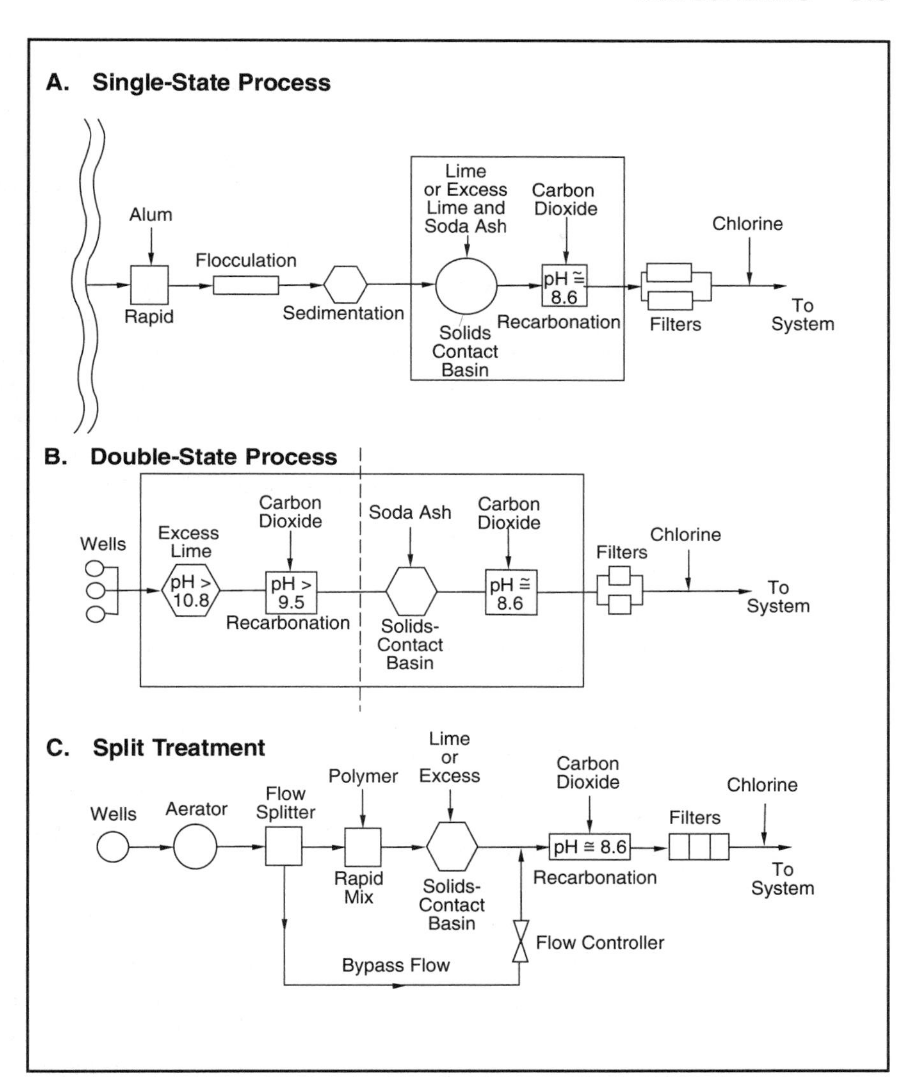

FIGURE 11-2 Types of lime–soda ash softening processes

Excess-Lime treatment

It is usually desirable to reduce the magnesium hardness of water when it exceeds about 40 mg/L. To reduce the magnesium hardness, more lime must be added to the water than is used in the conventional lime–soda ash process. The extra lime will raise the pH above 10.6 so that magnesium hydroxide will precipitate out of the water.

When this treatment process is used, soda ash is added to remove noncarbonate hardness, and recarbonation must be used to reduce the pH of the water before it enters the distribution system. Excess-lime treatment can be performed in either a single-stage process or a double-stage process. The double-stage process (Figure 11-2B) allows greater removal of magnesium hardness and more control over the quality of the treated water.

Split Treatment

Split treatment is a modification of the excess-lime process and is used to reduce the amount of chemicals required. As shown in Figure 11-2C, only a portion of the water is treated with excess lime. A smaller remaining portion bypasses the treatment process, and the two are recombined. The amount of water bypassing lime treatment depends on the quality of the raw water and the desired quality of the finished water. The carbon dioxide and bicarbonate alkalinity in the untreated portion of water help stabilize the treated portion of water, which minimizes or eliminates the need for adding carbon dioxide for recarbonation. Systems that must meet requirements of the Surface Water Treatment Rule would probably not be able to use this process because it requires that part of the water bypass coagulation treatment.

Caustic-Soda Treatment

Caustic soda (sodium hydroxide, NaOH) can be used in place of lime and soda ash for softening and will remove both carbonate and noncarbonate hardness. The process is generally more expensive than lime or lime and soda ash treatment, and it also increases the total dissolved solids in the treated water. Some advantages are that it produces less sludge, and the chemical is much easier to store, feed, and handle than lime.

Coagulant Use in Softening

The addition of coagulants is often necessary in the lime–soda ash softening process to help gather the fine calcium and magnesium precipitates into particles that will readily settle out of the water. If sedimentation is not efficiently accomplished, the result will be shorter filter runs, less efficient turbidity removal, and scale formation on the filter media. Small amounts of alum, ferric sulfate, or ferric chloride are often used for this process. Polymers have also been found to be effective and have the additional advantages of creating less sludge, being effective over a much broader pH range, and creating sludge that dewaters more easily.

Chemical Reactions

The following equations represent the basic reactions involved in the lime–soda ash process. The down arrow (↓) indicates precipitates formed by the reactions.

First, although carbon dioxide does not cause hardness, it reacts with and consumes the lime added to remove hardness. Therefore, it must be considered when the lime dosage is being determined. Sufficient lime must be added to convert carbon dioxide (CO_2) to calcium carbonate ($CaCO_3$), as shown in Eq 11-1 below. This reaction is complete when a pH of 8.3 is reached. Some systems use aeration to remove some of the CO_2, thereby reducing the lime requirement.

$$\underset{\text{carbon dioxide}}{CO_2} + \underset{\text{lime}}{Ca(OH)_2} \rightarrow \underset{\text{calcium carbonate}}{CaCO_3\downarrow} + \underset{\text{water}}{H_2O} \tag{11-1}$$

The minerals that cause carbonate hardness are then precipitated out of the water either as calcium carbonate (Eq 11-2) or as magnesium hydroxide (Eq 11-3 and Eq 11-4). In Eq 11-2, enough lime is added to raise the pH to 9.4, at which point the calcium precipitate is formed. A similar process is shown in Eq 11-3 and Eq 11-4, where excess lime is added to elevate the pH above 10.6 to form the magnesium hydroxide precipitate, $Mg(OH)_2$.

$$\underset{\text{calcium bicarbonate}}{Ca(HCO_3)_2} + \underset{\text{lime}}{Ca(OH)_2} \rightarrow \underset{\text{calcium carbonate}}{2CaCO_3\downarrow} + \underset{\text{water}}{2H_2O} \tag{11-2}$$

$$\underset{\text{magnesium bicarbonate}}{Mg(HCO_3)_2} + \underset{\text{lime}}{Ca(OH)_2} \rightarrow \underset{\text{calcium carbonate}}{CaCO_3\downarrow} + \underset{\text{magnesium carbonate}}{MgCO_3} + \underset{\text{water}}{2H_2O} \tag{11-3}$$

$$\underset{\text{magnesium carbonate}}{MgCO_3} + \underset{\text{lime}}{Ca(OH)_2} \rightarrow \underset{\text{calcium carbonate}}{CaCO_3\downarrow} + \underset{\text{magnesium hydroxide}}{Mg(OH)_2\downarrow} \tag{11-4}$$

The minerals that cause noncarbonate calcium hardness are precipitated out of the water by the addition of soda ash. The chemical reactions differ slightly, based on the types of noncarbonate calcium compounds causing the hardness. Two examples of these chemical reactions are

$$\underset{\text{calcium sulfate}}{CaSO_4} + \underset{\text{soda ash}}{Na_2CO_3} \rightarrow \underset{\text{calcium carbonate}}{CaCO_3\downarrow} + \underset{\text{sodium sulfate}}{Na_2SO_4} \tag{11-5}$$

$$\underset{\text{calcium chloride}}{CaCl_2} + \underset{\text{soda ash}}{Na_2CO_3} \rightarrow \underset{\text{calcium carbonate}}{CaCO_3\downarrow} + \underset{\text{sodium chloride}}{2NaCl} \tag{11-6}$$

The minerals that cause noncarbonate magnesium hardness are precipitated out of the water by the addition of lime (Eq 11-7 and Eq 11-9). However, this process also results in the formation of noncarbonate salts (such as $CaSO_4$ and $CaCl_2$) that cause noncarbonate hardness. Therefore, soda ash must be added to the water to react with these salts to form $CaCO_3$ (Eq 11-8 and Eq 11-10).

$$\underset{\text{magnesium chloride}}{MgCl_2} + \underset{\text{lime}}{Ca(OH)_2} \rightarrow \underset{\text{magnesium hydroxide}}{Mg(OH)_2\downarrow} + \underset{\text{calcium chloride}}{CaCl_2} \tag{11-7}$$

$$\underset{\text{calcium chloride}}{CaCl_2} + \underset{\text{soda ash}}{Na_2CO_3} \rightarrow \underset{\text{calcium carbonate}}{CaCO_3\downarrow} + \underset{\text{sodium chloride}}{2NaCl} \tag{11-8}$$

$$\underset{\text{magnesium sulfate}}{MgSO_4} + \underset{\text{lime}}{Ca(OH)_2} \rightarrow \underset{\text{magnesium hydroxide}}{Mg(OH)_2\downarrow} + \underset{\text{calcium sulfate}}{CaSO_4} \tag{11-9}$$

$$\underset{\text{calcium sulfate}}{CaSO_4} + \underset{\text{soda ash}}{Na_2CO_3} \rightarrow \underset{\text{calcium carbonate}}{CaCO_3\downarrow} + \underset{\text{sodium sulfate}}{Na_2SO_4} \tag{11-10}$$

Because the softened water has a pH close to 11 and a high concentration of $CaCO_3$, it must be stabilized so that the $CaCO_3$ will not precipitate on the filter media, on filter underdrains, or in the distribution system. The water is stabilized through a process called recarbonation, which involves adding carbon dioxide to the softened water (Eq 11-11). When carbon dioxide is added, soluble calcium bicarbonate is formed (resulting in a small amount of hardness), and the pH is reduced to about 8.6, or to a level at which the water is stabilized to prevent scale formation or corrosion.

Eq 11-12 shows that any residual $Mg(OH)_2$ will also be converted to a soluble compound, $MgCO_3$.

$CaCO_3$	+	CO_2	+	H_2O	→	$Ca(HCO_3)_2$	(11-11)
calcium carbonate		carbon dioxide		water		calcium bicarbonate	

$Mg(OH)_2$	+	CO_2	→	$MgCO_3\downarrow$	+	H_2O	(11-12)
magnesium hydroxide		carbon dioxide		magnesium carbonate		water	

Sulfuric acid is sometimes used to stabilize water. However, this process adds noncarbonate hardness ($CaSO_4$), and the acid is more difficult to handle. Therefore, carbon dioxide is more frequently used for stabilizing softened water.

Softening Facilities

The lime–soda ash softening process involves the following components:

- chemical storage facilities
- chemical feed facilities
- rapid-mix basins
- flocculation basins
- sedimentation basins and related equipment
- solids-contact basins (often used in place of the rapid-mix, flocculation, and sedimentation basins)
- pellet reactors (may be used in place of rapid-mix, flocculation, and sedimentation basins)
- sludge recirculation, dewatering, and disposal equipment
- recarbonation facilities
- filtration facilities

This section describes these components. A later section discusses operational concerns associated with them.

Chemical Storage Facilities

The size and number of facilities used for storing chemicals depend on the type and quantity of chemicals used. Lime can be purchased in two forms:

1. Calcium hydroxide, $CaOH_2$, which is called either hydrated lime or slaked lime.
2. Calcium oxide, CaO, which is called either quicklime or unslaked lime.

Both $CaOH_2$ and CaO are dry chemicals, varying in texture from a light powder to pebbles (Figure 11-3). Although calcium oxide requires special feeding equipment, large treatment plants typically use it because it is less expensive. Smaller plants, however, usually find calcium hydroxide less expensive. Soda ash (sodium carbonate, Na_2CO_3) is available in either powdered or granular form. All of these chemicals are available in bags, drums, or in bulk truck or train carloads.

Large quantities of chemicals are used in the water-softening process. For example, in order to reduce the hardness of 1 mgd (4 ML/d) of water from 200 mg/L to 50–85 mg/L hardness, more than 1,200 lb (540 kg) of soda ash is required. Chemical storage bins (Figure 11-4) are used to store the large quantities of chemicals that are required.

FIGURE 11-3
Pebble quicklime

Courtesy of the National Lime Association, Arlington, Va.

FIGURE 11-4 Chemical storage bins

Courtesy of the National Lime Association, Arlington, Va.

It is generally good practice to store a minimum of one month's supply of each chemical, so chemical storage space requirements for a plant must be carefully calculated. Calcium oxide and soda ash are most often delivered by hopper-bottom trucks and are transferred by mechanical or pneumatic conveyors to storage bins (Figure 11-5). Calcium hydroxide is usually

FIGURE 11-5 Conveyor used to transport calcium oxide and soda ash

Courtesy of the National Lime Association, Arlington, Va.

delivered and stored in bags. When liquid caustic soda (50 percent NaOH) is used, it is stored and fed from lined-steel or plastic tanks.

Chemical Feed Facilities

Slaked lime and soda ash can conveniently be fed by conventional dry feeders, which are usually located directly beneath the storage bins. Dispensing calcium oxide is a more complex process. In addition to a dry feeder, a lime slaker (Figure 11-6) and a solution feeder are required. The dry feeder dispenses the chemical into the slaker, where the calcium oxide and water are mixed together to form a slurry known as milk of lime [$CaO + H_2O \rightarrow Ca(OH)_2$]. The milk-of-lime slurry is then fed into the treatment system by a solution feeder.

The slurry will cake on any surface, so all feed equipment must be cleaned regularly. Open troughs or troughs that have removable covers are normally used to transport the slurry to allow for easy cleaning.

Tremendous heat is generated when lime is slaked, so adequate ventilation must be provided to the area where slakers are located. Provisions must also be made to remove the grit produced during slaking. The grit consists of undissolved lime and impurities. The quantity of grit produced depends on the amount of lime slaked. Small plants can often collect the grit in a wheelbarrow, but larger plants are usually designed to load the grit directly to dump trucks, which haul it to a disposal site daily.

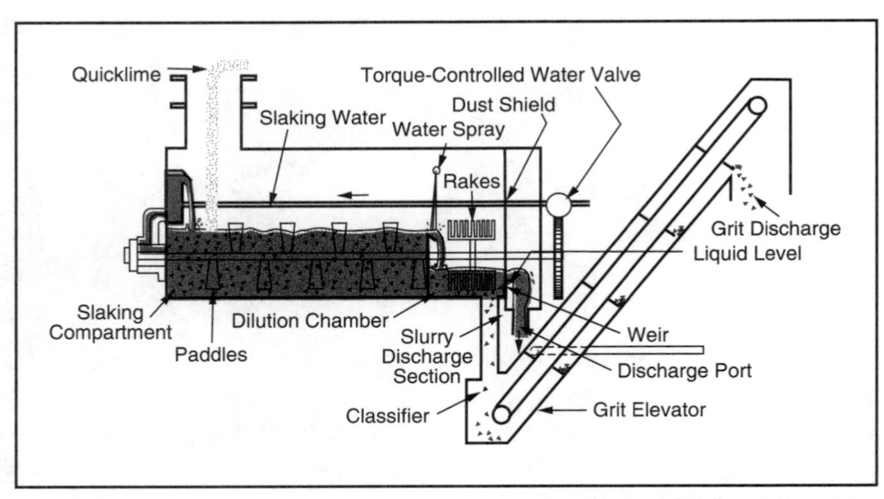

FIGURE 11-6
Lime slaker

Courtesy of Wallace & Tiernan, Inc.

Conventional dry feeders and solution feeders are also used to dispense coagulants (such as alum) or coagulant aids (such as organic polymers) if necessary. Chemical feed equipment is discussed in more detail in chapter 4.

Whenever lime, soda ash, or other dry powdered chemicals are used, chemical dust can pose a health hazard. Therefore, dust control equipment should always be used while working with the chemicals. Dust control equipment is discussed in more detail in chapter 8.

Rapid-Mix Basins

Rapid-mix basins used in lime–soda ash softening are similar to those described in chapter 4 for coagulation and flocculation. They serve the following two purposes:

- to thoroughly mix lime and soda ash with water
- to maintain mixing long enough for the chemicals to dissolve — especially critical for lime because it dissolves rather slowly

Rapid-mix basins with high-speed, mechanical mixers are preferred over baffled basins because the mixing rate can be controlled independently of the flow rate. This makes it possible to mix the chemicals more vigorously. Some rapid-mix basins are also equipped to receive sludge that is recirculated from the settling basin in order to speed up the chemical reactions.

Although coagulants can be added to rapid-mix basins, better results are usually obtained if they are added upstream in a separate rapid-mix basin.

Flocculation Basins

Flocculation is the process by which impurities are gathered together into solid particles, known as floc, that form and grow after coagulants have been added to the water. Flocculation basins used in the lime–soda ash softening process are similar to those used in conventional water treatment. They are equipped with gentle mixing devices, such as rotating paddle wheels, that control the mixing of chemicals. Variable-speed drives on the mixers provide a means of varying the mixing speed required for the most effective flocculation under the varying conditions of raw-water quality and flow rate.

Sedimentation Basins

Sedimentation is the process by which suspended solid particles of floc settle out of water. Sedimentation basins used in the lime–soda ash softening process can be identical to those described in chapter 5. However, because of the larger amount of sludge formed, mechanical equipment to collect and remove the sludge in the basins is essential.

Solids-Contact Clarifiers

The solids-contact basin (or clarifier), described in chapter 5, is sometimes used in the conventional treatment process. The basin is a single unit in which the processes of coagulation–flocculation, sedimentation, sludge collection, and sludge recirculation are performed (Figure 11-7). Several designs of solids-contact basins are available. All have two major zones:

- A *mixing zone,* which includes the reaction flocculation area in the basin. In this zone, lime (or lime with soda ash) is mixed with water and also with some of the previously formed slurry, which is recirculated to enhance the chemical reactions and formation of precipitates.
- A *settling or clarification zone* where the precipitates are allowed to settle as the clarified water passes upward through the sludge blanket to the effluent troughs.

A portion of the settling solids is returned to the mixing zone. The remainder is allowed to settle completely and is periodically withdrawn to maintain a relatively fixed slurry concentration in the basin.

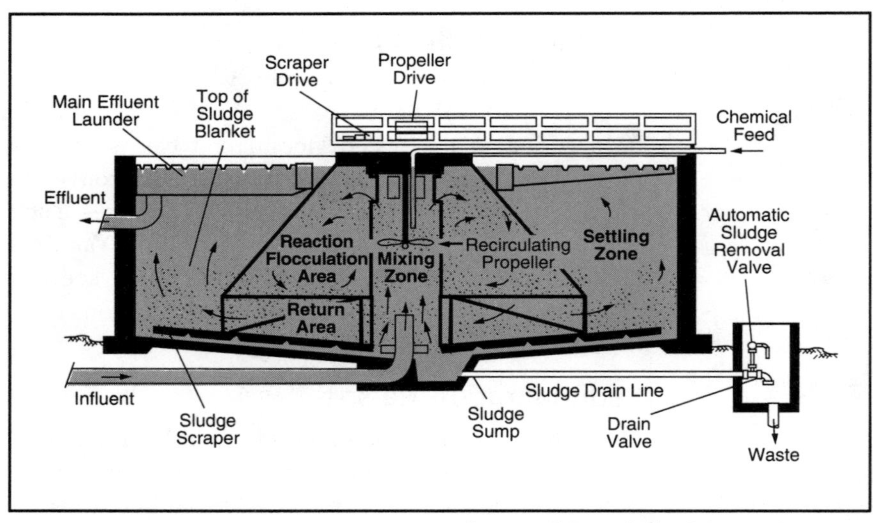

Courtesy of General Filter Company, Ames, Iowa

FIGURE 11-7 Solids-contact basin

The major advantage of the solids-contact basin is that it saves on construction costs by handling many softening processes in one component.

Pellet Reactors

A pellet softener is a conical tank in which the softening reactions take place quite rapidly as the water passes upward through the unit (Figure 11-8). In operation, the tank is half-filled with fine granules of calcium carbonate. Raw water enters the bottom through a nozzle placed at an angle so that an upward swirling action is produced. At the same time, a dose of lime or lime and soda ash is introduced through another opening.

The chemicals mix with the water as it moves upward, and the softening action takes place. The calcium carbonate granules gradually grow in size, with the heaviest sinking to the bottom, where they are periodically drawn off. At the same time, additional finer particles are added to make up for the loss. One advantage of the process is that the equipment does not require very much space, and the initial cost is relatively low. The detention time is only about 5 to 10 minutes. In addition, the "waste" from the unit is a

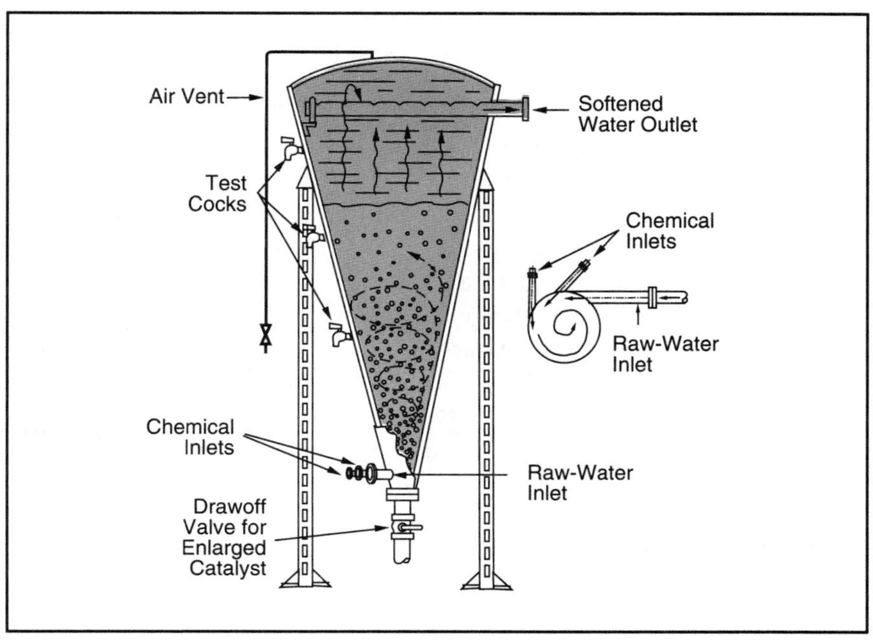

FIGURE 11-8 Design of a pellet reactor

Courtesy of the National Lime Association, Arlington, Va.

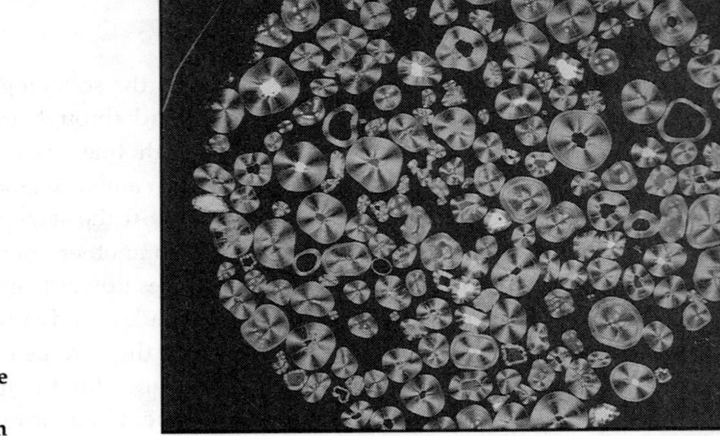

Photograph by DJ. Wiersma, The Netherlands

FIGURE 11-9 Thin section of water softening pellets from a pellet reactor; concentric buildup of calcium carbonate on the seed particles is shown

granular material (Figure 11-9), which is much easier to handle than conventional lime-softening sludge.

A principal disadvantage is that magnesium removal is difficult in the units, so pellet reactors should not be considered if the raw-water magnesium content is high.

Sludge Pumps

The flocculation and settling of calcium carbonate and magnesium hydroxide precipitates occur more quickly in the presence of previously formed floc particles. For this reason, some settled sludge is commonly recirculated to a rapid-mix basin. Whether or not sludge recirculation is practiced, sludge must be removed from the settling basin. Although sludge removal can be accomplished by gravity drains, it is rare that the plant can be set up for this to be possible. As a rule, special sludge-handling pumps must be provided for both sludge recirculation and removal. The handling and disposal of sludge are discussed further in chapter 5.

Pumps for handling sludge are usually located in rooms adjacent to the settling basins. Centrifugal pumps with open impellers designed to prevent plugging are generally used for sludge disposal. Use of slow-speed centrifugal pumps or positive-displacement pumps minimizes the breakup of the flocculated solids, which helps the sludge retain good settling characteristics.

Sludge recirculation equipment also includes flow-measuring devices, such as a magnetic or ultrasonic meter that will not become clogged. A progressing-cavity, variable-speed metering pump is also sometimes used to both pump and measure the returned sludge.

Sludge Dewatering and Disposal

The large amount of sludge produced during lime–soda ash softening must be either reclaimed or disposed of in an environmentally acceptable manner. The best method for a treatment plant to use depends on many factors, such as the amount and type of sludge and the availability of disposal sites.

The process of sludge dewatering reduces the amount of water in the sludge, thereby reducing the bulk and making it easier to handle. Most plants dispose of sludge by trucking it to a land disposal site. In general, the drier the sludge is, the lower the disposal cost will be. The more commonly used sludge-dewatering devices include

- drying beds
- lagoons
- sludge thickeners
- vacuum filters
- belt filters
- filter presses
- centrifuges

Drying Beds

A drying bed is a layer of sand placed over graded gravel or stones and an underdrain system. The sand is usually 4–9 in. (100–230 mm) deep. It is placed on top of 8–18 in. (200–460 mm) of graded aggregate.

The aggregate rests on top of an underdrain system consisting of perforated pipe or pipe laid with open joints. The underdrain collects the water and either returns it to the water treatment plant or pipes it to a waste disposal site. Uncovered drying beds perform best in regions having clear sunny days, warm weather, and little rain and snow.

Lagoons

Lagoons are large holding ponds that generally provide the most economical way to dewater sludge if sufficient land area is available within a reasonable distance of the treatment plant. They are usually about 10 ft (3 m) deep and can be 1 acre (0.4 ha) or more in size. Sludge is piped directly into the lagoon, where the solids settle out. The water on top (referred to as decant) is usually piped back to the treatment plant. More than one lagoon is always required, so that as the settled solids are being cleaned from one, the sludge from the plant can be diverted to the other.

Sludge Thickeners

Sludge thickeners (Figure 11-10) are similar to circular sedimentation basins. The rake arms rotate like the sludge collector on a solids-contact clarifier and gently stir the sludge blanket to allow water to escape to the surface. Sludge is often thickened first before a mechanical dewatering process such as vacuum filtration or centrifugation is used (Figure 11-11), so that these processes will work more effectively. The sludge drawn from a thickener is usually comprised of between 20 and 40 percent solid particles.

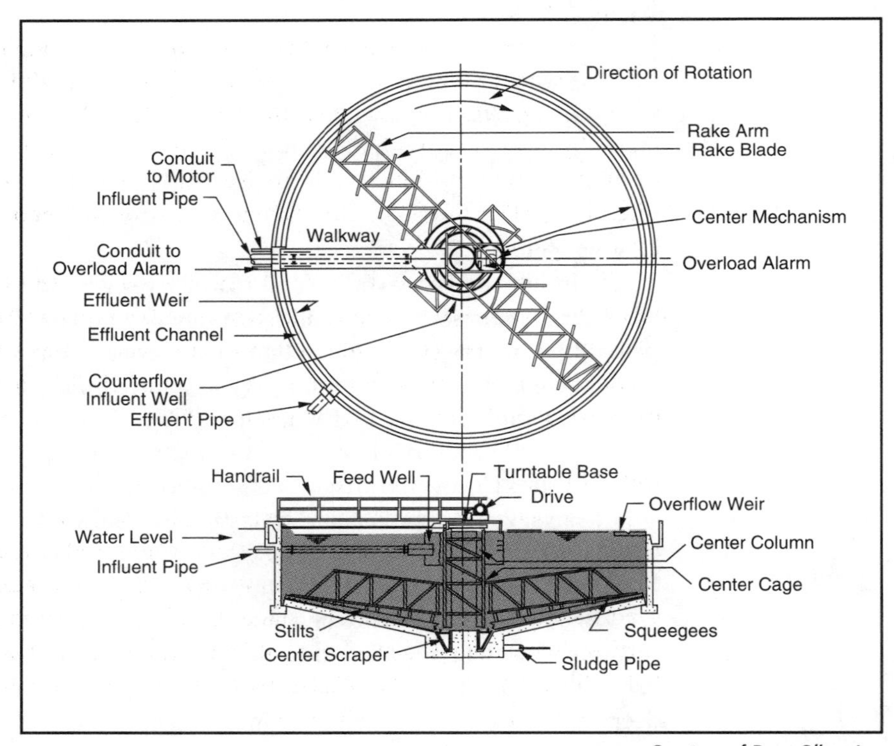

FIGURE 11-10
Sludge thickener

Courtesy of Dorr–Oliver Inc.

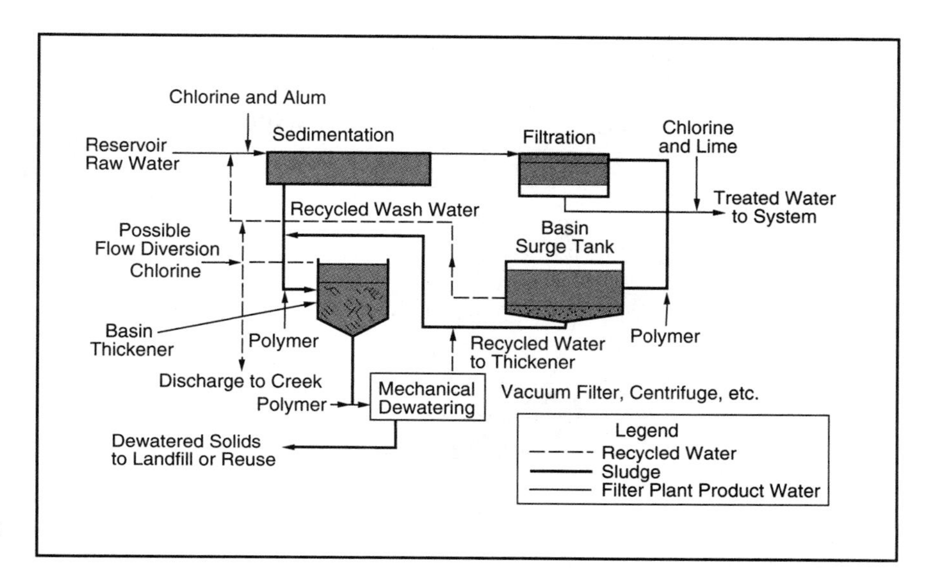

FIGURE 11-11 Mechanical sludge-dewatering system

Vacuum Filters

A vacuum filter (Figure 11-12) consists of a cylindrical filter drum covered with a porous fabric woven from fine metal wire or from fiber such as cotton or nylon. Sludge fills the feed tank and is pulled onto the filter fabric by a vacuum within the filter drum. This vacuum pulls water away from the sludge and into the drum, leaving a thin, dry cake of sludge on the filter fabric (Figure 11-13). This cake is continuously removed from the fabric by a scraper.

The filter cake is hauled to a waste disposal site, and the water removed from the sludge is returned to the thickener. Vacuum filters are normally not efficient for processing alum sludge unless precoating is used. The process works better on lime-softening sludge, but it is difficult to achieve a filter cake dry enough for landfill disposal.

Belt Filters

Belt filters use permeable belts to compress and shear the sludge to remove water. In the first step of the process, the sludge is drained by gravity to a nonfluid consistency. In the second stage, pressure is applied and gradually increased. In the last stage, the cake is sheared from the belt and further dewatered. The sludge must be conditioned so that it will stay on the belt. This is usually accomplished by the addition of a polymer.

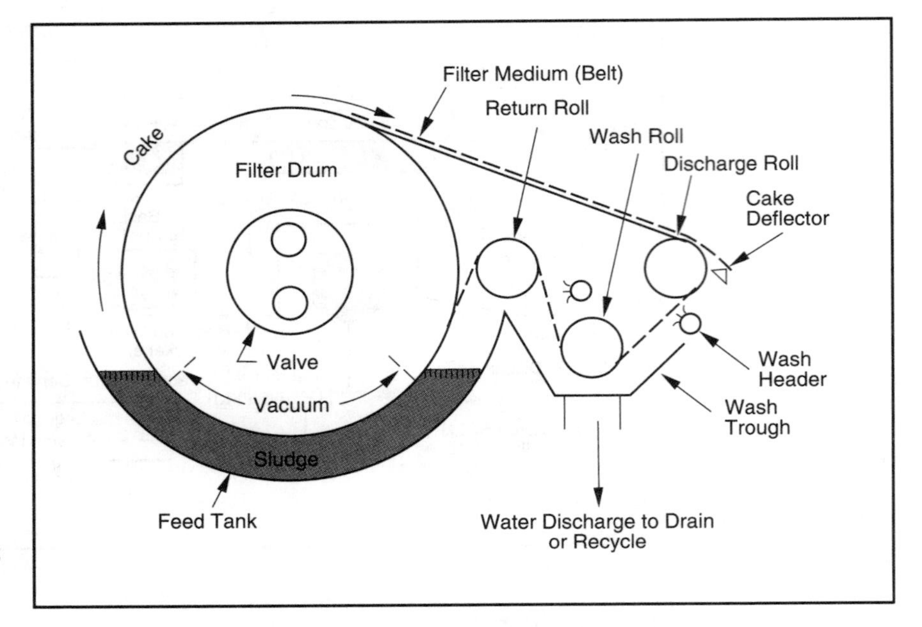

FIGURE 11-12 Sludge-dewatering vacuum filter (belt type)

FIGURE 11-13 Dry sludge on filter fabric

Courtesy of Komline–Sanderson

Filter Presses

A filter press can also be used to dewater sludge for final land disposal. The sludge must first be conditioned, usually by a polymer. It is then forced into contact with the filter cloth. The older types of units were called plate and frame filters, in which the sludge was forced through the filter under pressure. With the newer diaphragm filters, sludge is initially filtered through the cloth for about 20 minutes, and then compressed air is applied to squeeze water out of the sludge. The cake is then dislodged by shaking or rotating the cloth.

Centrifuges

A centrifuge is a sedimentation bowl that rotates at high speeds to help separate sludge solids from the water by centrifugal force. As shown in Figure 11-14, sludge enters the rotating bowl through a stationary feed pipe. The rotating bowl causes the solids in the sludge to be thrown outward to the wall at a force equal to 3,000 to 10,000 times the force of gravity.

The solids that settle against the wall of the bowl are scraped forward by a screw-type rotating conveyor and are discharged as shown in the figure. Clear water, called centrate, discharges through controlled outlets and is returned to the thickener. The sludge cake produced by a centrifuge is 50 to 65 percent solid particles.

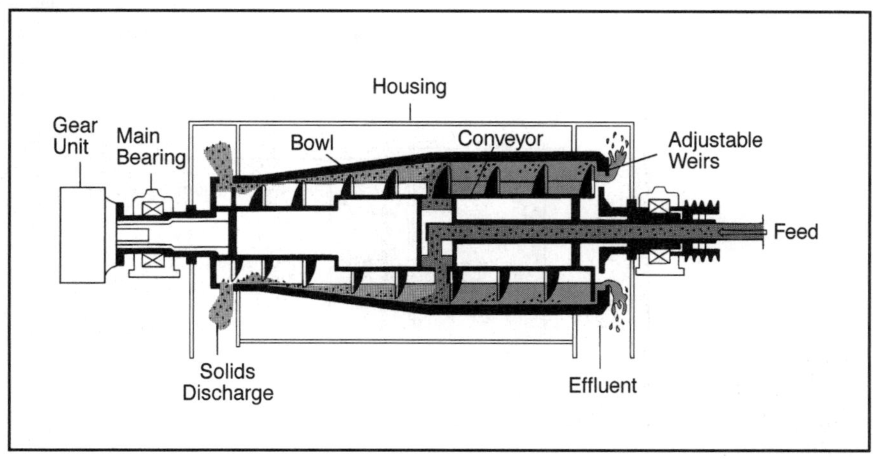

FIGURE 11-14 Sludge-dewatering centrifuge

Courtesy of Bird Machine Company, Inc.

Recarbonation Facilities

Recarbonation is usually accomplished by adding carbon dioxide gas to softened water to stabilize the water before it enters the distribution system. Recarbonation is performed in a basin that normally provides 15–30 minutes of detention time.

There are several ways to obtain carbon dioxide gas. In large, older plants, it is collected from furnace exhaust gases. It is then scrubbed by spraying water through it to remove impurities, moved by a compressor to the recarbonation basin, and diffused into the water. This technique is no longer recommended because sulfur and phenolic materials in the gas may cause taste-and-odor problems in the finished water. In addition, the equipment required for this system requires considerable maintenance because moist carbon dioxide is very corrosive.

In large, newer plants, carbon dioxide is produced by (1) burning a mixture of natural gas and air in a submerged combustion chamber (Figure 11-15), or (2) burning a mixture of natural gas and air in a forced-draft

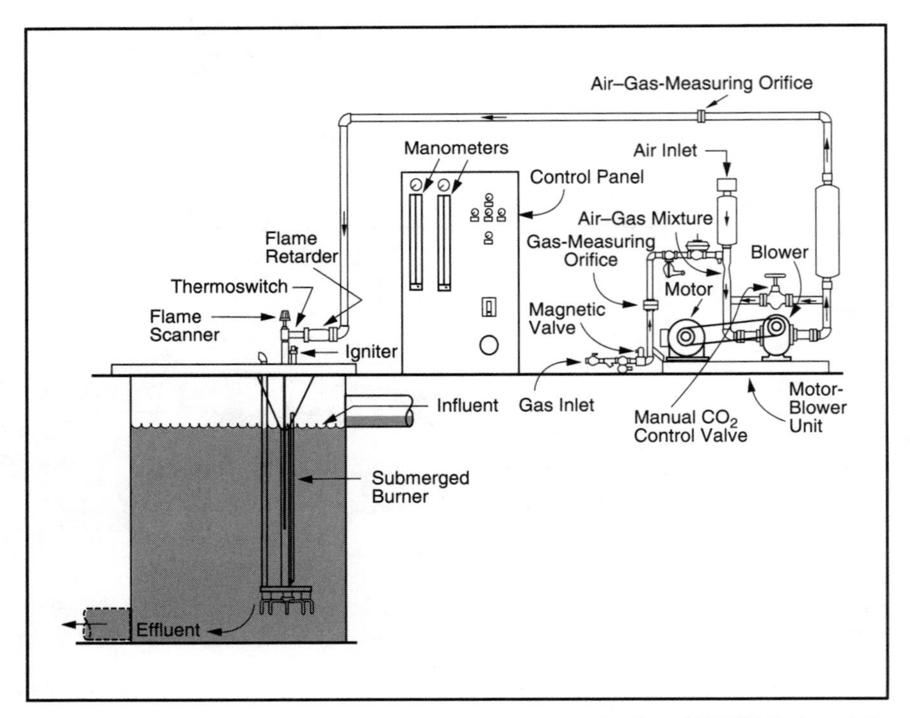

FIGURE 11-15 Submerged-combustion recarbonation system

Courtesy of TOMCO Equipment Co.

generator on the surface. Natural gas is used in these processes because it has relatively few impurities. The CO_2 produced is released near the bottom of the recarbonation basin, and the amount of gas generated is controlled by regulating the amount of gas burned. This equipment also produces carbon monoxide, so good ventilation is essential to prevent the level from reaching dangerous concentrations in low areas around the basin.

Small treatment plants and plants requiring only small amounts of carbon dioxide can purchase CO_2 in the form of either dry ice or liquid. Some principal advantages to purchasing the gas are the elimination of the costs of purchasing and maintaining combustion equipment and compressors and the danger of carbon monoxide being present in the plant.

If CO_2 is purchased in the form of dry ice or liquid, an evaporator is used to convert it to a gas. The gas then flows through a pressure-regulating valve, and the flow is controlled by a rotameter, similar to the type used on gas chlorinators. The gas is then fed into the recarbonation basin through diffusers. Because the gas generated from dry ice or liquid carbon dioxide is pure CO_2, pipes for the diffusers can be much smaller than those needed for CO_2 generated on-site by combustion.

Carbon dioxide gas is heavier than air. A leak can result in a hazardous buildup of the gas in low areas of the treatment plant. Therefore, a good ventilation system is needed wherever carbon dioxide is used.

Filtration Facilities

The filters used in a lime–soda ash softening plant are identical to those used in a conventional rapid sand or multimedia filter plant. Special operational considerations are discussed later in this chapter.

Regulations

There are no federal regulations that require the softening of water. It should be noted that addition of lime–soda ash softening could affect a system's compliance with some regulations. Additional details on federal drinking water regulations are provided in *Water Quality*, also part of this series.

Trihalomethane Control

The formation of trihalomethanes (THMs) or other disinfection by-products can be increased by softening because of the high pH that is required to operate the process. If this is found to be a problem, other methods can be considered for reducing the THMs sufficiently to meet the requirements. One method is to simply change disinfectants (see chapter 7). If this is not sufficient, a preliminary treatment step may have to be added to remove precursors before the softening process. Although THMs can also be removed after other treatment is completed by use of granular activated carbon, this is a rather expensive alternative.

Meeting Lead and Copper Rule Requirements

The federal Lead and Copper Rule requires that samples from customers' taps be analyzed for the presence of lead and copper. Excessive lead and copper are normally the result of corrosion in lead and copper piping. Soft water is generally more aggressive than hard water, so systems that have softening treatment may find it necessary to take additional steps to stabilize their water to meet the Lead and Copper Rule requirements (see chapter 9 for additional details).

Meeting Surface Water Treatment Rule Requirements

The Surface Water Treatment Rule (SWTR) requires that all water systems using a surface source, as well as certain groundwater systems designated by the state, must maintain specified removal and disinfection levels to ensure microbiological safety of the finished water (see *Water Quality*, part of this series, for details). Although the lime–soda ash process is known to be effective in removing viruses, the amount of "credit" that should be given to a softening plant for overall microbiological removal or control has not yet been determined. Systems required to comply with SWTR requirements should consult with state authorities on how the requirements must be met by their system.

Operation of the Lime–Soda Ash Process

Operation of the lime–soda ash process involves storing and feeding large amounts of chemicals. It also involves processes to perform the actual softening, as well as modifications to the operation of other processes. Finally, careful calculation of chemical dosages is required.

Storing and Feeding Chemicals

Storing Chemicals

Calcium oxide (quicklime) is usually purchased in pebble form to avoid the dust problems associated with the powdered form. However, some dust is created when any form of calcium oxide is used, and operators should always wear protective equipment when handling this chemical to avoid injury.

Calcium oxide is shipped in bulk form by covered railroad hopper cars or by truck. Unloading can be done by screw conveyors, covered belt conveyors, bucket elevators, or pneumatic conveyors. The dry chemical is not corrosive and can be stored in concrete or steel silos or bins.

Calcium hydroxide (hydrated lime) is available in 50- and 100-lb (23- and 45-kg) bags. The bags must be kept intact because the chemical will adsorb carbon dioxide from moisture in the air and change to calcium carbonate.

Soda ash can be purchased in bulk, bags, or barrels. It can be unloaded in bulk by mechanical or pneumatic conveyors, but it must be properly handled because of the dust created by the fine powder. Soda ash is noncorrosive, so it may be stored in ordinary steel or concrete bins.

It is good practice to keep a 30-day supply of lime and soda ash on hand at all times to avoid running out in the event of unexpected shipping problems. Many systems provide a storage capacity of 60 days and take delivery of a 30-day supply every month. Because chemicals lose effectiveness over time, it is important to operate storage facilities in a way that will always use the oldest chemicals first.

Feeding Chemicals

The specific methods of operating chemical feed facilities will depend on the chemicals and feed equipment used. Instructions from chemical suppliers should be followed to ensure proper operation of chemical feed facilities.

As discussed earlier, calcium oxide (unslaked lime) is fed by a gravimetric feeder into a lime slaker. Two types of slakers are used: the detention type and the paste type.

The detention-type lime slaker is the older of the two types and requires a water-to-lime ratio (by weight) of 4:1 to 5:1. It ordinarily takes from 20 to 30 minutes to slake lime. The water ratio is adjustable and is controlled by the operator. Heat generated during slaking is transferred by a heat exchanger to the incoming water, which accelerates the slaking process. A slaking temperature of 160°F (71°C) or higher should be maintained.

The paste-type lime slaker requires a water-to-lime ratio (by weight) of 2:1 to 2.5:1, and slaking is performed at near-boiling temperatures. These higher temperatures accelerate the slaking process. Paste-type slakers take 10 to 15 minutes to slake lime. The slaked lime is then diluted to approximately 10 percent $Ca(OH)_2$ in a solution chamber, forming a milk-of-lime slurry.

Calcium hydroxide (slaked lime) is fed directly into a solution chamber by a gravimetric feeder. The resulting slurry is similar to the milk-of-lime slurry.

The process of feeding lime slurries presents many problems because the slurries will deposit (or cake) on any surface. Valveless proportioning pumps can provide a satisfactory means of pumping the slurry. The slaker should be located as close to the point of feed as possible, and the slurry is best fed through an open trough that can easily be cleaned when necessary.

Operation of Softening Facilities

Rapid Mixing

A high-energy, rapid mixing process called flash mixing is used to dissolve and thoroughly mix lime and soda ash into the water being treated. The optimal mixing speed is that which produces the best chemical reactions; it can be determined only through operating experience. Mixers are operated at varying speeds to produce precipitates that have the best settling characteristics.

A detention time of 5 to 10 minutes is required when lime and soda ash are mixed with water because lime is slow to dissolve. All mixing equipment and the settling basin should be cleaned frequently to remove precipitates. The concentration of precipitates can be improved by recirculating the sludge from the settling basin. This process allows precipitation to take place on the previously formed solids rather than in the basin.

Flocculation

Flocculation requires slow, gentle mixing to allow the chemical reactions to be completed and to promote the growth and precipitation of the floc. Consequently, flocculation basins usually require a detention time of 40 to 60 minutes. The speed of the mixing paddles must be adjusted to obtain the best floc. Once the floc has formed, local eddy currents and flow surges must be avoided to prevent breaking up the solid floc particles.

Sedimentation Basins

The operation of sedimentation basins is discussed in more detail in chapter 5. To soften water, detention times of 2 to 4 hours are usually adequate. Surface overflow rates vary, but they are generally higher than the rates in a conventional water treatment plant because the floc settles more readily than alum floc. Because of the large amounts of sludge produced in the softening process, continuous mechanical collection of sludge is essential.

The sludge drawoff lines should be backflushed periodically to remove any clogs or caking.

Solids-Contact Basins

In a solids-contact basin (Figure 11-7), the water passes upward through a blanket of flocculated material called the sludge blanket, which entraps slowly settling particles that would otherwise escape the basin and end up in the filters. The mixing speed and the level of the sludge blanket are key operational factors for the system. The mixing speed determines how well the chemicals and water are mixed. Increasing the speed shortens the detention time of the initial mix, which can cause inefficient treatment because of incomplete chemical reactions. Inadequate mixing speed, however, can cause poor mixing of the chemicals and water. Settling tests and visual observations must be used to determine the best mixing speed. Many plants run the mixer continuously in order to keep the sludge blanket in a state of suspension at all times, including periods when the plant is shut down.

In addition to proper dosage and mixing, successful treatment in a solids-contact basin depends on keeping the sludge blanket at desirable levels. If it drops below the "skirt" of the flocculation compartment, water will not pass through a full bed of solids, and some particles could pass through to the filters.

At the same time, the sludge blanket must not be allowed to build up or rise to levels so high that solids will pass over into the main effluent launder and discharge into the filters. The top of the sludge blanket normally should be 4 to 8 ft (1.2 to 2.4 m) below the top of the overflow weirs. The height of the sludge blanket is adjusted according to the frequency, duration, and rate of sludge removal (sludge blowdown). Blowdown valves are operated intermittently in response to a timer or a water meter that signals when a given quantity of water has passed through the basin.

Sampling lines should be provided at various levels in the settling zone to permit the operator to determine with accuracy the height of the sludge blanket and the degree of solids present in the water. In the lower level of the blanket, the concentration of solids, by weight, should range from 5 to 15 percent. Maintaining a high concentration of solids, while keeping the sludge blanket at a level low enough to avoid a loss of solids over the weir, minimizes the amount of chemicals needed.

Solids-contact basins are usually designed for a flow rate of 1–1.75 gpm/ft^2 (0.7–1.2 mm/s). Higher flow rates are sometimes possible, depending on the settling characteristics of the floc. Where multiple basins are used, flow must be balanced between the basins so that no individual basin will be overloaded. This balance can be achieved either by flowmeters or by adjusting overflow weirs to provide flow rates that are proportional to the sizes of the basins.

Solids-contact basins must be drained periodically for cleaning. The cleaning must be done once every 3 to 12 months, depending on the types and amounts of chemicals used. Sludge withdrawal lines and sampling lines should be vigorously flushed and cleaned routinely to remove sludge residues and scale buildup. An alum or chlorine solution will make cleaning of these lines more effective.

Sludge Collection and Disposal Equipment

Sludge collectors normally operate at a slow constant speed of about one revolution every 20 to 30 minutes. The operator does not usually have to vary the speed. Once the sludge has been scraped into the sludge hopper in the sedimentation basin, the hopper is emptied by pumping or gravity. A portion of the removed sludge is then recirculated, and the remainder is discarded.

Constant-speed centrifugal sludge pumps may have a valve on the discharge for the operator to control the pumping rate. If a variable-speed, positive-displacement pump is used, the operator can control sludge flow by varying the motor speed.

The frequency at which sludge is pumped varies significantly based on the amount of hardness removed and the concentration of sludge solids. As a rule, for each pound of lime used, 2.5 lb (1.1 kg) of sludge will be formed. Sludge concentrations can vary from less than 5 percent to more than 15 percent solids by weight.

The operating procedures for dewatering sludge vary depending on the type of process used. If drying beds are used, the sludge should be spread to a depth of 8 to 12 in. (200 to 300 mm) over the sand. In good weather, the sludge will dry and crack in about a week. The dried sludge contains about 50 to 60 percent solids; it can be removed by hand raking or using a front-end loader. The dried sludge is then hauled to a burial or refuse site. The success of the drying-bed process for dewatering sludge depends on clear, dry, warm weather. In cold, wet climates, the process can be used if the beds are covered like greenhouses and have adequate heat and ventilation.

If lagoons are used to dewater sludge, the sludge is pumped directly to the lagoons and allowed to settle. It is best to fill a lagoon at one end and decant the water from the other end. When the depth of sludge reaches 3 to 5 ft (0.9 to 1.5 m), use of the lagoons should be discontinued and the water level should be kept as low as possible. After a period of time, the dewatered sludge should contain about 50 percent solids. It is then usually removed by a dragline or clamshell crane.

The sludge may be spread on the ground near the lagoon to dry further before final disposal. In some cases, the sludge solids are left to accumulate in the lagoon. When the lagoon is filled, it is abandoned, and a new lagoon is excavated. In cold climates, freezing temperatures help dewater the sludge by separating water from the sludge solids. After thawing, the solids are small granular particles.

In softening plants that dewater sludge by using a vacuum filter, belt filter, filter press, or centrifuge, sludge is usually pumped into a sludge thickener before proceeding to the mechanical dewatering device. Sludge entering the thickener with 5 percent solids, for example, can be thickened to 15 to 20 percent solids. This reduced volume of sludge is easier and less costly to treat mechanically. The dried sludge produced by a vacuum filter or centrifuge is a fairly dry, thin, cake-like sludge that has a solids concentration of 40 percent or more.

Final disposal of sludge is usually achieved either by burying it in a sanitary landfill or by spreading it on farmland to serve as a soil conditioner. If the sludge is incinerated, lime can be reclaimed and reused. However, this procedure is economical only for large treatment plants.

Recarbonation Basins

Recarbonation basins are used to stabilize water after softening. In order to provide adequate time for the chemical reactions to take place, recarbonation basins usually provide 15 to 30 minutes of detention time. If carbon dioxide is being injected by diffusers, the operator should routinely check the diffusers to see that they are not plugged. The ventilation system in the area should also periodically be checked for proper operation, and the flow of gas should be monitored at least every 8 hours. Fluctuations in the flow rate of gas in the basin can cause excessive scale to form downstream. Control of the recarbonation process should be based on (1) meeting water stabilization goals as determined by the Langelier saturation index or other stability index, and (2) using coupons in the distribution system (see chapter 9).

Filters

The filters used for the softening process are essentially the same as those for conventional filtration, except for one difference that can cause serious operating problems. Because the water being filtered after softening contains calcium carbonate and, in some cases, magnesium hydroxide, precipitates of these chemicals can form on the filter media, causing the grains of media to stick together. If this happens, the effectiveness of the filtration process will be destroyed. To avoid this problem, the filter media should be inspected frequently for scale buildup. If the grains are coated with white (calcium carbonate) or black (magnesium hydroxide) scale, recarbonation is probably inadequate. The presence of scale can be verified by the addition of a small quantity of hydrochloric acid to a small quantity of filter media. If scale is present, the acid will react with it and will boil and foam.

Scale can sometimes be removed from filter media by increasing the carbon dioxide feed rate. Care must be taken during this operation to maintain water quality and to avoid damaging metallic underdrain components by prolonged exposure to corrosive water.

Determination of Chemical Dosages

The amount of lime and soda ash required to soften water depends on (1) the hardness of the water, and (2) how much of the hardness is to be removed. The required dosage of lime is based on the combined amount of carbonate hardness, noncarbonate hardness, and carbon dioxide to be removed. The required dosage of soda ash is based only on the amount of noncarbonate hardness to be removed.

There are two methods for calculating dosages: the conventional-dosage method and the conversion-factor method. Initial dosages for lime and soda ash can be computed based on these two methods. (Information on chemical feed calculations for lime–soda ash softening are included in *Basic Science Concepts and Applications*, also part of this series.) However, it is important to note that either sludge recirculation or the use of solids-contact basins will allow dosages to be reduced. Adding excessive amounts of chemicals may not increase the amount of hardness removed from the water. An operator should start by using the computed figures as a base, and then experiment with various dosages in the same range to find the most effective rates for the water being treated.

The use of a softening curve can be helpful in establishing a base dosage. This curve can be developed by using jar tests, followed by process testing in the plant. Jar tests are used to perform laboratory experimentation with

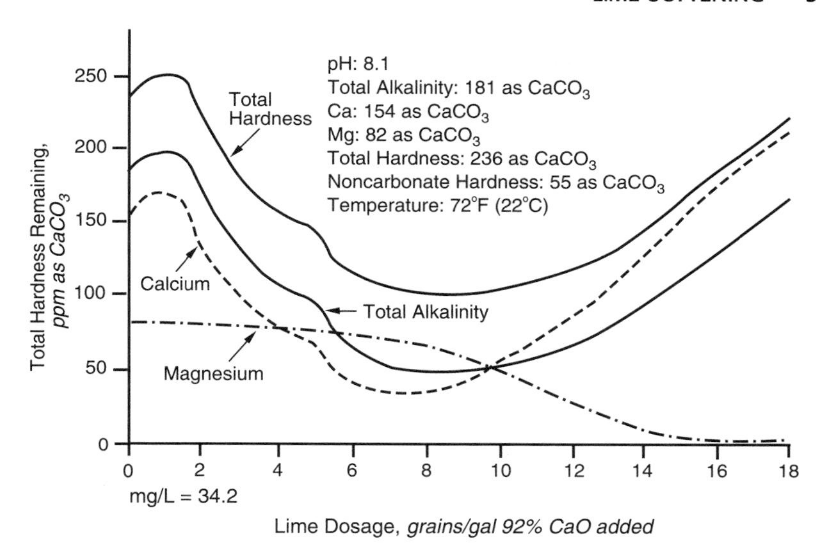

FIGURE 11-16 Softening curve used to establish most effective lime dosage

various chemical dosages and mixing rates and to determine the most effective treatment. See *Water Quality*, part of this series, for more information on jar tests. As shown in Figure 11-16, a softening curve is developed by plotting the lime dosage against the concentrations of carbonate and noncarbonate hardness remaining in the water. At each point on the curve, pH should also be recorded. In the example shown in the figure, a total hardness of 100 mg/L can be achieved with a lime dosage ranging from 7.4 to 10 grains (127 to 171 mg/L). If the lower dosage is used, a substantial amount of lime and money can be saved.

The dosage of soda ash needed to remove noncarbonate hardness should be based on calculations for either the conventional-dosage method or the conversion-factor method. Monitoring the finished water's noncarbonate hardness and alkalinity allows the dosage of soda ash to be adjusted to the most efficient level.

Calculations for the theoretical feed rate for carbon dioxide in the recarbonation process are based on the amount of alkalinity that must be removed in order to stabilize the water. Any adjustments in the carbon dioxide dosage should be based on both the Langelier index and the results of coupon testing in the distribution system.

Operational Control Tests

The lime–soda ash softening process is complex and requires the following control tests:

- alkalinity tests
- total hardness tests
- carbon dioxide tests
- pH tests
- jar tests
- Langelier saturation index determination
- monitoring of coupons in the distribution system

Most of these tests are described in *Water Quality*, part of this series. The key to successful control of the softening process is the correct use of these test results.

Alkalinity Tests

The results of an alkalinity test are used in calculating dosages for both lime and soda ash. The three specific types of alkalinity are

- hydroxide alkalinity
- carbonate alkalinity
- bicarbonate alkalinity

Details of the alkalinity tests can be found in *Basic Science Concepts and Applications*, also part of this series.

Total Hardness Tests

The total hardness test indicates the need for softening and provides information vital to the proper operation of the softening process. The test determines the total hardness concentration of a water. This information, together with total alkalinity, is used to calculate carbonate and noncarbonate hardness (Table 11-2).

If noncarbonate hardness is present, the total hardness concentration and the bicarbonate alkalinity are used to calculate the dosage of soda ash needed to remove the noncarbonate hardness. If noncarbonate hardness is absent, no soda ash is needed, and the total hardness concentration is used in place of bicarbonate alkalinity to calculate the correct dosage.

The total hardness test should also be conducted routinely on raw and finished waters to determine how effectively the softening processes are removing hardness from the water.

TABLE 11-2 Carbonate and noncarbonate hardness determination

Laboratory Results	Noncarbonate Hardness (Lime and Soda Ash Used)	Carbonate Hardness (Lime Only Used)
1. TH < TA	0	TH
2. TH = TA	0	TH
3. TH > TA	TH – TA	TA

NOTE: TH = total hardness, TA = total alkalinity.

Magnesium Tests

The magnesium test is used to determine how much magnesium hardness is in the water. This amount, in turn, is used in determining lime dosages. Because the results of magnesium tests indicate the magnesium hardness present, they can be used to calculate calcium hardness when the total hardness is also known.

The magnesium test also provides valuable insight into whether or not the water being treated contains enough magnesium to cause scale in water heaters and boilers. Whenever magnesium hardness exceeds 40 mg/L (as $CaCO_3$), there is a strong possibility that scale will form in hot water heaters and boilers that are operated at temperatures over 140°F (60°C). If magnesium hardness is removed to levels at or below 40 mg/L, scale formation should not be a problem.

Carbon Dioxide Tests

The carbon dioxide test is used to determine the concentration of free carbon dioxide in the water. In turn, this value is used to calculate lime dosages. Although carbon dioxide does not cause hardness, it reacts with lime, precipitating as calcium carbonate. A sufficient amount of lime must be added to the water to allow this chemical reaction to occur and still have enough lime remaining in the water to complete the softening process.

pH Tests

Measurement of the water's pH should be made before, during, and following lime–soda ash softening. During the softening process, pH measurements provide valuable information about the success or completeness of the chemical reactions that are occurring. For example, when excess-lime treatment is used, the lime dosage is designed to raise the pH of the water to 10.6, which is the best pH for magnesium hydroxide precipitation. The pH is then reduced by recarbonation to 9.4, which is the best pH for calcium carbonate precipitation.

After soda ash is added and softening reactions have been completed, the final pH is adjusted for recarbonation to whatever pH is required to achieve stabilization (usually about 8.6). This process converts any residual calcium carbonate into soluble calcium bicarbonate, thus preventing after-precipitation either on the filters or in the distribution system mains.

By monitoring pH with an accurate, calibrated pH meter throughout the softening process, an operator can determine whether or not softening and recarbonation dosages are working successfully. In addition, pH and total alkalinity can be used to determine the concentration of carbon dioxide in the water.

Jar Tests

The jar test is a good way to test the performance of lime and soda ash dosages before they are used in the plant. A dosage correctly calculated only from the results of the alkalinity test, total hardness test, and carbon dioxide test is not necessarily the best dosage. Numerous other factors, not easily defined or tested, can greatly influence the selection of the "best" dosage.

By performing the jar test, an operator can check the overall performance of the lime and soda ash dosages, such as the size and condition of floc particles, the ease and speed with which floc particles settle, and the resultant final pH. A softening curve can also be developed to help identify the lime dosage that will soften water to the desired point at the least cost.

The Langelier Saturation Index and Coupons in the Distribution System

The stability of softened water should be monitored to ensure that the water will neither cause corrosion nor form scale. The Langelier saturation index should be determined daily at the plant, and it should be evaluated in connection with inspection of coupons placed in the distribution system. (The Langelier saturation index and coupons are described in chapter 9.)

Operating Problems

The following are several potential problems commonly associated with the operation of the lime–soda ash softening process:

- excess calcium carbonate
- magnesium hydroxide scale
- after-precipitation
- carryover of sludge solids
- unstable water
- interference with other treatment processes

Unless recognized and eliminated, these conditions can lead to high maintenance costs, failure of filtration processes, loss of distribution capacity, inefficient operation of water heaters and boilers, and ineffective treatment.

Excess Calcium Carbonate

Extremely small, almost colloidal particles of calcium carbonate can be formed during the coagulation–flocculation process. These particles can pass through the entire softening process to the filters, where they adhere very tightly to the individual grains of filter media. The filter grains can become coated with the calcium carbonate and cement themselves together. In a similar way, calcium carbonate can pass through the media and plug the filter underdrains. Clogged underdrains will usually cause visibly uneven backwash patterns. If these conditions are not corrected, they can cause problems serious enough to require complete replacement of the filter media and underdrains.

In addition, excess calcium carbonate can also precipitate in fine particles on the walls of pipelines. As shown in Figure 11-17, the particles can eventually build up in thickness, reducing the pipeline capacity. The restricted pipe capacity makes pumps work harder, which increases operating costs and shortens equipment life.

Excess calcium carbonate is also the cause of scale deposits on filter walls and wash troughs. Such deposits are both unsightly and costly to remove. These deposits may be caused by inadequate mixing, short detention times, or interference from organic contaminants in the water during chemical precipitation.

The best solution to the problem is to eliminate whatever is causing the incomplete coagulation–flocculation process by improving the mixing, increasing detention times, or pretreating the water to remove organic contaminants. If the fine particles of calcium carbonate cannot be eliminated

FIGURE 11-17 Calcium carbonate buildup removed from pipe

Courtesy of Girard Industries

at the source, they must be either removed by recarbonation or controlled by applying a recommended dosage of 0.25 to 1.0 mg/L of sodium hexametaphosphate (a sequestering agent). This will usually hold the calcium carbonate in solution and prevent precipitation onto the filter media.

Magnesium Hydroxide Scale

When water softened by the lime–soda ash process contains magnesium hardness in excess of 40 mg/L, magnesium hydroxide scale can form inside boilers and household water heaters. To prevent scale buildup, magnesium hardness should be reduced to 40 mg/L or less through excess-lime treatment.

After-Precipitation

Immediately after the lime–soda ash process is performed, the pH of the treated water is about 9.4. If the pH is left at this level, calcium carbonate will continue to precipitate out of the treated water, even after the water leaves the sedimentation basin or upflow basin. This effect, called after-precipitation, can cause the same problems that occur when excess calcium carbonate precipitates out of treated water because of inadequate coagulation–flocculation. To prevent after-precipitation, carbon dioxide should be added to the softened, settled water in order to decrease the pH to about 8.7, where it will be chemically stable. This procedure converts any unsettled calcium carbonate to soluble calcium bicarbonate. Encrustation can also be prevented by adding sodium hexametaphosphate.

Carryover of Sludge Solids

The primary reasons that sludge solids and properly formed floc particles are carried over the weirs of a sedimentation basin are (1) improper hydraulic conditions, and (2) sudden changes in water quality.

Carryover is indicated by the presence of unclear water above the sludge blanket, by sludge solids appearing in the basin effluent, and by short filter runs. The specific causes of carryover are listed in chapter 5.

Carryover of sludge solids is often most severe when a considerable portion of the settleable solids is composed of magnesium hydroxide floc particles, which are very light. To prevent carryover, it is necessary either to reduce the hydraulic loading on the sedimentation or upflow basin or to improve the settling characteristics of the floc. Recirculation of previously formed calcium carbonate sludge is effective in improving the settling characteristics of the floc.

Unstable Water

If softened water is not properly stabilized, one of two major operating problems will result: scale deposits or corrosiveness. Recarbonation and other stabilization techniques are used to control these operating problems. These techniques are discussed in chapter 9.

Interference With Other Treatment Processes

Disinfection of water is essential to make water safe for consumption. Although the high pH associated with lime–soda ash softening contributes somewhat to disinfection, it also results in the formation of a chlorine residual primarily in the form of hypochlorite (OCl^-) when the water is chlorinated. The hypochlorite residual has considerably less disinfecting power than hypochlorous acid ($HOCl$), which exists at lower pH values. Therefore, higher dosages of chlorine or longer contact times may be required to disinfect water softened by the lime–soda ash process.

If the lime–soda ash process is used to treat surface water, the processes for removing tastes and odors (particularly adsorption) may be hindered as a result of the increased solubility of taste-and-odor-causing compounds at elevated pH values. If this problem occurs, it can usually be controlled by removing these compounds during pretreatment.

Many surface water treatment plants must treat raw water that has a high color concentration due to organic compounds. For the most part, color must be removed prior to softening because a pH in the range of 4.0 to 5.5 is often required for alum coagulation to remove color effectively.

The rate of formation of trihalomethanes (THMs) increases at pH levels required for the softening process. If THM problems exist, pretreatment processes may need to be improved or disinfection processes may need to be modified. Regulatory agencies or technically qualified consultants should be consulted before a plan to minimize THM formation is adopted.

Safety Precautions

The lime–soda ash softening process exposes operators to daily contact with large quantities of lime and soda ash. These chemicals pose hazards if not handled properly. As shown in Figure 11-18, operators should be adequately protected when handling chemicals.

Lime Handling Safety

Quicklime is a strong, caustic chemical. Handling the chemical creates dust that can be irritating to the eyes, mucous membranes, and lungs. Prolonged or repeated inhalation can cause lung damage. Operators exposed to the dust should wear goggles and a suitable dust mask.

Skin contact with the dust can cause dermatitis or skin burns, particularly when the dust mixes with perspiration on the skin. Operators should wear heavy cotton clothing with long sleeves, gauntlet gloves, hat, and trousers tied around the shoe tops when handling the chemical. If clothing becomes covered with dust or slurry, it should be laundered immediately.

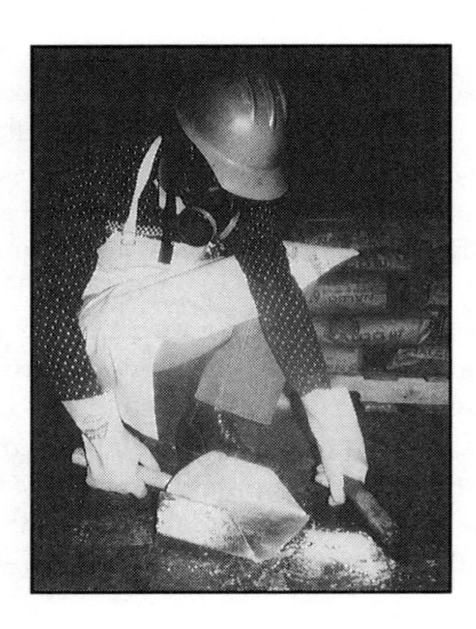

FIGURE 11-18 Worker properly dressed for handling chemicals

Operators should always shower immediately after handling quicklime, even if the dust is not visibly present. Dust collection equipment should frequently be checked for proper operation. Chemical spills and dust accumulation should be cleaned up with a dry vacuum, not with a broom.

Calcium oxide is strongly attracted to water. When the two come into contact, the heat generated can be sufficient to start a fire if there are flammable materials close by. The chemical must therefore be stored in a dry place.

Dry calcium oxide and dry alum should never be mixed. The heat that results when they are mixed can generate hydrogen and cause an explosion. To avoid accidental contact, these two chemicals should be loaded, handled, and stored in completely separate locations.

Lime slakers should be equipped with temperature override devices that will stop the lime feed before dangerous temperatures are reached.

Calcium hydroxide (hydrated lime) is less caustic and less irritating to the skin than calcium oxide, but it is still a dangerous chemical to handle. It is particularly dangerous to the eyes. In general, the same handling precautions should be exercised for calcium hydroxide as for calcium oxide.

Soda Ash Handling Safety

Moisture will cause soda ash to cake, making it difficult and hazardous to handle. To prevent this problem, soda ash must be stored in a cool, dry place. All soda ash equipment should be fitted with a dust collection system.

Operators working with soda ash should wear protective equipment, including safety goggles for eye protection, a close-fitting dust respirator, and clothing that protects against skin contact. Skin areas that will be exposed unavoidably should be covered with a protective cream or petroleum jelly to minimize the harmful effects of the dust. Some brands of soda ash contain ammonia, and operators who are allergic to ammonia should be particularly careful to protect themselves.

Pumps and equipment that handle soda ash solutions should be equipped with spray or splash guards to protect personnel working in the area. Potential danger areas and storage rooms should all be marked with warning signs.

Carbon Dioxide Handling Safety

Carbon dioxide is a colorless, odorless gas that is heavier than air. If it accumulates, it will displace the air. If it is inhaled, it can result in oxygen deficiency, causing asphyxiation.

Because carbon dioxide is often generated on-site, preventive maintenance of the generating equipment is essential to prevent leaks and to ensure that safety devices and alarms will function properly. The generating equipment should be located in a well-ventilated area. In addition, explosion-proof lighting should be installed, and smoking should be strictly prohibited when carbon dioxide is being used.

Generation of carbon dioxide can also produce carbon monoxide, another highly toxic, odorless, and colorless gas. Good ventilation of the carbonation basin is essential, particularly if submerged-combustion generating units are used. Extreme care should be taken during the cleaning of carbonation basins, particularly if they are enclosed. They should be thoroughly ventilated by blowers before and during the cleaning operation. A carbon monoxide tester should be used to make sure the air is safe before anyone enters the basin.

First Aid for Chemical Exposure

Soda ash dust and the mist from soda ash solutions are very irritating to the respiratory system, mucous membranes, and eyes. Prolonged exposure can damage the nasal passages.

If soda ash dust or solution or lime slurry gets into the eyes, flush the eyes with warm water for at least 15 minutes. Consult a physician immediately. If skin contact with any chemical solutions occurs, flush the area with large quantities of water. In the event that the dust or mist is inhaled, consult a physician immediately. Gargling or spraying the nasal passages and throat with warm water will reduce mild irritation.

Record Keeping

The following information should be recorded daily for the lime–soda ash process:

- quantity of lime, soda ash, and carbon dioxide fed (in pounds or kilograms)
- chemical feeder settings
- amount of water treated (in million gallons per day or megaliters per day)
- hardness, alkalinity, and magnesium–calcium concentration in raw and treated water
- free carbon dioxide in raw water

- the pH of
  - —raw water
  - —treated water
  - —stages of the process that are critical to chemical reactions
- amount of sludge pumped to disposal
- dosage calculations and results of jar tests (softening curves)
- depth of the sludge blanket and results of settling tests if a solids-contact basin is used

Selected Supplementary Readings

Briggs, J.C., and J.F. Ficke. 1977. ***Quality of Rivers in the United States, 1975 Water Year.*** USGS Open-File Rpt. 78-200. Reston, Va.

Logsdon, G.S., M.M. Frey, T.D. Stefamich, S.L. Johnson, D.E. Feeley, J.B. Rose, and M. Sobsey. 1994. ***The Removal and Disinfection Efficiency of Lime Softening Processes for* Giardia *and Viruses.*** Denver, Colo.: American Water Works Association Research Foundation and American Water Works Association.

Manual M3, Safety Practices for Water Utilities. 1990. Denver, Colo.: American Water Works Association.

Manual of Instruction for Water Treatment Plant Operators. 1975. Albany, N.Y.: New York State Department of Health.

Manual of Water Utility Operations. 8th ed. 1988. Austin, Texas: Texas Water Utilities Association.

Peters, G.H., E.R. Baumann, and M.A. Larson. 1989. Effects of Various Parameters on the Thickening of Softening Sludge. ***Jour. AWWA,*** 81(3):74.

Recommended Standards for Water Works. 1992. Albany, N.Y.: Health Education Services.

Sawyer, C.N. 1960. ***Chemistry for Sanitary Engineers.*** New York: McGraw-Hill.

Sludge: Handling and Disposal. 1989. Denver, Colo.: American Water Works Association.

Water Quality and Treatment. 4th ed. 1990. New York: McGraw-Hill and American Water Works Association (available from AWWA).

Water Supply and Treatment. Arlington, Va.: National Lime Association.

Water Treatment Plant Design. 2nd ed. 1990. New York: McGraw-Hill and American Water Works Association (available from AWWA).

CHAPTER 12

Ion Exchange Processes

Ion exchange processes are a common alternative to the use of lime and soda ash for softening water. Water from all natural sources contains dissolved minerals that dissociate in water to form charged particles called ions. Calcium, magnesium, and sodium are the positively charged ions of principal concern, and bicarbonate, sulfate, and chloride are the normal negatively charged ions of concern. An ion exchange medium, called resin, is a material that will exchange a hardness-causing ion for another one that does not cause hardness, hold the new ion temporarily, and then release it when a regenerating solution is passed over the resin.

The early types of resins used for ion exchange softening were zeolites, which are silica compounds. The softening process is still commonly referred to as "zeolite" softening.

New organic exchange materials are now available with improved properties and expanded uses. The ion exchange process is most frequently used in the sodium cycle, in which calcium and magnesium are removed by replacement with sodium. In this case, a saltwater solution is used for regeneration (see formulas later in this chapter). Ion exchange can also take place through the hydrogen cycle, in which calcium, magnesium, and bases such as sodium are replaced with hydrogen; sulfuric acid is normally used for regeneration.

The problems associated with hard water are discussed in the first section of chapter 11, so they will not be repeated here.

The Ion Exchange Process

Some of the advantages, disadvantages, and operating points of the ion exchange process are as follows:

- Ion exchange has a much lower initial cost and smaller space requirements than those for lime softening.
- Ion exchange units can easily be operated by automatic control; they require minimal operating staff, as opposed to lime–soda ash plants, which require constant monitoring and a high level of operating skill.
- In cases where wells are spaced widely apart throughout a system, it is usually impractical to install small lime-softening treatment equipment at each location. The process will normally be feasible only if all raw water is piped to a central location for treatment. However, it is often feasible to install small ion exchange units at these multiple sites.
- The only chemical used in most ion exchange applications is salt, which is safe and easy to handle. There is relatively little danger of serious contamination of the water supply through equipment failure or improper operation if salt is used.
- The problem of disposing of this regeneration waste from ion exchange units is highly variable. In some instances, disposal is very simple, and at other locations it can be very difficult. If ion exchange is being used for radium removal, radium will be concentrated in the waste, which may create a disposal problem.
- Lime–soda ash softening is usually less expensive to operate for larger systems where all of the raw water is available at a single location. An exception is when a large percentage of the hardness in a water source is in noncarbonate form, in which case the ion exchange process might be less expensive to operate.
- The ion exchange process is not frequently used to treat surface waters. If it is, full conventional treatment must be used first to prevent turbidity and algae from fouling the resin.
- The ion exchange process removes all hardness from the water, so by blending a proportion of completely soft water with hard water, a water having any desired hardness can be produced.

Chemistry of Ion Exchange Softening

The following chemical reactions describe the chemistry of ion exchange softening. In the equations, X represents the ion exchange material.

Carbonate hardness:

$$\underset{\text{calcium bicarbonate}}{Ca(HCO_3)_2} + Na_2X \rightarrow CaX + \underset{\text{sodium bicarbonate}}{2NaHCO_3} \quad (12\text{-}1)$$

$$\underset{\text{magnesium bicarbonate}}{Mg(HCO_3)_2} + Na_2X \rightarrow MgX + \underset{\text{sodium bicarbonate}}{2NaHCO_3} \quad (12\text{-}12)$$

Noncarbonate hardness:

$$\underset{\text{calcium sulfate}}{CaSO_4} + Na_2X \rightarrow CaX + \underset{\text{sodium sulfate}}{Na_2SO_4} \quad (12\text{-}3)$$

$$\underset{\text{calcium chloride}}{CaCl_2} + Na_2X \rightarrow CaX + \underset{\text{sodium chloride}}{2NaCl} \quad (12\text{-}4)$$

$$\underset{\text{magnesium sulfate}}{MgSO_4} + Na_2X \rightarrow MgX + \underset{\text{sodium sulfate}}{Na_2SO_4} \quad (12\text{-}5)$$

$$\underset{\text{magnesium chloride}}{MgCl_2} + Na_2X \rightarrow MgX + \underset{\text{sodium chloride}}{2NaCl} \quad (12\text{-}6)$$

These chemical reactions represent cation exchange because the positive cations Ca^+, Mg^+, and Na^+ are involved. To replenish the sodium ions, the ion exchange materials must periodically be backwashed with brine (salt solution). This process, called regeneration, allows the ion exchange materials to be used over and over again. The regeneration process involves the following chemical reactions:

$$CaX + 2NaCl \rightarrow CaCl_2 + Na_2X \quad (12\text{-}7)$$

$$MgX + 2NaCl \rightarrow MgCl_2 + Na_2X \quad (12\text{-}8)$$

The ion exchange softening process does not alter the pH or alkalinity of the water. However, the stability of the water is altered by the removal of calcium and by an increase in total dissolved solids (TDS). For each 1 mg/L of calcium removed and replaced with sodium, the TDS increases 0.15 mg/L. For each 1 mg/L of magnesium removed and replaced with sodium, the TDS increases by 0.88 mg/L.

The measurements commonly used to express water hardness in the ion exchange softening process are different from those in the lime–soda ash process. Hardness in the lime–soda ash process is commonly expressed in terms of milligrams per liter as calcium carbonate ($CaCO_3$). In the ion exchange process, hardness is expressed in grains per gallon, which is usually simply referred to as *grains*. The following conversion factors show the relationship between mg/L and grains:

1 grain = 17.12 mg/L

1 grain = 0.142 lb per 1,000 gal

7,000 grains = 1 lb per gal

Removal of Barium, Radium, and Nitrate

Three inorganic chemicals for which maximum contaminant levels (MCLs) have been set by the US Environmental Protection Agency (USEPA) can be removed by the ion exchange process. Barium and radium are found frequently in groundwater in several locations in the United States, and they exceed the MCL in quite a few wells. Nitrate is most often found in shallow wells in rural areas, primarily as a result of agricultural contamination, and is also present in excess of the MCL in some surface sources.

Barium and radium are effectively removed by ion exchange softening. The removal action of these chemicals takes longer than hardness removal. Anion exchange is currently the simplest and lowest-cost method for removing nitrate from contaminated groundwater.

Water systems that have raw-water sources contaminated with excessive levels of these chemicals typically first investigate the possibility of using another water source that has little or no contamination. The new source could be used to replace the old source entirely, or it could be mixed with the water having higher contamination to dilute the water furnished to the public to below the MCL. Where this is not possible, treatment for contaminant removal must be installed.

The use of ion exchange for removing nitrate and barium has been limited, and several different types of resins and operating procedures have been used experimentally. Systems investigating ion exchange for this purpose should seek professional assistance in obtaining the latest technology available.

Health Concerns Over Water-Softening Processes

A number of studies have associated high levels of sodium intake with high blood pressure and heart disease in those people who are considered "at risk" for hypertension. In addition, certain diseases are aggravated by a high salt intake, including congestive heart failure, cirrhosis, and renal disease. There has been some concern because the sodium level of water softened by the ion exchange process is somewhat higher than it was before softening.

The greatest intake of sodium for the average person comes from food. The typical intake for a normal adult is between 1,100 and 3,300 mg/day, so the amount added by softening is extremely small proportionally. However, persons who have been placed on a severely restricted sodium diet by their physician may find that they cannot tolerate the amount in the water in addition to the unavoidable amount in their food. If the amount of sodium in the drinking water is substantial, these people may have to obtain low-sodium drinking water elsewhere.

The USEPA has not proposed a primary MCL for sodium. Current data showing that it significantly contributes to hypertension in the general population are insufficient. In addition, the USEPA notes that the level is normally minor compared with that from dietary intake. The USEPA has suggested that for protection of the at-risk population, the level of sodium in drinking water be limited to 20 mg/L, as recommended by the American Heart Association.

In addition, a number of studies have indicated a possible relationship between the hardness of drinking water and the incidence of cardiovascular disease. There are theories that soft water promotes the problem, or that the various minerals in hard water help to prevent it. Studies are continuing, but information is not sufficient at this time to recommend any changes in the softening treatment process for public health reasons.

Ion Exchange Softening Facilities

The ion exchange process requires the following basic components:

- ion exchange materials (resins)
- ion exchange units
- salt storage tanks
- brine-feeding equipment
- devices for blending hard and soft water

Ion Exchange Resins

A number of materials, including some types of soil, can act as cation exchangers for softening water. For example, a natural greensand called glauconite has very good ion exchange capabilities and was once widely used for water softening. It is composed of sodium aluminum silicate and was most commonly called zeolite.

Synthetic zeolites and organic polymers, known as polystyrene resins, have now replaced natural zeolites because their quality can more easily be controlled and they have higher ion exchange capacities.

Polystyrene resins are most commonly used today because they have three to six times the ion exchange capacities of other materials. As shown in Figure 12-1, beads of the resin look like small BBs. The resin is not totally depleted in the process; it instead serves as a kind of "parking lot" in which ions are traded. In practice, a small amount of the resin is lost in the backwashing process as the resin particles rub against each other.

Ion Exchange Units

The tanks holding the ion exchange resins (Figure 12-2) resemble pressure filters. The one major difference is that the interior of the tanks must be coated with a special lining to protect them from corrosion from the brine used in regeneration.

Figure 12-3 is a cutaway view of a vertical-downflow ion exchange unit. A unit is usually provided with the following components:

- hard-water inlet
- soft-water outlet
- wash-water inlet and collector

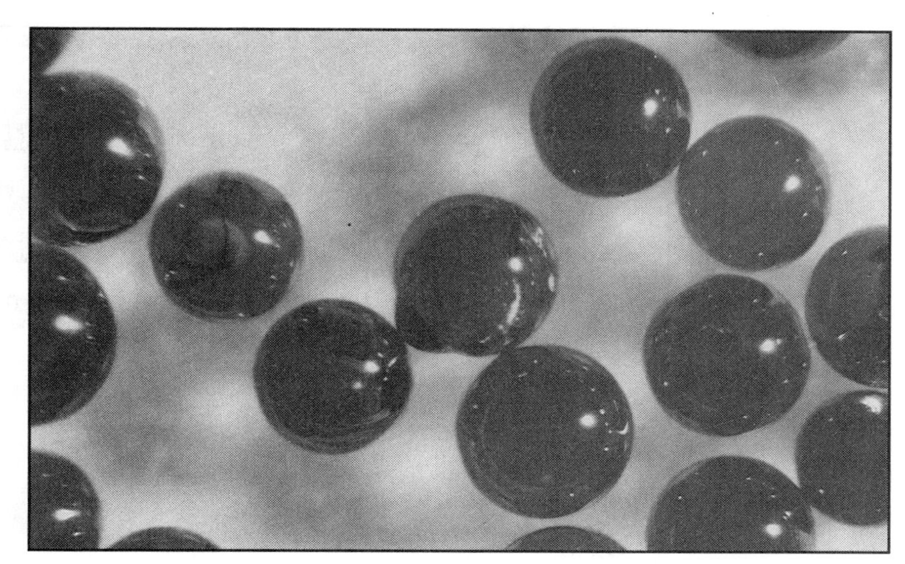

FIGURE 12-1 Beads of polystyrene resin

Courtesy of U.S. Filter/Permutit

FIGURE 12-2 Ion exchange pressure tanks

Courtesy of Ionics, Incorporated, Water Systems Division

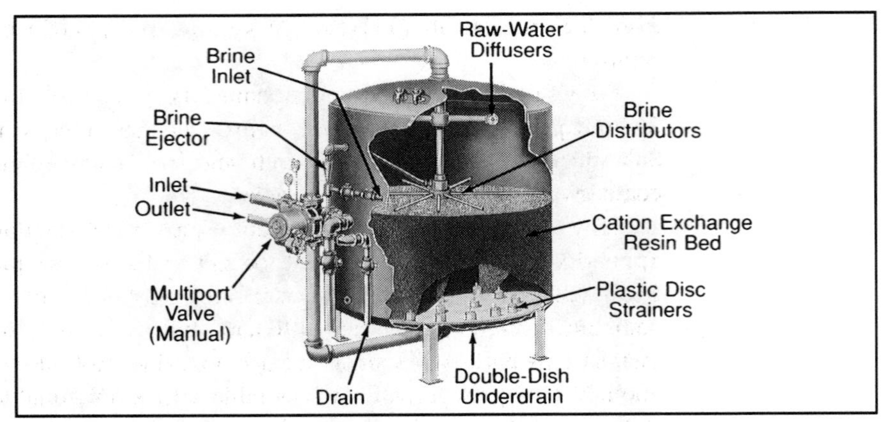

FIGURE 12-3 Vertical-downflow ion exchange unit

Courtesy of U.S. Filter/Permutit

- brine inlet and distribution system
- brine and rinse-water outlet
- rate-of-flow controllers
- sampling taps
- underdrain system, which also serves to distribute backwash water
- graded gravel to support the ion exchange resins

Both upflow and downflow ion exchange units are available. Units designed for the water to flow downward are more commonly used because they remove some sediment as they soften the water. All types of units can be equipped with automatic controls.

The size of the ion exchange unit and the volume of resin needed to soften water are determined by the hardness of the raw water and the desired length of time between regenerations. The minimum recommended depth for resin is 24 in. (0.6 m). The resin is supported either by an underdrain system or by 15–18 in. (0.40–0.45 m) of graded gravel.

In downflow units, a brine distribution system is used to direct the flow of brine downward and evenly through the unit during regeneration so that all resin comes into contact with it. The rinse water can also be distributed through the same system.

Salt Storage Tanks

Salt is used in the water-softening process to form brine, which regenerates the resins. The amount of salt used in creating the brine ranges from 0.25 to 0.45 lb (0.11 to 0.20 kg) for every 1,000 grains of hardness removed.

The salt is usually stored in tanks large enough to hold brine for a 24-hour period of operation or for three regenerations, whichever is greater. Salt attacks and wears away concrete and steel, so tanks must be made of, or coated with, a salt-resistant material.

The salt used for resin regeneration must meet standards for purity (see appendix A). Rock salt or pellet-type salt is the best for preparing the brine. Road salt is not suitable because it often contains impurities. Block salt is sometimes used for home softeners, but it is not suitable for larger installations because its small surface area does not allow it to dissolve fast enough. Fine-grained salt, such as table salt, is not suitable because it packs tightly and does not dissolve easily.

Salt storage tanks should be covered to prevent dirt and foreign material from entering. The access holes in the cover should have a raised lip so that any dirty water on the cover does not flow into the hole and contaminate the salt.

To fill a salt storage tank and prepare the brine, water is added to the tank and the tank is filled with rock salt. More water is then added to submerge the salt so that it will dissolve. The water fill line must allow for an air gap above the top of the tank to prevent the possibility of brine siphoning back into the water supply. An excess of undissolved salt should be kept in the tank to ensure that a concentrated solution is achieved.

Because brine is heavier than water, the highest concentration of brine will be in the lower part of the storage tank. The brine used for regeneration is therefore usually pumped from the bottom of the tank.

Brine-Feeding Equipment

Concentrated brine contains about 25 percent salt, but to be effective the brine should be diluted to contain 10 percent salt. A metering pump or hydraulic ejector is used to dilute the concentrated brine as it is applied to the resin bed. The brine is very corrosive, and special plastic or other salt-resistant materials must be used for pumps and piping.

The solubility of salt decreases at low temperatures, forcing salt out of solution. The water that remains after the salt has separated out of solution is subject to freezing, so the brine tanks and piping should be protected.

Blending Hard and Soft Water

A properly operated ion exchange unit will produce water with zero hardness. Most water systems try to supply water to the system with a hardness of 85 to 100 mg/L because it is a level acceptable to customers and is relatively cost effective. If zero-hardness water were furnished to the system, it would be very corrosive.

To dilute the softened water, a bypass or some other device is used to blend hard and soft water at the discharge of the softener. Water meters are commonly installed to determine the proportions of blended water. An automatic softener bypass valve is illustrated in Figure 12-4.

Regulations

There are no federal regulations requiring the softening of water. However, the softening provided by a system could affect compliance with the Lead and Copper Rule. Additional information on that regulation is given in chapter 9, and in *Water Quality*, also part of this series.

Operation of Ion Exchange Processes

There are four basic cycles in the ion exchange water-softening process. As shown in Figure 12-5, these are

- softening
- backwash

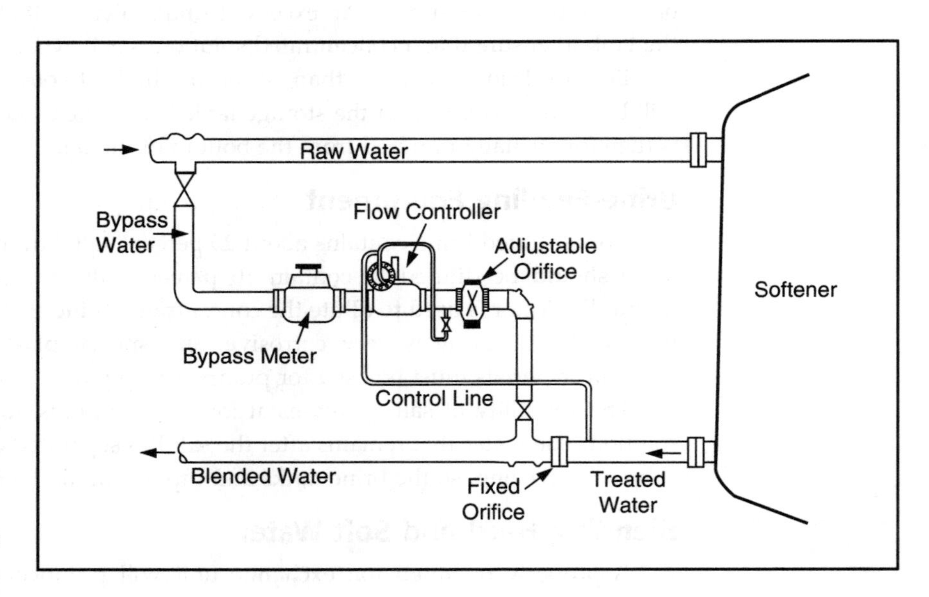

FIGURE 12-4 Illustration of automatic softener bypass valve

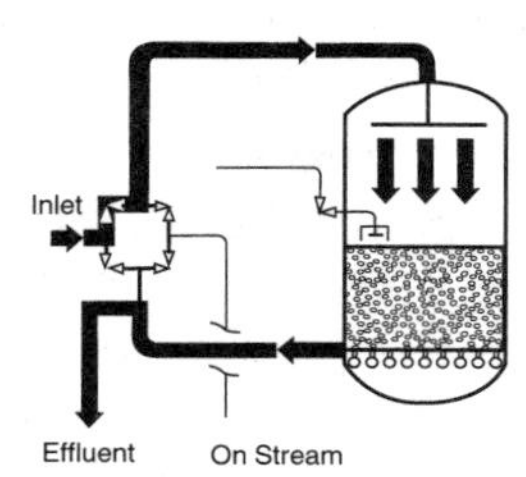

Softening. Influent water passes downward through the bed of ion exchange material to the effluent.

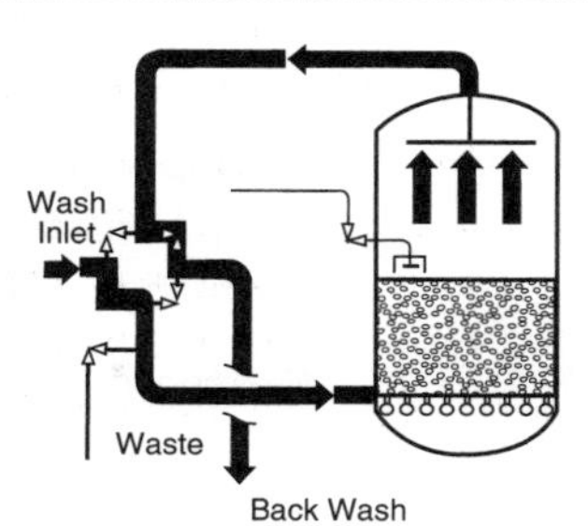

Backwash. Influent water is passed upward through the bed of ion exchange material to loosen the bed and remove suspended solids that may have been deposited in the bed during operation.

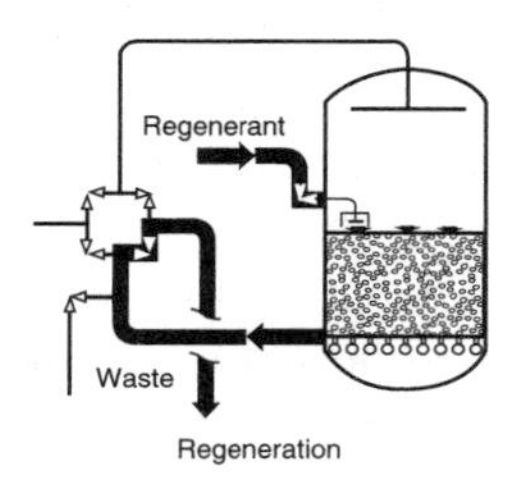

Regeneration. Regenerant solution is passed through the bed to waste at a controlled concentration and flow.

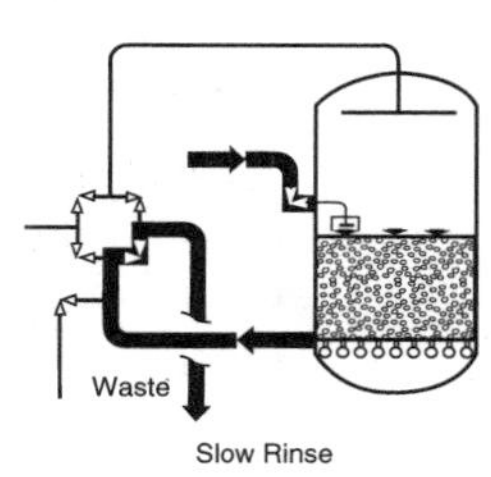

Slow rinse. Water is passed through the bed to displace the regenerant solution to waste.

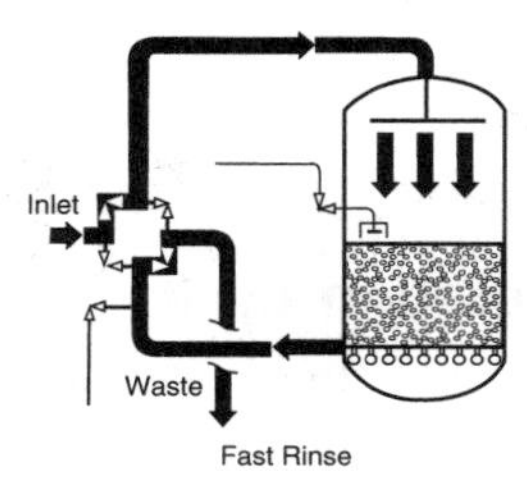

Fast rinse. Influent water is passed through the bed to waste to remove the last traces of regenerant chemicals.

FIGURE 12-5
Four cycles in the ion exchange process

Courtesy of Infilco Degremont, Inc., Richmond, Va.

- regeneration
- slow and fast rinse

Operating procedures will vary somewhat depending on the type of equipment used, so it is important to follow the manufacturer's specific instructions. General guidelines for the process operation follow.

Softening Cycle

The softening cycle involves feeding hard water into the ion exchange unit until the unit can no longer produce water with a hardness near zero. The cycle is normally completed when the effluent hardness reaches 1 to 5 mg/L. Ion exchange units are generally furnished with a meter for measuring water flow. This meter can be adjusted to sound an alarm when the volume of water corresponding to the unit's rated capacity has passed through the unit. The loading rate for units that use polystyrene resin is in the range of 10 to 15 gpm/ft^2 (6.8 to 10.2 mm/s).

Backwash Cycle

Once hardness "breaks through," the ion exchange unit must be removed from service and the resin regenerated. In vertical-downflow units, the resin must be backwashed with previously treated water before being regenerated. This loosens the resin that had compacted during the softening cycle, randomly mixes the resin, and removes foreign material that has been filtered or precipitated from the water. The backwash rate is usually 6–8 gpm/ft^2 (4.1–5.4 mm/s), which is the flow rate needed to expand the resin bed by at least 50 percent of its depth. The flow rate is usually controlled by a float-operated butterfly valve calibrated to obtain the desired rate for each operation.

Distributor pipes at the top of the unit provide uniform distribution of raw water and uniform collection of backwash water. An underdrain collector–distributor maintains a uniform distribution of flow.

Regeneration Cycle

To regenerate ion exchange resins, concentrated brine is drawn from the salt storage basin, diluted by an ejector to a solution containing 10 percent salt by weight, and passed slowly and continuously through the resin bed. Regeneration usually requires a contact time of 20 to 30 minutes for the brine to pass through the resin bed at application rates of about 1 gpm/ft^3 (0.7 mm/s). Adequate brine contact time is extremely important in order to return proper ion exchange capacity to the resin. During the contact time, sodium ions from the brine are exchanged for calcium and magnesium ions in the resin bed.

Rinse Cycle

The regeneration cycle must be followed by a thorough rinsing cycle to remove the unused salt. Rinsing is accomplished by first running unsoftened water through the brine distribution line. Rinse water (previously treated water) is then added to the unit at a rate of about 2 gpm/ft^2 (1.4 mm/s) through the raw-water distribution line until the water leaving the unit has a chloride concentration nearly equal to that of the water entering the unit. As shown in Figure 12-5, some ion exchange units use a slow rinse to remove the bulk of the brine, followed by a fast rinse to remove the last remaining traces of brine. The amount of water needed to perform the rinse cycle is about 20 to 35 gal/ft^3 (2.6 to 4.7 kL/m^3) of resin.

Disposal of Wastewater

A potential problem with the ion exchange process is the disposal of the wastewater from the backwash, regeneration, and rinse cycles. Proper disposal techniques must be thoroughly evaluated before the ion exchange process is selected for use. The costs involved in disposing of the brine might make other softening processes more cost effective.

The total amount of spent brine will usually vary from 1.5 to 7 percent of the amount of water softened. The wastewater contains calcium chloride, magnesium chloride, and sodium chloride. If there is radium in the source water, there will also be a high concentration of radium in the wastewater. Even with the dilution provided by the backwash and rinse water, the total concentration of dissolved solids in the wastewater is usually between 35,000 and 45,000 mg/L.

Wastewater with this concentration of dissolved solids can cause serious corrosion of piping and, if discharged to a sanitary sewer, may upset the sewage treatment process. If the waste is discharged to a waterway where there is not adequate dilution, it can seriously harm aquatic life. If it is discharged to a land surface, it can make the soil unusable for agricultural purposes. And if it is discharged to a seepage bed, it could contaminate an underground aquifer to the extent that it becomes unusable as a water source.

If the waste is considered to be sufficiently diluted, it may be accepted by a sewage treatment authority, or the state may allow it to be discharged to a flowing stream or a large body of water. The method allowed will depend primarily on the requirements of the state pollution control agency.

Operational Control Tests

The ion exchange process is relatively simple and requires only the following control tests:

- total hardness tests
- Langelier saturation index determination for water in the distribution system
- chloride tests (required periodically to check that rinsing after regeneration is adequate)

These tests are described in more detail in *Water Quality*, also part of this series. Some details of the total hardness test are included in chapter 11.

Operating Problems

Several conditions cause operating problems in the ion exchange softening process, including

- resin breakdown
- iron fouling
- turbidity, organic color, and bacterial-slime fouling
- unstable water

Resin Breakdown

Polystyrene resins generally have an operational life of 15 to 20 years before needing replacement. However, certain conditions can cause the resin to degrade sooner. Oxidation by chlorine is the primary reason for resin breakdown. The chlorine dosages normally used in disinfecting water have only a minimal effect on resin. However, high chlorine residuals, such as those that result when chlorine is used for removing iron from water, can cause severe resin breakdown. When high chlorine levels are used for pretreatment, the excess chlorine should be removed before the water enters the ion exchange softener.

Iron Fouling

The presence of iron in raw water can seriously affect the capacity of resin to exchange sodium ions for calcium and magnesium ions. Ferrous iron is oxidized and precipitates as iron oxide within the resin bed. After this occurs, no amount of brine will remove the iron oxide, and the condition can worsen until the bed completely loses its ion exchange capacity.

If iron oxide is formed before the raw water enters the unit, it is deposited on the resin bed, but it will generally be removed during normal backwashing. A resin bed that has become contaminated by iron develops a dark red color. Raw water containing iron should generally be pretreated for iron removal before the water enters ion exchange units.

Turbidity, Organic Color, and Bacterial-Slime Fouling

Turbidity, color caused by humic substances, and bacterial slimes in raw water will coat ion exchange resin, which causes loss of ion exchange capacity and excessive head loss. Although backwashing helps to remove some of these materials, it will not remove large volumes of it because the materials become tightly held on the resins.

If these materials are present in high enough quantities, it is best to treat the water by conventional coagulation–flocculation, sedimentation, and filtration prior to softening. If this water is not pretreated, the resin will frequently have to be replaced or restored, which can become quite costly.

Unstable Water

Water that has been softened by the ion exchange process is usually corrosive. Blending raw and softened water so that the finished water is only nominally soft usually helps. If the water is still corrosive, chemicals may have to be added to provide corrosion control. Additional information on corrosion control is included in chapter 9.

Record Keeping

The following information should be recorded daily concerning the operation of the ion exchange process:

- hardness, alkalinity, magnesium and calcium concentrations, and pH of the raw and treated water
- amount of water treated, in million gallons per day (megaliters per day)
- amount of water treated in each softening cycle, in gallons (liters) per day
- amount of backwash water, rinse water, and brine used, in gallons (liters) per day
- amount of salt added to the storage tank, in pounds (kilograms) per day

These records will be useful for keeping track of the softening process and for ordering chemicals. However, each operator should develop additional records as necessary for the particular treatment equipment.

Selected Supplementary Readings

Controlling Radionuclides and Other Contaminants in Drinking Water Supplies: A Workbook for Small Systems. 1991. Denver, Colo.: American Water Works Association.

Recommended Standards for Water Works. 1992. Albany, N.Y.: Health Education Services.

Subramonian, S., D. Clifford, and W. Vijjeswarapu. 1990. Evaluating Ion Exchange for Removing Radium From Groundwater. ***Jour. AWWA,*** 82(5):61.

Water Quality and Treatment. 4th ed. 1990. New York: McGraw-Hill and American Water Works Association (available from AWWA).

Water Supply and Treatment. Arlington, Va.: National Lime Association.

CHAPTER 13

Adsorption

Essentially all natural water contains varying amounts of carbon-containing substances that are dissolved from soil and vegetation. These substances are usually referred to as natural organic matter.

In addition, there are thousands of synthetic organic substances, and new ones are invented every day. These are generally termed synthetic organic chemicals (SOCs). They include a wide range of pesticides, industrial chemicals, oils, and other manufactured chemicals. Well over 1,000 SOCs have been identified in drinking water at one location or another. Organic chemicals can also appear at an objectionable level in finished water as a result of reactions that take place in the water treatment process. The term most commonly used in referring to organic compounds from all of the different sources is *organics*.

The organics of interest in water treatment generally fall into the following categories:

- class I: organic compounds that cause taste, odor, and color problems
- class II: synthetic organic chemicals that must be limited in drinking water because of concern about adverse health effects. Some are regulated by established maximum contaminant levels (MCLs), and many more have suggested limitations reported in health advisories
- class III: precursors, principally humic and fulvic acids, that react with disinfectants to produce disinfection by-products (DBPs)

- class IV: DBPs that have been formed in the treatment process, such as trihalomethanes (THMs)

Additional information on organic substances may be found in *Water Quality*, part of this series.

Process Description

The adsorption process is used in the water works industry primarily for the removal of organic substances, so it is important to understand where the organics originate and why they are objectionable.

Organic Substance Occurrence and Concerns

Organics in Surface Water

Essentially all surface water contains some natural organic material. Many water systems that use surface sources have bothersome tastes, odors, or color in the source water at some time or another. Organic substances in the source water are also precursors for the development of THMs and other DBPs.

In addition, low levels of SOCs have been identified at times in most surface sources. They are usually so diluted and their presence so variable that they have not yet been regulated. Some water systems, however, have installed special organics removal systems because the overall level of organics is consistently quite high.

Organics in Groundwater

Water from deeper wells usually contains relatively low levels of organics, but water from shallow wells, on occasion, has relatively high levels of humic substances drawn from the surrounding aquifer. In these cases, the water systems face the same operating problems as surface systems.

In addition, tests of groundwater sources in recent years have revealed that an alarmingly high percentage of wells are contaminated by detectable amounts of SOCs. Thousands of manufactured chemicals (such as pesticides and industrial solvents) have, over the years, been leaked, spilled, and disposed of near wells and have filtered down into the aquifer. In many cases, the chemicals have been identified as toxic or carcinogenic to humans at relatively low concentrations.

The concentrations detected in most contaminated wells amount to only a trace. In a great number of locations, however, the concentration is high enough that use of the water has been discontinued, or special treatment has been provided for SOC removal when no other water source was available.

Organic Chemical Removal

Removal at the Source

The best place to control organics in drinking water is at the source. To control the organics level of raw water, surface water systems might locate the raw-water intake at a more favorable site, restrict certain human activities in the watershed, or control algae growth. Groundwater systems might establish restrictions on land use in the recharge zone or construct barriers to prevent contaminants from flowing toward a well in the aquifer. This is discussed in greater detail in *Water Sources*, which is also part of this series.

Removal by Other Treatment Processes

When organic chemicals cannot adequately be controlled at the source, they can often be removed by standard treatment processes. For instance, many taste-and-odor-causing substances are sufficiently oxidized by chlorine to make the water quality acceptable. When chlorine is ineffective, permanganate, chlorine dioxide, or ozone may work better. In addition, aeration and the coagulation, sedimentation, and filtration processes work well for reducing the level of many organic substances.

Removal by Adsorption

At times the conventional treatment processes are not effective enough to remove organics. In such instances, adsorption might be used instead of other treatment methods because it can be implemented quickly. It can also be more cost-effective than other treatments because of the size or location of the treatment site.

The Principle of Adsorption

Adsorption works on the principle of adhesion. In the case of water treatment, organic contaminants are attracted to the adsorbing material. They adhere to its surface by a combination of complex physical forces and chemical action.

For adsorption to be effective, the adsorbent must provide an extremely large surface area on which the contaminant chemicals can adhere. If the process is to be economical to build and operate, the total surface area of adsorbent required must be contained in a tank of reasonable size.

Porous adsorbing materials help achieve these objectives. Activated carbon is an excellent adsorbent because it has a vast network of pores of varying size to accept both large and small contaminant molecules. These pores give activated carbon a very large surface area. Just 1 lb (0.45 kg) of activated carbon has a total surface area of about 150 acres (60 ha). This can, for example, trap and hold over 0.55 lb (0.25 kg) of carbon tetrachloride.

Figure 13-1 is a photograph, taken through an electron microscope, showing the large number of pores in a grain of activated carbon. These pores are created during the manufacturing process by exposing the carbon to very high heat in the presence of steam. This is known as *activating* the carbon. Activation oxidizes all particles on the surface of the carbon, leaving the surfaces free to attract and hold organic substances. Figure 13-2 shows the structure of the carbon after activation and indicates that different sizes of chemical molecules can be adsorbed within the pores.

Once the surface of the pores is covered by adsorbed material, the carbon loses its ability to adsorb. The spent carbon can then be reactivated by essentially the same process as the original activation, or it can be discarded and replaced with fresh carbon.

Adsorption Facilities

Activated carbon can be made from a variety of materials such as wood, nutshells, coal, peat, and petroleum residues. One of the principal sources of activated carbon used in water treatment is bituminous or lignite coal. The coal is slowly heated in a furnace without oxygen so that it will not burn, thus converting the coal to carbon. The carbon is then activated through

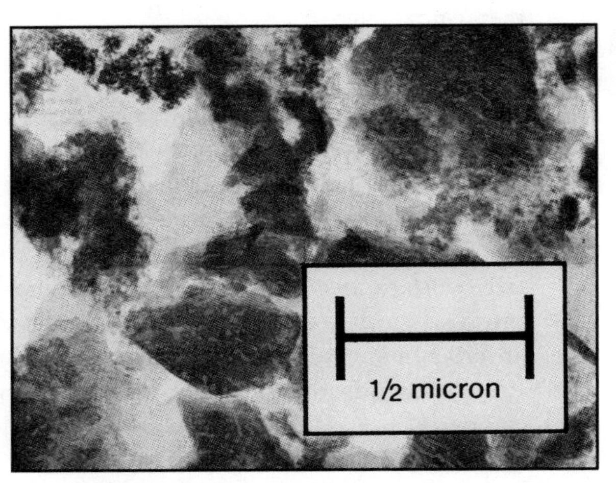

Photograph supplied by Activated Carbon Division, Calgon Corp.

FIGURE 13-1 Details of the fine structure of activated carbon

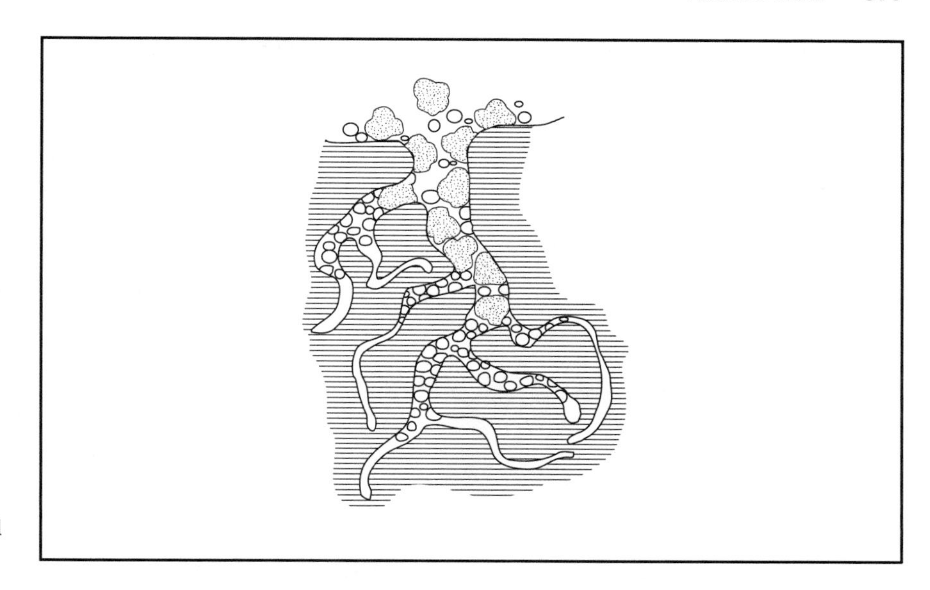

FIGURE 13-2 Carbon structure after activation showing small and large adsorbed chemical molecules

exposure to a steam–air mixture. The activated carbon is then crushed and screened to obtain the desired particle size.

The two common forms of activated carbon used in water treatment are powdered activated carbon (PAC) and granular activated carbon (GAC). Table 13-1 compares the general properties of types of PAC and GAC. Figure 13-3 shows the difference in particle size.

Powdered Activated Carbon

Powdered activated carbon is typically available in 50-lb (23-kg) bags or in bulk truck or railroad car shipments. It can be fed in dry form or as a slurry with water.

Dry Feed Systems

Treatment plants that only periodically need to use PAC normally use the dry feed method. The PAC can be stored in bags or in bulk form in steel tanks. The tanks should be located so that they can feed directly into the hoppers of dry feeders by using gravity only.

When PAC is used in dry form, chemical feeders specifically designed to handle carbon should be used. The feeders must be capable of operating over a wide range of feed rates. In some instances, only a very light application may be desired to eliminate slight tastes or odors in the water. In cases, very

TABLE 13-1 Comparison of GAC and PAC properties

Property	Types of Carbon GAC	PAC
Density, lb/ft^3	26–30	20–45
(g/cm^3)	(0.42–0.48)	(0.32–0.72)
Surface area, m^2/g	650–1,150	500–600
Mean particle diameter, mm	1.2–1.6	less than 0.1

FIGURE 13-3 Difference in particle size between GAC and PAC

heavy dosages may be required if there are serious tastes or odors or chemical contamination of the raw water.

The feeder must feed to a tank or ejector where the carbon will quickly be mixed with water to form a thin slurry and then discharged to the application point (Figure 13-4). The hopper above the feeder must have walls slanting at a 60° angle, as well as a vibrator mechanism to keep the carbon flowing and to prevent it from arching.

Systems that feed PAC only occasionally find that a relatively low-cost method is to use a helix-type feeder positioned over an eductor, as illustrated in Figure 13-5. Although the feed rate may not be very accurate, the dosage of carbon required during a temporary taste or odor episode is generally an estimate based on past experience, so precise feed is only a secondary consideration.

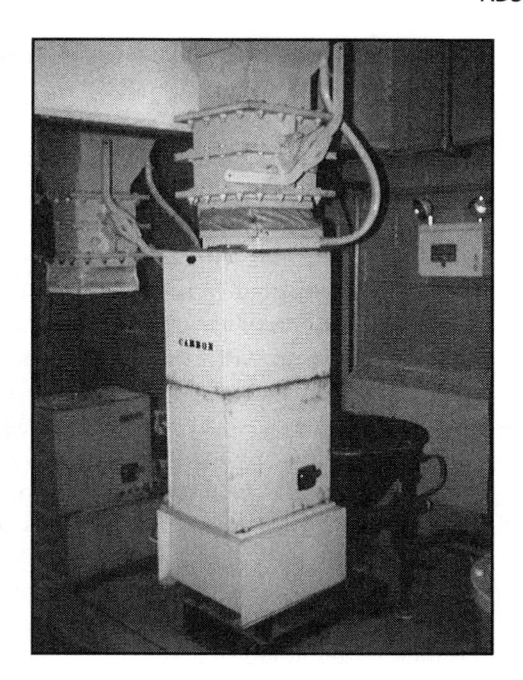

FIGURE 13-4 Dry carbon feeder

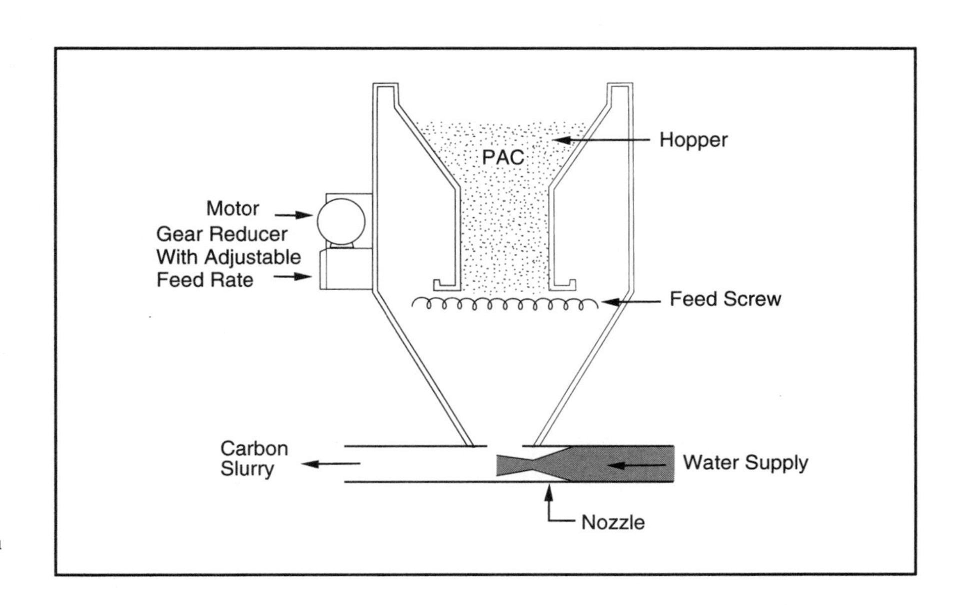

FIGURE 13-5 PAC feed using a helix feeder

Carbon dust is extremely fine and light. If not carefully confined and handled, it will easily float throughout a plant and cover floors and surfaces with a fine, black coating. To keep dust to a minimum, dry carbon feeders should be located in a confined room of the treatment plant.

Slurry Feed Systems

Because of the handling problems encountered with dry PAC, many plants that consistently use PAC have a slurry feed system of the type shown in Figure 13-6. If delivered in bulk form, the PAC is removed from the truck or rail car via an eductor or pneumatic system directly to a storage tank. Plants usually have two storage tanks so that a shipment can be placed in one before the other is empty. The tanks are generally made of steel or concrete and have a capacity about 20 percent greater than the maximum load that will be delivered. The tanks should have an epoxy or other type of lining to protect them from corrosion. If PAC is delivered in bags, it can be added directly to the storage tank.

The tanks are generally provided with two-speed mixers. If dry carbon is added directly to the water in the tank, the mixer must initially be operated at high speed to create a slurry — otherwise, the carbon just tends to lie on

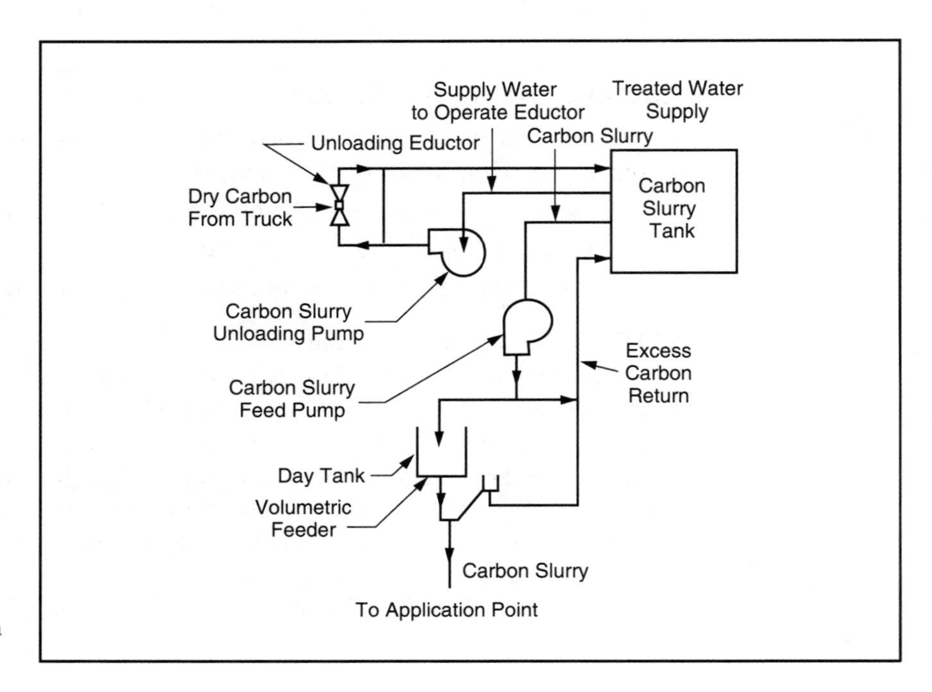

FIGURE 13-6
Slurry feed system for handling PAC

top of the water. After all of the carbon has been thoroughly wetted to form a slurry, the mixer can be set to a slow speed to maintain the proper slurry condition. If power to the mixer is ever turned off, perhaps because of an electrical outage, it may be necessary to operate the mixer at the high speed again for a period of time to reestablish a uniform slurry.

The slurry is usually pumped to a day tank that holds the quantity to be used for a day or shift. The tank should be plastic or lined steel and equipped with a mixer to keep the slurry in suspension. The slurry is then fed from the day tank by a volumetric feeder.

Unless the feeder is located at exactly the feed point, there is a danger that the carbon will settle out in the feed line because of the low flow velocity. To prevent this, the feeder can discharge onto an eductor that maintains a high flow velocity in the feed line. The feed line should slope downward all the way to the application point, and provisions should be made for flushing any carbon that may settle out and clog the pipe. The type of pipe used for the feed line must be both corrosion- and erosion-resistant material, such as rubber, plastic, or stainless steel.

PAC is usually added to the water before the normal coagulation–flocculation step. The carbon must be removed from the water to prevent it from moving through the plant into the distribution system. Most of the PAC becomes part of the sedimentation basin sludge, so it is not practical to recover and reuse it.

Granular Activated Carbon

Granular activated carbon is used principally where continuous removal of organics is required. Typical uses include

- removal of organics following surface water treatment
- removal of organics from groundwater
- special-purpose removal of certain inorganics or radionuclides

GAC Used as a Filter Medium

Where the total organics load of a water source is not too high but there is a continuing need to remove organics, GAC can be installed in place of sand or anthracite in open filters (Figures 13-7 and 13-8). The GAC can be either a capping layer over other media or a complete replacement. The material then acts as both an adsorbent and a filtering medium.

The decision to use this approach is practical must be based on a study of how long the adsorption qualities of the GAC will last, how much it will cost to remove exhausted material, and how much it will cost to have the old

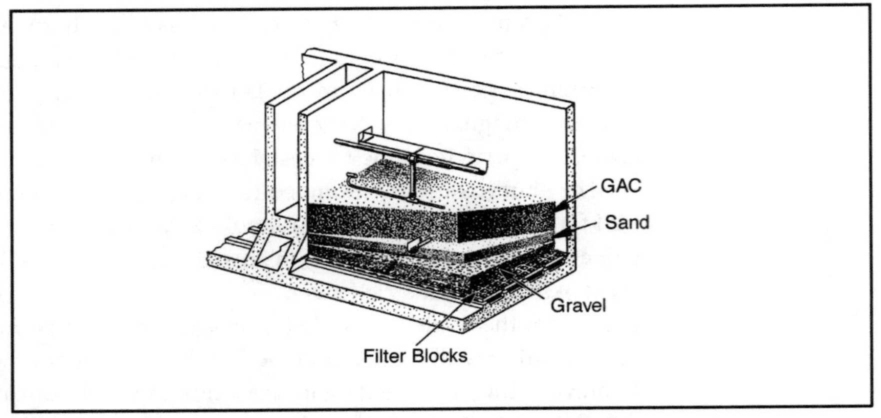

FIGURE 13-7 Partial replacement with GAC

Artwork supplied by Activated Carbon Division, Calgon Corp.

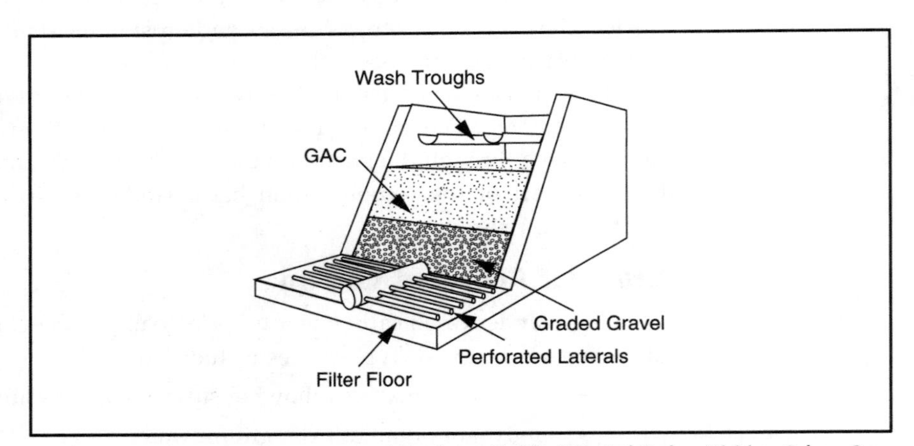

FIGURE 13-8 Complete replacement with GAC

Artwork supplied by Activated Carbon Division, Calgon Corp.

material either reactivated or replaced with new material. The effective life of the GAC in typical installations has been anywhere from a few months to 3 or 4 years, depending not only on the organics concentration, but also on the types of substances being removed. For instance, GAC will generally last much longer for removal of tastes and odors than it will for removal of SOCs.

It is recommended that the GAC bed be at least 24-in. (610-mm) thick, but the deeper it is, the longer the contact time will be, and the longer the bed will last before it must be replaced. On existing installations, the wash-water troughs may have to be raised to provide adequate bed depth.

GAC Contactors

GAC contained in closed pressure tanks (GAC contactors) has been used to treat the effluent of surface water systems after full conventional filtration treatment (Figure 13-9). This treatment may be provided for taste-and-odor control or to reduce the level of SOCs not removed by other treatment processes. Where the total organics load of the raw-water source is relatively high, using GAC after other treatment is typically found to be more economical than using it as a filter medium.

GAC contactors can also be used when a concentration of THMs or other DBPs cannot be prevented from forming by an adjustment of the disinfectants or other treatment processes. In these cases, GAC can be used as a final treatment to remove the DBPs to an acceptable level before the finished water enters the distribution system.

The principal use of GAC contactors on groundwater systems has been for removal of SOCs from wells that have been found to be contaminated. In most cases where contamination of groundwater has been identified, it is most cost effective to abandon the well if another good-quality source is available. The second choice is typically to use aeration if the problem

FIGURE 13-9 GAC contactors installed as the final treatment at a large surface water treatment plant

Courtesy of the Cincinnati Water Works

contaminants are volatile, which means they diffuse freely from the water. Where there are no other choices, GAC will normally work for chemical removal, but the operating cost will probably be relatively high. Groundwater has also been treated by passing the water through open beds, rather than tanks, of GAC. This has a few advantages, such as ease in monitoring and replacement of the GAC, but enclosed tank contactors are usually preferred.

GAC contactors have also been used extensively as an emergency method for treating a wide variety of water contamination problems. Truck or skid-mounted GAC contactors are available and can be set up quickly at a site to treat a contaminated water supply on an interim basis (Figure 13-10). In these cases, the capacity of the GAC unit is usually sufficient to give the water system enough time to examine alternatives and to design either a change in water source or a permanent treatment system before the GAC is exhausted.

New GAC is available in 60-lb (27-kg) bags or in bulk form delivered by trucks or railcars. GAC is usually placed in filters or contactors in a slurry form by an eductor, both to facilitate handling and to reduce dust.

GAC Regeneration

Depending on the type and concentration of organic compounds in the water being treated, GAC gradually loses its adsorption ability over a period

FIGURE 13-10 Skid-mounted GAC contactors

Photograph supplied by Activated Carbon Division, Calgon Corp.

of time, ranging from a few months to several years. The old GAC must then be removed and replaced with fresh carbon. The used GAC can be reactivated and reused, as illustrated in Figure 13-11. Reactivation consists of passing the spent carbon through a regeneration furnace, where it is heated to 1,500–1,700°F (820–930°C) in a controlled atmosphere that oxidizes the adsorbed impurities. About 5 percent of the carbon is lost during the process, so more new carbon must be added when the GAC is replaced in the filter or contactor.

Reactivation can be done on-site, but it is generally practical only for a relatively large installation. It is usually not cost-effective to ship exhausted GAC to a distant location to be reactivated and then return it to the treatment plant.

Regulations

The use of activated carbon in water treatment has increased in recent years as a result of new regulations and health advisories limiting public exposure to various SOCs. Most of the uses have involved groundwater systems where well contamination exceeding an MCL has been identified. These systems are usually directed by state authorities to stop using the well or to add treatment for contaminant removal.

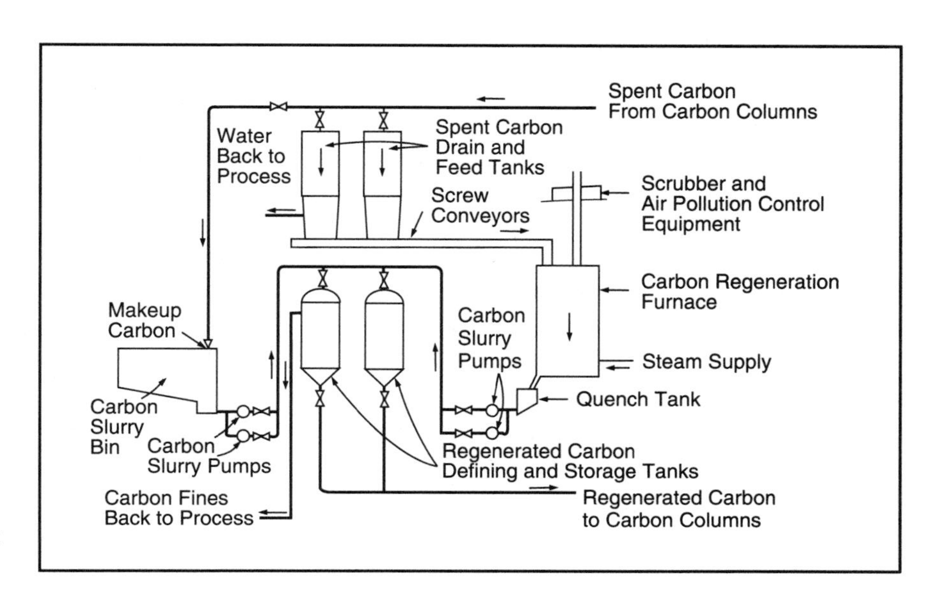

FIGURE 13-11
Schematic of GAC reactivation

In many cases, GAC is used as an emergency treatment measure because it will effectively remove almost all SOCs of concern; treatment units can be delivered to the site and made operational on very short notice.

Operating Procedures for Adsorption

The operating procedures for adsorption differ depending on whether PAC or GAC is used.

Application of PAC

PAC is used primarily to help control those organic compounds responsible for tastes and odors. It has also been used to remove compounds that will form THMs or other DBPs. In some instances massive doses of PAC have been used for temporary treatment of a water source that has been contaminated by a chemical spill.

In addition to adsorption of organic compounds, the carbon particles will often aid coagulation by providing a center, or nuclei, on which floc will form. This is a benefit where a water source has naturally low turbidity.

PAC can be applied in the treatment plant at almost any point before filtration, but the following considerations should be made in the selection of an application point:

- The contact time between the PAC and the organics is important and depends on the ability of the carbon to remain in suspension. At least 15 minutes of contact time is generally considered advisable.
- The surfaces of the PAC particles lose their capacity to adsorb if coated with coagulants or other water treatment chemicals.
- PAC will adsorb chlorine. If PAC and chlorine are fed at the same time, the chlorine will reduce the effectiveness of the PAC. At the same time, more chlorine than normal must be fed in order to provide the required disinfection.

The above factors indicate that the raw-water intake is the most advantageous point for PAC application if adequate mixing and retention facilities are available. This allows organic compounds to be removed before the application of chlorine. When chlorine reacts with organic compounds, new compounds are often produced that may not be as easily removed by the carbon.

If the raw-water intake is not a possible application point, PAC can be used at other points, but the dosages normally must be higher to account for shorter contact times and the interference by other chemicals.

If PAC can be fed only to the effluent from the sedimentation basins or as the water enters the filters, particular care must be taken in filter operation because finely divided carbon particles can pass through filters. If is does pass through the filter, some of the PAC will settle to the bottom of clearwells, where it can be stirred up and pumped to the distribution system at a time when the pumping rate is unusually high. The system operator will then most certainly receive "black water" complaints from customers. If PAC does pass into the clearwells, the clearwells should be cleaned as soon as possible.

A particularly effective way of applying PAC is to use two or more application points. Some carbon is added to the raw water, and smaller doses are added before filtration to remove any remaining taste-and-odor-causing compounds.

The PAC dosage depends on the types and concentrations of organic compounds that are present. Common dosages range from 2 to 20 mg/L, but they can also be as high as 100 mg/L to handle severe problems.

Application of GAC in Conventional Filters

When GAC is used as a medium in a conventional filter, an eductor like the one shown in Figure 13-12 is typically used to remove the carbon when it has become exhausted. If the GAC has been placed over a layer of sand, it is difficult to remove the GAC without also removing some of the sand.

Except for very large installations, it is generally not cost-effective to regenerate the GAC when it has become exhausted. The old GAC is therefore discarded and replaced with new material. It is usually easiest and least troublesome in terms of dust to replace the new material in the filter as a slurry, as shown in Figure 13-13.

Once the carbon has been placed, the filters should be backwashed to remove trapped air and small particles of carbon, or carbon fines. It will probably be noticeable for several weeks that fines are still being washed out during backwashing.

Contact Time

The time that water is in contact with GAC as it passes through the bed is usually termed the empty bed contact time (EBCT). This is equal to the volume of the bed, which is based on the medium's depth and the dimensions of the filter compartment, divided by the flow rate through the

FIGURE 13-12 Eductor used for filter media removal

FIGURE 13-13 Placement of GAC slurry

Photograph supplied by Activated Carbon Division, Calgon Corp.

bed. Filters with GAC as a medium are normally operated in the same manner as regular rapid sand filters, with a filtration rate of about 2 gpm/ft^2 (1.4 mm/s). This provides an EBCT of about 7.5 to 9 minutes, which is sufficient to remove many organic compounds.

Backwashing

A filter bed containing GAC is backwashed using the same general procedures as for conventional filters. The backwash rate depends to some extent on the GAC particle diameter and density, the water temperature, and the amount of bed expansion desired. A 50 percent bed expansion is recommended for GAC. Operational curves are available from GAC manufacturers to help select the proper backwash rate. It is also recommended that surface washers be used to ensure adequate cleaning of the carbon layer and to prevent the formation of mudballs within the bed.

It is important to have precise control of the backwash rate in order to achieve good cleaning of the bed without losing too much of the filter medium over the wash-water troughs. The particle density of GAC is about 1.4 g/cm^3, which is much less than that of sand, which is about 2.65 g/cm^3. An appropriate adjustment must therefore be made in the backwash rate to prevent loss of the medium if GAC is used to replace sand.

Carbon Loss

The major causes of carbon loss are entrapment within the bed and excessive backwash rates. Prior to backwashing, the filter should be drained to a minimal distance beneath the backwash troughs. Surface washers should be turned on for about 2 minutes to break up any mat that may have formed on the surface and to clean floc particles from the carbon. If polymers are used as a coagulant or filter aid, it is particularly important to provide thorough surface washing because the floc will be very sticky and hard to remove. The agitation created by the surface washers also helps to remove any air entrapped within the bed.

Backwashing should begin at 2–3 gpm/ft^2 (1.4–2.0 mm/s) to allow any remaining air to be released slowly, so that it does not cause carbon loss into the wash-water troughs. The rate should then be increased to 5–6 gpm/ft^2 (3.4–4 mm/s) for about 1 minute, and then gradually increased to the rate necessary to achieve the desired bed expansion.

Backwashing should continue until the flow into the wash-water troughs is clear. The time required for backwashing and the lengths of filter runs depend primarily on the amount of floc and other suspended matter in the water passing through the filters.

Some loss of carbon will occur during backwashing. It is important to keep track of the bed depth; this way, more GAC can be added when the depth becomes insufficient. One method of monitoring bed depth is to take routine measurements of the distance from the top of the wash-water troughs to the top of the carbon bed. Another method is to place a marker, such as a stainless steel plate, on the filter walls to indicate the proper carbon level. It is not unusual for about 1 in. (25 mm) of carbon to be lost from filters over a period of a year.

Carbon Life

The life of a GAC bed depends primarily on the concentrations and types of organic compounds being removed. For typical taste-and-odor-causing compounds, bed life may be as long as three years. However, the bed life for removing organic compounds such as chloroforms can be as short as one month. Generally, as the influent concentration increases, the bed life decreases.

One major advantage of GAC beds is that they are not all used up at once. Breakthrough of the contaminants takes place over a long period of time, rather than suddenly. Figure 13-14 indicates a typical breakthrough pattern for a 30-in. (760-mm) GAC bed. This pattern is important because it

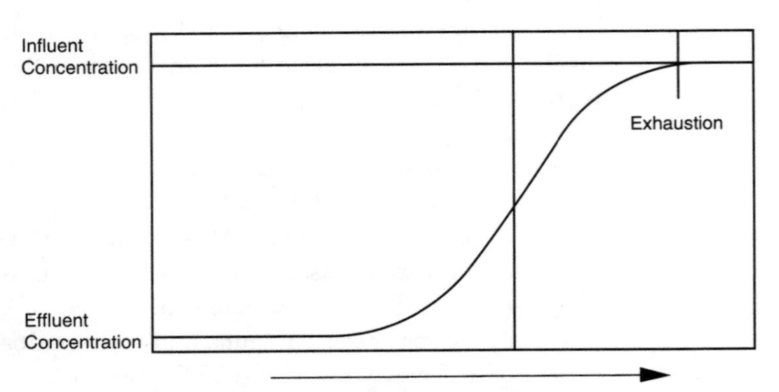

FIGURE 13-14 Typical breakthrough pattern for a GAC bed

allows the replacement of the GAC to be phased, so that only one filter needs to be out of service at a given time.

Application of GAC in Contactors

When GAC is used in closed tanks, the containers are known as contactors or adsorbers. An advantage of contactors is that they can be manufactured to the size necessary to provide the desired EBCT. For example, if the contact time of GAC in a filter bed is not sufficient, there is little that can be done about it. However, if a contactor is used, the size and flow rate can be designed to provide any contact time desired. Table 13-2 shows the effect of these factors in a hypothetical water treatment situation. The design of contactors for a water system is generally based on pilot tests of the water to be treated.

In most installations, two or more contactors are operated in parallel. The units are started up at different times so that the beds will not all be exhausted at the same time. The beds are usually backwashed periodically to remove suspended matter and carbon fines. Contactors are backwashed when there is a noticeable pressure drop between the inlet and outlet sides of the unit.

Samples of the effluent from each contactor must be collected and analyzed periodically for the presence of the organic chemicals that are to be removed. When monitoring indicates that the chemicals are passing through the contactor, the GAC is exhausted and must be replaced.

TABLE 13-2 Example of GAC adsorption characteristics for various bed depths

Bed Depth,			Average Influent Chloroform Concentration,	Time to Bed Exhaustion,*
ft	*(m)*	EBCT, *min*	*mg/L*	*weeks*
2.5	(0.8)	6.2	67	3.4
5.0	(1.5)	12	67	7.0
7.5	(2.3)	19	67	10.9
10.0	(3.0)	25	67	14.0

*Bed is exhausted when effluent concentration is equal to influent concentration.

Operating Problems

Powdered Activated Carbon

The most common operating problem with PAC is handling it. It is a fine powder, so dust can be a major problem. If it is used continuously or if large quantities are used periodically, a slurry system should be considered.

PAC passing through the filters and entering the distribution system can result in complaints of "black water." Carbon passing through the filters is usually caused by inadequate coagulation–sedimentation, resulting in carbon carryover to the filters. It can also be caused by adding heavy doses of PAC just before the filters.

An occasional problem is that taste-and-odor problems persist, or even worsen, regardless of the PAC dosage. This problem is usually caused by adding the PAC and chlorine in stages too close to one another. The chlorine reacts with the organic compounds to produce additional compounds that are more difficult to adsorb. One solution to this problem is to add the PAC so that it has at least 15 minutes contact time before any chlorine is added.

Granular Activated Carbon

When GAC is used as a filter medium, it is important that the coagulation, flocculation, and sedimentation processes be operated continuously to maximize the removal of suspended matter going to the filters. The adsorption capacity of the carbon can quickly be lost if the carbon becomes coated with floc, resulting in shorter bed life and higher operating costs.

In addition, filter rates should be carefully controlled so that they do not fluctuate rapidly. Fluctuations can drive previously deposited material through the filter and can form channels through the bed. The channels will reduce the filtering capacity of the bed and greatly reduce the contact time of water passing through the filter medium.

Proper backwashing is essential for effective filtration and adsorption. The rate must be sufficient to achieve a bed expansion of about 50 percent for proper cleaning. If the bed is not adequately cleaned, both filtration and adsorption capacity will be lost, and mudballs will begin to form. This is particularly a problem where polymers are used as coagulants and filter aids.

GAC is lighter than other filter media, so it can easily be washed away during backwashing. If an excessive amount of GAC is being lost, backwash rates may have to be reduced. The water pressure to surface washers may also have to be decreased if it is found that they are contributing to excessive GAC loss. If carbon loss is over 2 in. (50 mm) per year, the backwashing procedures should be reviewed to determine how the loss can be reduced. If the height of the wash-water troughs above the filter beds is not adequate, the troughs may have to be raised to prevent further carbon loss.

A special problem with GAC filters is the rapid growth of bacteria within the bed. The carbon removes and holds organic compounds, and bacteria can feed and thrive on this material. The carbon also adsorbs chlorine, so any chlorine added ahead of the filters to control bacterial growth is ineffective. As a result, the bacterial concentration in the filter effluent may be thousands of times higher than in the filter influent.

Research has indicated that the high bacterial populations in the bed actually enhance organics removal because the bacteria break down complex compounds to simpler products that are more easily adsorbed by the GAC. Some treatment plants backwash with heavily chlorinated water to help control the bacterial levels. In general, it is not good practice to allow high concentrations of bacteria to pass through the filters. The final chlorination must be closely controlled to ensure that the bacteria are destroyed before they enter the distribution system.

If GAC contactors are used after conventional filters, it is essential that only a minimum of suspended material comes through the filters. If the length of time between backwashing a contactor begins to decrease, or if the bed life is much shorter than it should be, floc carryover from the filters may be a problem. The turbidity of filter effluent should be monitored continuously to prevent reducing the life of GAC in the contactors. Bacterial growth in contactors can also be a problem, but it can be handled in the same manner as that for GAC in conventional filters.

Control Tests

Testing of the effluent from carbon adsorption systems to determine if there is a breakthrough of organic chemicals requires sophisticated water quality analyses using a gas chromatograph. In general, only very large water systems have the equipment and expertise to run these analyses themselves. Other systems should contract with a convenient private laboratory that will perform the analyses.

Determining PAC Dosage Rates

The required dosage of PAC can be approximated by modifying the jar test. The stirring apparatus, as well as all glassware, should be cleaned with an unscented detergent and rinsed thoroughly with odor-free water. (The method for producing odor-free water is described in *Standard Methods for the*

Examination of Water and Wastewater. The standard jar test and taste-and-odor testing are described in *Water Quality*, a part of this series.)

One-liter samples of the raw water are then dosed with varying amounts of a well-shaken stock PAC solution — for example, 5, 10, 20, and 40 mL. The stock solution is prepared by adding 1 g of PAC to 1 L of odor-free water. Each milliliter of this solution, when added to a 1-L sample of raw water, is equivalent to a dosage of 1 mg/L.

The four dosed samples, and a fifth sample to which no PAC is added, are stirred for a period that approximates the contact time the PAC will have with the water as it passes through the plant. At the end of that time, each sample is filtered through glass wool or filter paper to remove the PAC. The first 200 mL of each sample through the filter is discarded, and the remainder is subjected to the threshold odor test to arrive at a threshold odor number (TON) for each sample.

The results can then be plotted, as illustrated in Figure 13-15, to obtain an optimal PAC dosage. In the example, the dosage needed to reduce the TON to 3 (which is recommended by the USEPA's secondary drinking water regulations) would be 29 mg/L. Experience has shown that plant-scale PAC application is more efficient than indicated by jar tests. However, a plant should begin with the dosage indicated by the jar-test result, and then gradually reduce it while sampling the plant effluent. To maintain the optimum PAC dosage, the threshold odor test should be conducted by each shift, or at least daily, on the raw and treated water while PAC is being fed.

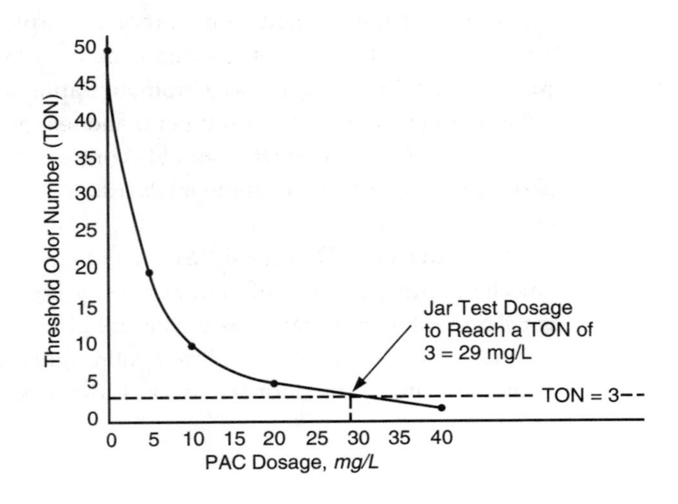

FIGURE 13-15 Example determination of optimal PAC dosage

If PAC is used to remove organic compounds other than those causing tastes and odors, different procedures are needed to determine the proper dosage.

Tests of GAC Used in Filters

GAC is used in a conventional gravity filter for both filtration and adsorption, so most of the control tests are the same as those used for filtration, as described in chapter 6. The condition of the filter bed should be checked frequently, particularly during and after backwashing, to determine if there are problems with cracks, shrinkage, channeling, or mudballs.

The distance between the top of the carbon and the top of the wash-water troughs (or other reference mark) should be measured at least every three months to determine the rate of carbon loss. Samples of backwash water can also be collected and tested to see if they contain excessive amounts of carbon. Excessive loss may indicate a need for changes in the backwashing procedures.

Core samples of the carbon bed should be taken at the time of installation and at least every six months to determine both the amount of bed life remaining and the condition of the medium. The sample should represent the carbon from top to bottom of the bed. GAC manufacturers can provide the necessary information on sampling, testing, and interpreting results.

If the filter is designed to remove taste-and-odor-causing compounds, the threshold odor test should be conducted routinely on the raw and finished water as a check on the effectiveness of the filter. If the filter is intended to remove other organic compounds, special analyses of samples will periodically have to be performed for these compounds. Taste-and-odor testing cannot be used as an indicator of the removal of other organic compounds. In general, the holding capacity of GAC for other organic compounds is much less than it is for taste-and-odor-causing compounds.

Because bacteria can thrive in GAC filters, standard plate count analyses should be performed each day on the filtered water and the water after final chlorination. (The standard plate count procedure is detailed in *Water Quality*, part of this series.) This will help determine the required final chlorination dosage and will indicate whether chlorination of the backwash is needed. The plate count should be kept well below 500 organisms/mL.

Tests of GAC Used in Contactors

Because GAC contactors are typically used for removing organic compounds other than those causing tastes and odors, more sophisticated control tests are usually needed. There is normally a specific compound or a group of compounds that are to be removed, so periodic analyses of the contactor effluent must be made to check on the effectiveness of the treatment and to determine when the carbon is exhausted.

Head loss through the contactors should also be monitored and recorded continuously so that backwashing can be performed at the proper time. Core samples of the carbon bed should be taken at least every three months so that the approximate remaining bed life can be determined. Turbidity of the effluent should also be monitored and recorded continuously to determine if carbon fines or other suspended matter is passing through the GAC bed. Standard plate count analyses of the contactor effluent before and after chlorination will also help determine what chlorine dosages are needed to control bacterial growth.

Safety Precautions

Bags of PAC and GAC should be stored on pallets in a clean, dry place so that air can circulate underneath. The bags should be stacked in single or double rows with access aisles around every stack to allow easy handling and fire inspection. They should never be stored in stacks over 6 ft (2 m) high.

Although activated carbon is not considered explosive, it will burn like charcoal without producing smoke or flame, and it glows with the release of intense heat. A carbon fire is difficult to detect and to extinguish. This is another advantage of using slurry storage rather than dry feed.

The storage area for dry carbon should be fireproof and have self-closing fire doors separating it from other storage areas. Storage bins for dry bulk carbon storage should be fireproof and equipped for fire control. Smoking must be prohibited during the handling and unloading of carbon and in the storage and feeding areas.

Burning carbon should not be doused with a large stream of water because this will only cause carbon particles to fly in all directions and spread the fire. A fine mist of spray from a hose or a chemical foam extinguisher is most effective in controlling a fire.

Activated carbon must never be stored near gasoline, mineral oils, or vegetable oils. When mixed with carbon, these substances will slowly oxidize until the ignition temperature of 600–800°F (316–427°C) is reached.

Stored carbon should also be kept well away from chlorine compounds and potassium permanganate because spontaneous combustion can occur when they are mixed. Carbon is an electrical conductor, so explosion-proof light fixtures and electrical wiring should be used in all carbon storage and feeding areas. As an added precaution, the electrical equipment should be cleaned frequently.

All tanks receiving dry carbon should be vented and provided with dust-control equipment, such as bag-type dust collectors. Even slurry storage tanks should be so equipped. If a PAC dry feed system is used, the hoppers and feeders should be enclosed in a separate room so that dust will be confined.

Oxygen is removed from air in the presence of wet activated carbon. As a result, slurry tanks or other enclosed spaces containing carbon may have seriously reduced oxygen levels. Personnel must be careful when entering these spaces to ensure that adequate oxygen is available. Devices to indicate the amount of oxygen present should be used before anyone enters this type of closed space. Personnel entering a tank or other enclosure should also have attached safety belts and another worker standing by to pull them from danger if necessary.

FIGURE 13-16 Safety clothing to be worn in the handling of carbon

Personnel who unload or handle carbon should be furnished with dust masks, face shields, gauntlets, and aprons (Figure 13-16). In addition, shower facilities should be provided for workers to use if they get unduly dirty from handling carbon.

Record Keeping

Good record keeping can help prevent taste-and-odor-causing compounds from reaching consumers, while preventing waste of PAC. Suggested records include

- the type of odor in the water (such as musty, septic, or rotten-egg odors) and the taste (such as sweet or bitter)
- dates on which PAC was added
- the calculated dosage based on jar tests
- the actual plant dosage that was effective
- the threshold odor number (TON) values for the raw and treated water

This information may give some clue to the source of the problems, how long they may last, how they may be prevented, and when they may occur again.

Records of GAC in filters should include

- hours of filter operation for computing the quantity of water processed
- monthly measurements of the carbon level to monitor GAC loss
- dates when GAC is replaced so that bed life can be monitored
- results of periodic threshold odor tests

Records of GAC contactor operations could include

- quantity of water treated
- times when the contactor is backwashed and the amount of backwash water used
- dates of GAC replacement for computing the bed life
- results of analyses of raw-water and finished-water organics concentrations

Selected Supplementary Readings

Activated Carbon for Water Treatment. Denver, Colo.: American Water Works Association.

Adams, J.Q., and R.M. Clark. 1989. Cost Estimates for GAC Treatment Systems. ***Jour. AWWA,*** 81(1):35.

Adams, J.Q., R.M. Clark, and R.J. Miltner. 1989. Controlling Organics With GAC: A Cost and Performance Analysis. ***Jour. AWWA,*** 81(4):132.

Chen, A.S.C., et al. 1989. Removing Dissolved Organic Compounds With Activated Alumina. ***Jour. AWWA,*** 81(1):53.

Controlling Radionuclides and Other Contaminants in Drinking Water Supplies: A Workbook for Small Systems. 1991. Denver, Colo.: American Water Works Association.

Crittendon, J.C., et al. 1991. Predicting GAC Performance With Rapid Small-Scale Column Tests. ***Jour. AWWA,*** 83(1):77.

Design and Use of Granular Activated Carbon: Practical Aspects. 1989. Denver, Colo.: American Water Works Association Research Foundation and American Water Works Association.

DiGiano, F.A., Kathryn Mallon, William Stringfellow, Nathan Cobb, James Moore, and J.C. Thompson. 1992. ***Microbial Activity on Filter-Adsorbers.*** Denver, Colo.: American Water Works Association Research Foundation and American Water Works Association.

Eaton, A.D., L.S. Clesceri, and A.E. Greenburg, eds. 1995. ***Standard Methods For the Examination of Water and Wastewater.*** 19th ed. American Public Health Association, American Water Works Association and Water Environment Federation.

Graese, S.L., V.L. Snoeyink, and R.G. Lee. ***GAC Filter Adsorbers.*** 1987. Denver, Colo.: American Water Works Association.

Hand, D.W., et al. 1989. Designing Fixed-Bed Adsorbers to Remove Mixtures of Organics. ***Jour. AWWA,*** 81(1):67.

Koffskey, W.E., and B.W. Lykins Jr. 1990. GAC Adsorption and Infrared Reaction: A Case Study. ***Jour. AWWA,*** 82(1):48.

Manual of Water Utility Operations. 8th ed. 1988. Austin, Texas: Texas Water Utilities Association.

McGuire, M.J., et al. 1991. Evaluating GAC for Trihalomethane Control. ***Jour. AWWA,*** 83(1):38.

McGuire, M.J., M.K. Davis, S. Liang, C.H. Tate, E.M. Aita, I.E. Wallace, D.R. Wilkes, J.C. Crittenden, and K. Vaith. 1989. ***Optimization and Economic Evaluation of Granular Activated Carbon for Organic Removal.*** Denver, Colo.: American Water Works Association Research Foundation and American Water Works Association.

McTigue, N.E., and D. Cornwell. 1994. ***The Hazardous Potential of Activated Carbons Use in Water Treatment.*** Denver, Colo.: American Water Works Association Research Foundation and American Water Works Association.

Najm, I.N., V.L. Snoeyink, T.L. Galvin, and Yves Richard. 1991. ***Control of Organic Compounds With Powdered Activated Carbon.*** Denver, Colo.: American Water Works Association Research Foundation and American Water Works Association.

Najm, I.N., et al. 1991. Effect of Initial Concentration of a SOC in Natural Water on Its Adsorption by Activated Carbon. ***Jour. AWWA,*** 83(8):57.

———. 1991. Using Powdered Activated Carbon: A Critical Review. ***Jour. AWWA,*** 83(1):65.

Organics Removal by Granular Activated Carbon. 1989. Denver, Colo.: American Water Works Association.

Oxenford, J.L., and B.W. Lykins Jr. 1991. Conference Summary: Practical Aspects of the Design and Use of GAC. ***Jour. AWWA,*** 83(1):58.

Pirbazari, M., et al. 1992. Evaluating GAC Adsorbers for the Removal of PCBs and Toxaphene. ***Jour. AWWA,*** 84(2):83.

Summers, R.S., L. Cummings, J. DeMarco, D.J. Hartman, D.H. Metz, E.W. Howe, B. MacLeod, and M. Simpson. 1992. ***Standardized Protocol for the Evaluation of GAC.*** Denver, Colo.: American Water Works Association Research Foundation and American Water Works Association.

VOCs and Unregulated Contaminants. 1990. Denver, Colo.: American Water Works Association.

Waer, M.A., et al. 1992. Carbon Regeneration: Dependence on Time and Temperature. ***Jour. AWWA,*** 84(3):82.

Water Quality and Treatment. 4th ed. 1990. New York: McGraw-Hill and American Water Works Association (available from AWWA).

CHAPTER 14

Aeration

Aeration is the process of bringing water and air into contact in order to remove dissolved gases, introduce oxygen to oxidize dissolved metals, and release volatile chemicals. When aeration is used, it is often the first process at a treatment plant. This way, certain constituents can be altered or removed before they can interfere with other treatment processes. In some systems, aeration may be the only treatment other than chlorination.

Process Description

A common example of aeration occurs naturally in a stream as water tumbles over rocks. This turbulence brings the air and water into contact, and the air dissolves into the water. The aeration process is also often referred to as desorption or air-stripping.

How Aeration Removes or Modifies Constituents

The primary uses of aeration in the water industry are

- to remove gases from solution and allow them to escape
- to oxidize metals that are in solution so that they can be removed as precipitates
- to remove volatile organic chemicals

Aeration is generally considered ineffective for the removal of most inorganic substances. In the few instances where it might be a usable technology, usually other treatment methods are more cost-effective or convenient.

The efficiency of the aeration process depends almost entirely on the amount of surface contact that can be achieved between air and water. This contact is controlled primarily by the size of the water drops or air bubbles that provide the contact area. For example, the cubic foot of water shown in Figure 14-1A has 6 ft^2 of surface area. When the same volume is divided into eight equal pieces ½ foot square as shown in Figure 14-1B, the exposed area is increased to 12 ft^2. If division of the cube is continued until each exposed face is 1⁄100 in., the surface area is increased to 7,200 ft^2. The greater the surface area that is created in an aeration process, the greater the oxygen transfer that can be achieved.

Constituents Affected by Aeration

Aeration is commonly used in water treatment for removal of the following substances:

- carbon dioxide
- hydrogen sulfide
- methane
- volatile organic chemicals
- radon
- iron and manganese (oxidation facilitates their removal)
- tastes and odors

It is also done to add dissolved oxygen to water.

Carbon Dioxide

Carbon dioxide (CO_2) is very soluble in water. Up to 1,700 mg/L can be dissolved in water at 68°F (20°C). Deep-well water usually contains less than

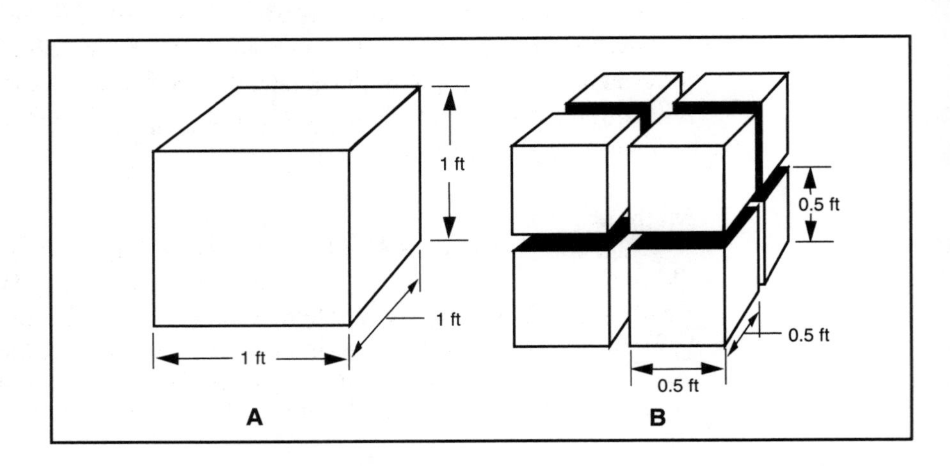

FIGURE 14-1 Increased surface area

50 mg/L, but shallow wells may have 50 to 300 mg/L. Surface water is usually low in CO_2, in the range of 0 to 5 mg/L. The exception is water drawn from a deep location in a lake that may have higher levels as a result of the respiration of microscopic animals and the lack of plant growth near the bottom.

When the CO_2 level in a water supply is above about 5 to 15 mg/L, it can cause the following operating problems:

- The acidity of the water is increased, making the water corrosive.
- It tends to keep iron in solution, making iron removal more difficult.
- If lime softening is used, the CO_2 reacts with the lime, increasing the cost of softening because of the additional lime that must be used.

Almost any type of aerator is able to remove CO_2. At normal temperatures, aeration can generally reduce the level to as little as 4.5 mg/L. Equilibrium between the CO_2 in the air and in the water prevents further removal.

Hydrogen Sulfide

Hydrogen sulfide (H_2S) occurs mainly in groundwater supplies and is the cause of the rotten-egg odor occasionally noticed in some well water. Even at concentrations as low as 0.05 mg/L, H_2S in drinking water will alter the taste of coffee, tea, ice cubes, and many foods. The gas is corrosive to piping, tanks, water heaters, and other plumbing. Another objection by customers is that silverware washed in water containing H_2S may turn black.

In addition, H_2S is a poisonous gas that can be dangerous if released in a treatment plant. Breathing concentrations as low as 0.1 percent by volume in air for less than 30 minutes can be fatal.

H_2S is very unstable and is easily removed from water by almost any method of aeration. However, care must be taken that there is sufficient air movement in the vicinity of the aerator so that the gas is carried away. This will both facilitate the aeration process and prevent the inhalation danger to treatment plant personnel.

Methane

Methane (CH_4) is commonly called "swamp gas." It is often found in groundwater supplies located near natural gas deposits. The gas alone is colorless, odorless, tasteless, and lighter than air, but when mixed with water, methane causes the water to taste like garlic. Methane is highly flammable and explosive when mixed with an appropriate proportion of air. It must be removed from water so that it does not accumulate in customers' homes where it could cause an explosion.

Methane is only slightly soluble in water, so it is easily removed by aeration. Extreme care must be taken to thoroughly dissipate the gas that is removed.

Volatile Organic Chemicals

"Manufactured" chemicals that have been found as contaminants in many groundwater supplies in recent years are called volatile organic chemicals (VOCs). The predominant VOCs found are listed in Table 14-1. VOCs regulated by the US Environmental Protection Agency (USEPA) are classified as known or suspected carcinogens, or as causing other adverse health effects.

When a well is found to be contaminated by VOCs in excess of the maximum contaminant levels (MCLs), and long-term removal must be used, aeration is usually found to be the most cost-effective method. Although VOCs can be removed by aeration, they are significantly less volatile than CO_2 and H_2S, so more sophisticated packed tower aeration equipment is required to achieve acceptable removal.

Radon

The radioactive gas radon is colorless and odorless and is present in unacceptable levels in many groundwater sources. Prolonged inhalation of concentrations of radon is considered a cause of lung cancer. Recent studies

TABLE 14-1 Volatile organic chemicals removable by air stripping

Benzene	Ethylene dibromide (EDB)
Carbon tetrachloride	Hexachlorocyclopentadiene
Di(2-ethylhexyl) adipate	Monochlorobenzene
Dibromochloropropane (DBCP)	Styrene
p-Dichlorobenzene	Tetrachloroethylene
o-Dichlorobenzene	Toluene
1,2-Dichloroethane	1,2,4-Trichlorobenzene
1,1-Dichloroethylene	1,1,1-Trichloroethane
cis-1,2-Dichloroethylene	1,1,2-Trichloroethane
trans-1,2-Dichloroethylene	Trichloroethylene
Dichloromethane (methylene chloride)	Vinyl chloride
1,2-Dichloropropane	Xylenes (total)
Ethylbenzene	

have indicated that radon is liberated from water in customers' homes from showers, washing machines, and other plumbing fixtures, and that the accumulation of radon in closed buildings presents an inhalation danger. USEPA regulations establishing limits on the allowable level of radon in drinking water will require numerous groundwater systems to install treatment for radon removal.

Radon is easily removed by aeration, so almost any method will be acceptable. Provisions must be made to dissipate the removed gas thoroughly so as not to present an inhalation danger to treatment plant personnel.

Iron and Manganese

Both iron and manganese are found in the dissolved form in many groundwater sources. They can also occasionally be a problem in surface water that is drawn from a stratified reservoir. The method generally used for removal of iron and manganese is to oxidize them to form a precipitate. This precipitate can then be removed by sedimentation and filtration. The processes are described in chapter 10.

Manganese is not oxidized very well by aeration, but iron generally responds very well to just about any type of aeration process.

Tastes and Odors

Volatile materials that cause tastes and odors are easily oxidized and can generally be removed by aeration. This includes a large percentage of the materials that cause tastes and odors in groundwater. Some materials, such as oils produced by some algae and some industrial chemicals, cause tastes and odors that cannot be removed adequately by aeration. These materials may respond to a chemical oxidant or may have to be removed by granular activated carbon, as detailed in chapters 7 and 13.

Dissolved Oxygen

Dissolved oxygen (DO) is introduced into water by the process of aeration. A certain amount of DO in drinking water is beneficial in that it increases palatability by removing the "flat" taste. However, too much DO can cause the water to be corrosive.

Aeration of water may either add or release DO. For example, water from the bottom of a lake is usually low in DO, so aeration will increase the level. Conversely, water from a source with a large algae concentration is often supersaturated with DO because of the oxygen given off by the algae, so aeration will usually reduce the DO level.

The amount of oxygen that can remain dissolved in water depends on the water's temperature. The colder the water, the higher the possible concentration of DO. The saturation levels for DO in water at various temperatures are given in Table 14-2.

TABLE 14-2 Oxygen saturation (or equilibrium) levels in water

Temperature, °C	Saturation Concentration, *mg/L*	Temperature, °C	Saturation Concentration, *mg/L*
0	14.621	26	8.113
1	14.216	27	7.968
2	13.829	28	7.827
3	13.460	29	7.691
4	13.107	30	7.559
5	12.770	31	7.430
6	12.447	32	7.305
7	12.139	33	7.183
8	11.843	34	7.065
9	11.559	35	6.950
10	11.288	36	6.837
11	11.027	37	6.727
12	10.777	38	6.620
13	10.537	39	6.515
14	10.306	40	6.412
15	10.084	41	6.312
16	9.870	42	6.213
17	9.665	43	6.116
18	9.467	44	6.021
19	9.276	45	5.927
20	9.092	46	5.835
21	8.915	47	5.744
22	8.743	48	5.654
23	8.578	49	5.565
24	8.418	50	5.477
25	8.263		

Source: Adapted from *Standard Methods for the Examination of Water and Wastewater*. 15th ed. 1981.

NOTE: Values given for atmospheric pressure (101.3 kPa).

Types of Aerators

The general categories of aerators are based on the two main aeration methods. Water-into-air aerators are designed to produce small drops of water that fall through air. Air-into-water aerators create small bubbles of air that rise through the water being aerated. Both categories of aerators are designed to create extensive contact area between the air and the water.

Within the two main categories, there is a wide variety of equipment. The more common types of aerators are listed and discussed below.

Water-into-air types:

- cascade
- cone
- slat and coke tray
- draft
- spray
- packed towers

Air-into-water types:

- diffuser
- draft tube

Combination types:

- mechanical
- pressure

Water-Into-Air Aerators

Cascade Aerators

A cascade aerator is a series of steps that may be designed like a stairway or stacked metal rings, as shown in Figures 14-2 and 14-3. In all cascade aerators, aeration occurs in the splash areas. Splash areas on the inclined cascade are created by placing riffle plates (blocks) across the incline. Cascade aerators can be used to oxidize iron and to partially reduce dissolved gases.

Cone Aerators

Cone aerators are similar to ring-type cascade aerators. Air portals draw air into stacked pans, mixing that air with the falling water. Water enters the top pan through a vertical center feed pipe. The water fills the top pan and begins cascading downward to the lower pans through specially designed,

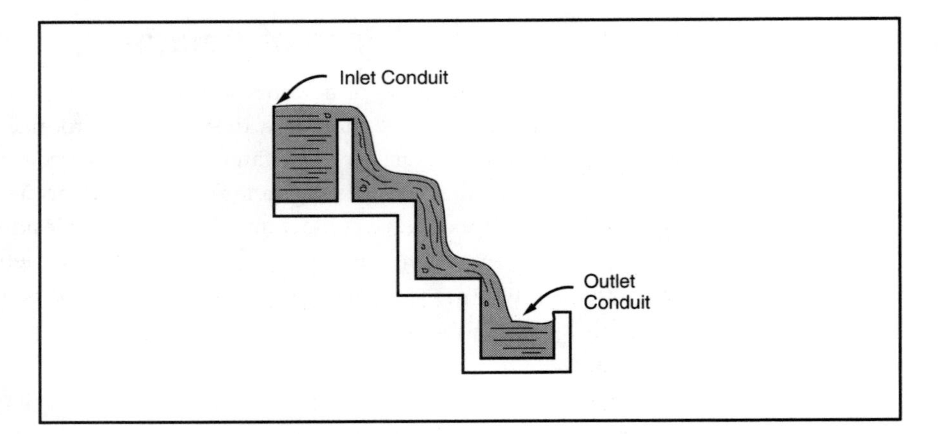

FIGURE 14-2 Stairway-type cascade aerator

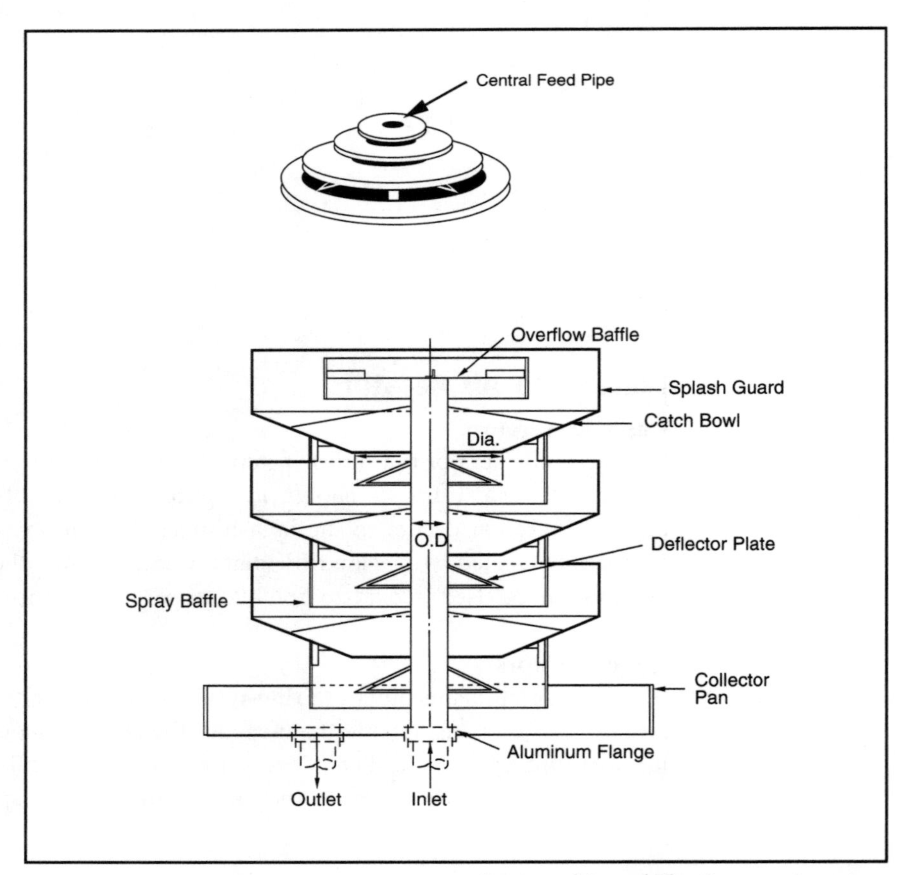

FIGURE 14-3 Ring-type cascade aerator

Courtesy of General Filter Company, Ames, Iowa

cone-shaped nozzles in the bottom of each pan, as shown in Figures 14-4 and 14-5. Cone aerators are used primarily to oxidize iron, although they are also used to partially reduce dissolved gases.

Slat-and-Coke-Tray Aerators

As shown in Figure 14-6, slat-and-coke-tray aerators usually consist of three to five stacked trays that have spaced wooden slats, usually made of redwood or cypress. Early models of this type of aerator were usually filled with about 6 in. (150 mm) of fist-sized pieces of coke. The media used in more recent years may also be rock, ceramic balls, limestone, or other materials.

The aerator is usually constructed to have sloping sides, called splash aprons, which are used to protect the splash from wind loss and freezing. Water is introduced into the top tray (the distributing tray) and moves down through the successive trays, splashing and taking in oxygen each time it hits a layer of media. This type of aerator is commonly used to oxidize iron and, to a limited extent, to lower the concentration of dissolved gases.

FIGURE 14-4
Cone aerator

Courtesy of Infilco Degremont, Inc., Richmond, Va.

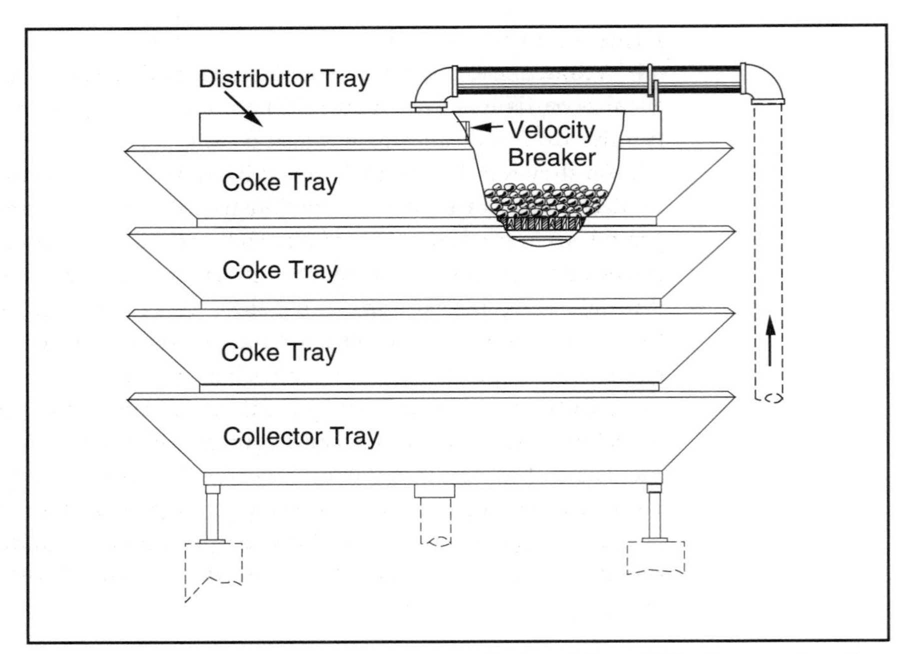

FIGURE 14-5 Schematic of a cone aerator

Courtesy of General Filter Company, Ames, Iowa

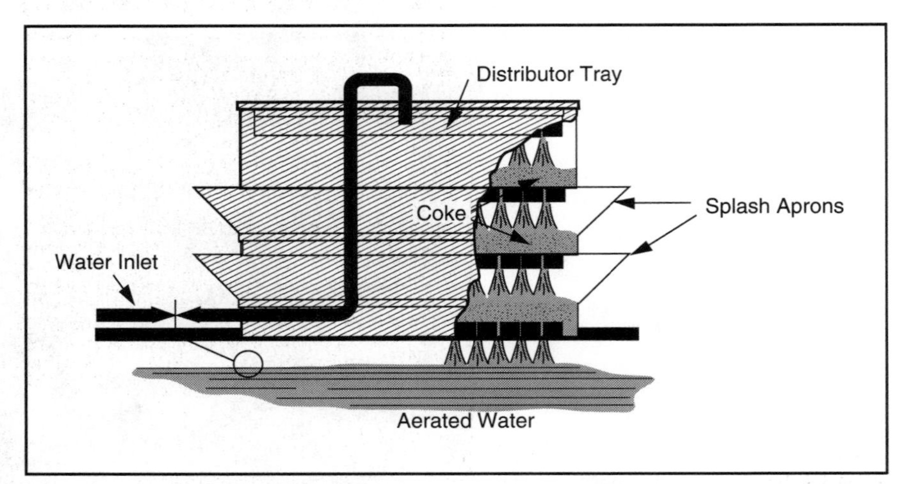

FIGURE 14-6 Slat-and-coke-tray aerator

Courtesy of Power Magazine, McGraw-Hill Inc.

Draft Aerators

A draft aerator is similar to those already discussed except that it also has an airflow created by a blower. There are two types of draft aerators: the positive draft type and the induced draft type.

As shown in Figure 14-7, a positive draft aerator is composed of a tower of tiered slats (wooden in the illustration) and an external blower that provides a continuous flow of air. The water is introduced at the top of the tower, and as it flows and splashes down through the slats, it is subjected to the high-velocity airstream from the blower. The impact caused by the free-falling droplets of water hitting the slats aids in releasing dissolved gases. The high-velocity airflow rushes past the water droplets, carrying away CO_2, methane, or hydrogen sulfide, and constantly renews the supply of DO needed to oxidize iron and manganese.

The induced draft aerator, shown in Figure 14-8, differs in that it has a top-mounted blower, which pulls an upward flow of air from vents located near the bottom of the tower. Draft aerators are much more efficient than the previously discussed types for dissolved gas removal and for oxidizing iron and manganese.

FIGURE 14-7 Positive draft aerator

Courtesy of General Filter Company, Ames, Iowa

FIGURE 14-8 Induced draft aerator

Courtesy of General Filter Company, Ames, Iowa

Spray Aerators

A spray aerator consists of one or more spray nozzles connected to a pipe manifold. Moving through the manifold under high pressure, the water leaves each nozzle in a fine spray and falls through the surrounding air, creating a fountain effect. When relatively large high-pressure nozzles are used, the resulting fountain effect can be quite attractive. This type of aeration device is sometimes located in a decorative setting at the entrance to the water treatment plant.

Spray towers (Figure 14-9) are also used for aeration because the structure protects the spray from windblown losses and reduces freezing problems. Spray aerators are also sometimes combined with cascade and draft aerators, as shown in Figure 14-10, to capture the best features of each, depending on the application.

In general, spray aeration is successful in oxidizing iron or manganese and is very successful in increasing the DO level of water.

Packed Towers

The use of packed towers (or air-strippers, as they are commonly called) for aeration of drinking water is a relatively new development. They have

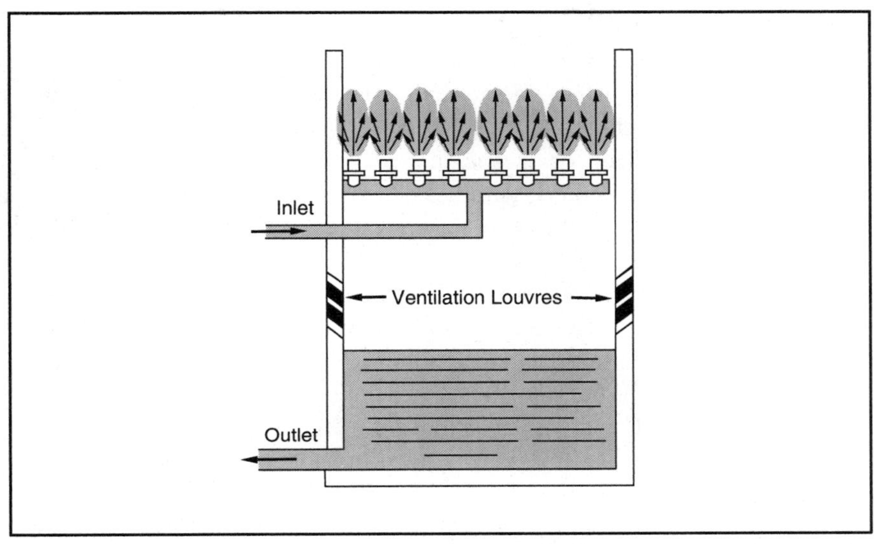

FIGURE 14-9
Spray tower

Courtesy of Infilco Degremont, Inc., Richmond, Va.

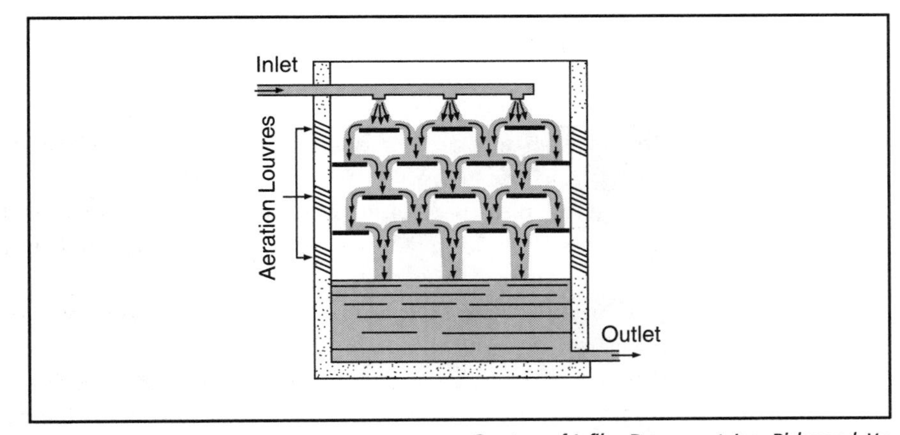

FIGURE 14-10
Spray aerator combined with cascade and draft aerators

Courtesy of Infilco Degremont, Inc., Richmond, Va.

been designed principally for removal of less volatile compounds, such as VOCs, from contaminated water.

A typical packed tower consists of a cylindrical tank containing a packing material. Water is usually distributed over the packing at the top, and air is forced in at the bottom. The very large surface area of the packing material provides for considerably more liquid–gas transfer compared with other aeration methods.

The packing material is of two general types. The type commonly referred to as *dumped packing* consists of shaped pieces of ceramic, stainless steel, or plastic that are randomly dumped into the tower. Figure 14-11 illustrates examples of commercially available packing pieces. Plastic packing is normally used in water treatment air-strippers. Fixed packing is also available as prefabricated sheets that are placed in the tower. The manufacturer can provide the operational transfer efficiency rates for these sheets.

The principal parts of a packed tower with *loose packing* are shown in Figures 14-12 and 14-13. The redistributors placed at intervals in the column direct water that is flowing along the column wall back toward the center. The demister at the top of the column removes moisture from the air as it leaves to prevent objectionable clouds of moisture from coming off the column.

Airflow is provided by a centrifugal blower driven by an electric motor. Small towers designed to remove relatively volatile compounds may require only a 5-hp (3,700-W) or smaller motor. Other columns require much larger and more powerful blowers. Formulas are available for computing the air-to-water ratio required for removing various volatile chemicals. Care

FIGURE 14-11 Typical plastic packing pieces

Source: US Environmental Protection Agency

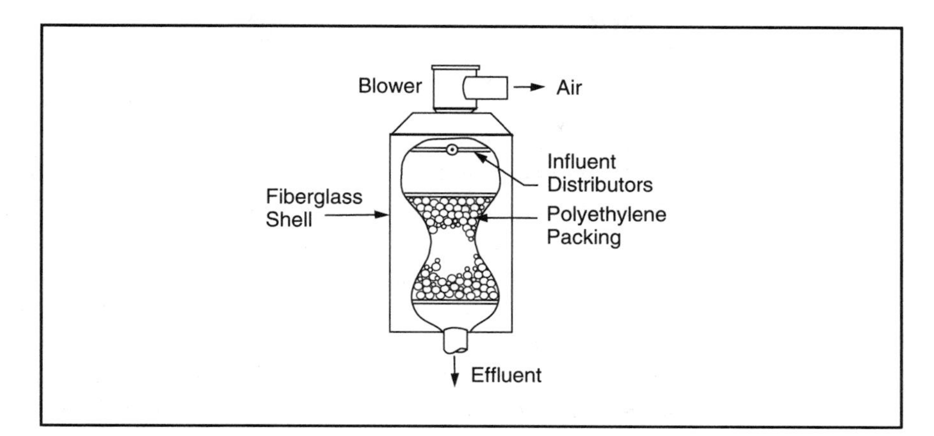

FIGURE 14-12 Cutaway view of interior of a packed tower

FIGURE 14-13 View of an installation of a packed tower

Source: US Environmental Protection Agency

must be taken in protecting the blower intake to prevent contaminants from being blown into the tower.

The relative volatility of a substance is expressed in terms of a factor known as Henry's constant. In general, compounds with a factor over 100 can feasibly be removed by a packed tower. For values from 10 to 100, removal may or may not be feasible, depending on what other contaminants are in the water and other factors. In most cases, the state will require that pilot plant tests be run on the water if there is any question of how well a contaminant would be removed.

Among the factors that must be considered in the design of a packed tower unit are

- the height and diameter of the unit
- the air-to-water ratio required
- the packing depth
- the surface loading rate

The minimum air-to-water ratio at peak flow in a packed tower should normally be at least 25:1, and the maximum should be no more than 80:1. The design of a tower should also make allowance for potential fouling of the packing by calcium carbonate, precipitated iron, and bacterial growth. If any of these are expected to be a problem, pretreatment may have to be provided. If the problem is not expected to be too serious, provisions can be made for periodic chemical backwashes to clean the packing.

Air-Into-Water Aerators

Diffuser Aerators

A typical diffuser aeration system (Figure 14-14) consists of an aeration basin or tank constructed of steel or concrete. The basin is equipped with compressed air piping, manifolds, and diffusers. The piping is usually steel

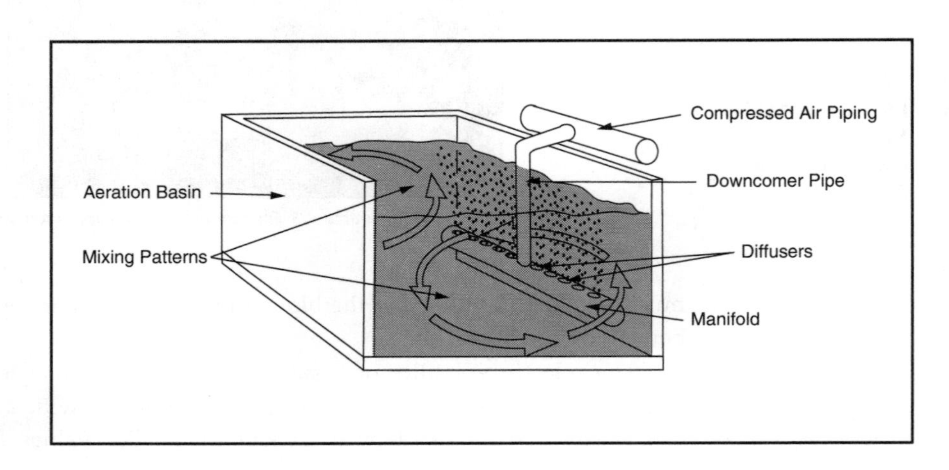

FIGURE 14-14 Diffuser aeration system

or plastic, and the individual diffusers may be plastic or metallic devices, porous ceramic plates, or simply holes drilled into the manifold pipe.

The air diffuser releases tiny bubbles of compressed air into the water, usually near the bottom of the aeration basin. The bubbles rise slowly but turbulently through the water, setting up a rolling-type mixing pattern. At the same time, each air bubble gives up some oxygen to the surrounding water. This form of aeration is used primarily to increase the DO content in order to prevent tastes and odors.

The essential piece of equipment in every diffused air system is the air compressor, or blower. There are two types of blowers: centrifugal blowers and positive-displacement blowers. A typical unit of the more common type, the centrifugal blower, is shown in Figure 14-15; the cutaway view in Figure 14-16 identifies the major component parts.

Draft-Tube Aerators

The draft-tube aerator is a submersible pump equipped with a draft tube (air intake pipe), as illustrated in Figure 14-17. A partial vacuum is created at the eye of the spinning turbine impeller, causing air to enter through the draft tube and water to enter through the water intake. The air and water are

FIGURE 14-15
Centrifugal blower

Courtesy of Hoffman Air & Filtration Systems, Syracuse, N.Y.

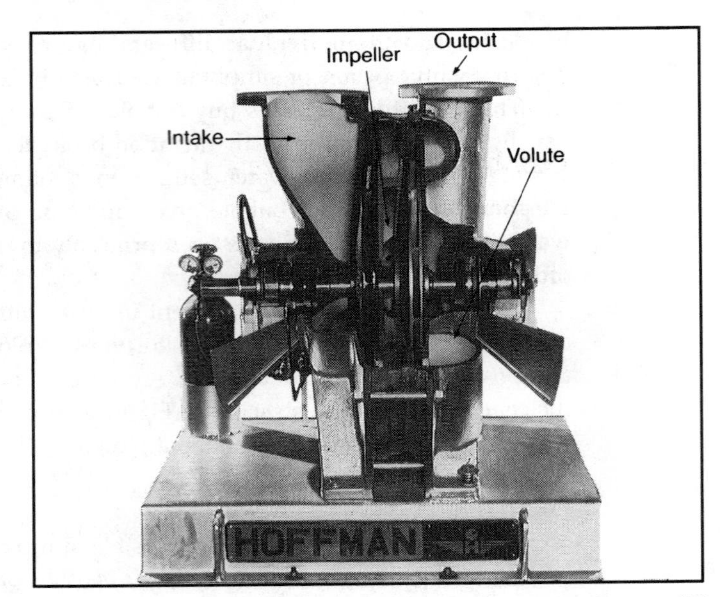

Courtesy of Hoffman Air & Filtration Systems, Syracuse, N.Y.

FIGURE 14-16
Cutaway view of a centrifugal blower

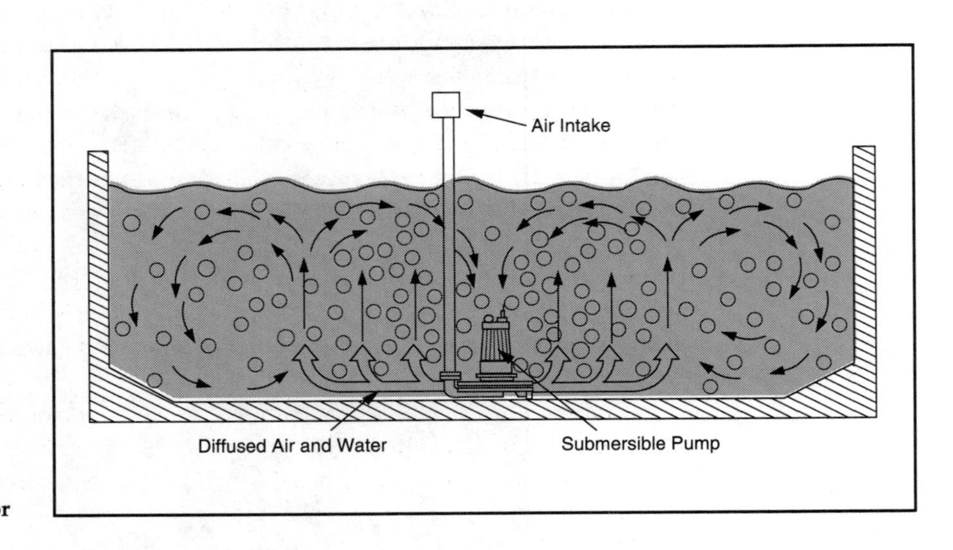

FIGURE 14-17
Draft-tube aerator

mixed by the turbine impeller and then discharged into the aeration basin. The draft-tube aerator is a convenient, low-cost method for adding aeration to an existing basin or tank.

Combination Aerators

Mechanical Aerators

A mechanical aerator consists of a propeller-like mixing blade mounted on the end of a vertical shaft driven by a motor. By a rapid rotation of the mixing blade in the water, air and water are violently mixed.

Figure 14-18 is a side-by-side comparison of the following four types of mechanical aerators:

- surface aerators (water-into-air type)
- submerged aerators (air-into-water type)
- combination mechanical aerators (combination type)
- draft-tube surface aerators (water-into-air type)

Surface aerators. Surface aerators draw water into the blade of the aerator and throw the water into the air in tiny droplets, so that the water can pick up oxygen. The mixing pattern of a typical surface aerator is shown in Figure 14-19. Violent mixing is necessary for efficient oxygen transfer and release of unwanted gases, tastes, and odors.

Submerged aerators. Submerged aerators usually consist of two components: a submerged air diffuser (called a sparger) and a submerged blade that mixes the air into the water. Figure 14-20 shows the mixing pattern of a submerged turbine aerator with air entering the water just below the propeller blade. Note that the mixing pattern is opposite to the surface aerator pattern. Figure 14-21 shows more clearly how the air is introduced by the sparger and how the submerged turbine distributes the air.

A submerged turbine aerator produces relatively calm water at the surface compared with surface aerators. Because of this relative calm, submerged aerators are best used to increase DO levels, rather than surface aerators, whose greater turbulence removes unwanted gases and incidentally acts as a very effective mixer.

Combination mechanical aerators. Combination mechanical aerators, like the one shown in Figure 14-18C, offer the features of both the surface and submerged types. They can be used to oxidize iron and manganese and to remove unwanted gases, tastes, and odors.

Draft-tube aerators. Draft-tube aerators, like the one shown in Figure 14-18D, are used to ensure better mixing when surface aerators are installed

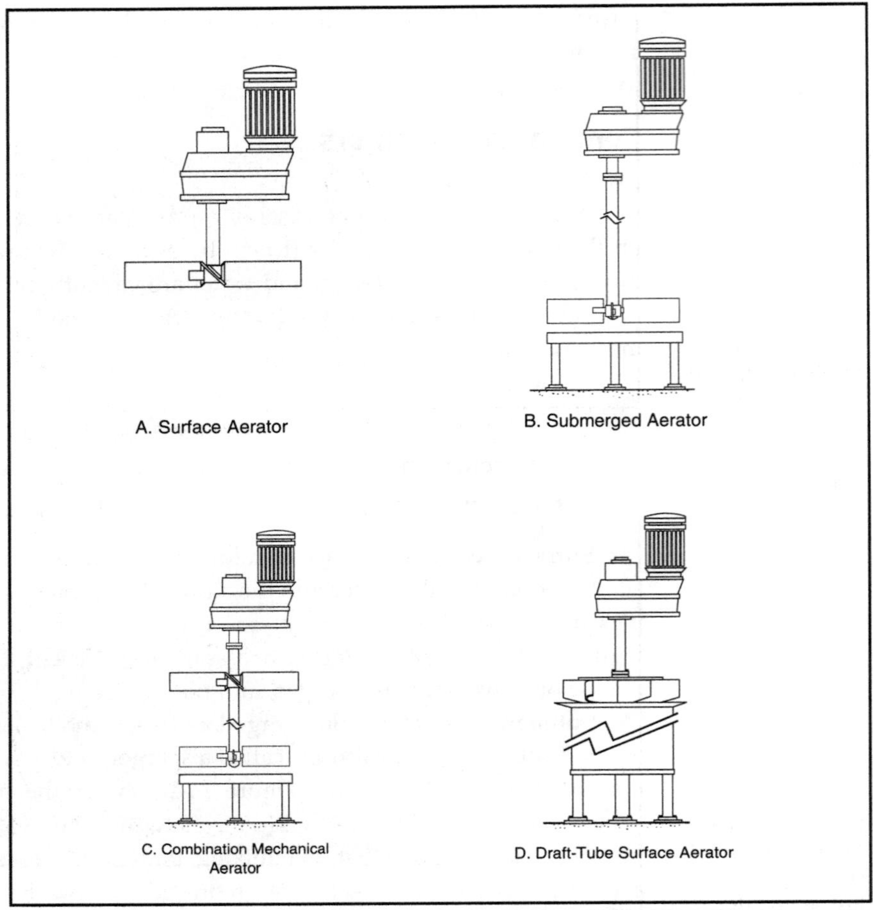

FIGURE 14-18 Four types of mechanical aerators

Courtesy of Philadelphia Mixers Corporation

in deep aeration basins. The draft tube is open at both ends. When the surface aerator is spinning, water is drawn from the very bottom of the basin into the bottom of the draft tube. The water rises up the tube into the impeller and is thrown out onto the water surface. In this way, the draft tube improves the bottom-to-top mixing and turnover.

Pressure Aerators

There are two basic types of pressure aerators. The type diagrammed in Figure 14-22 consists of a closed tank continuously supplied with air under pressure. The water to be treated is sprayed into the high-pressure air,

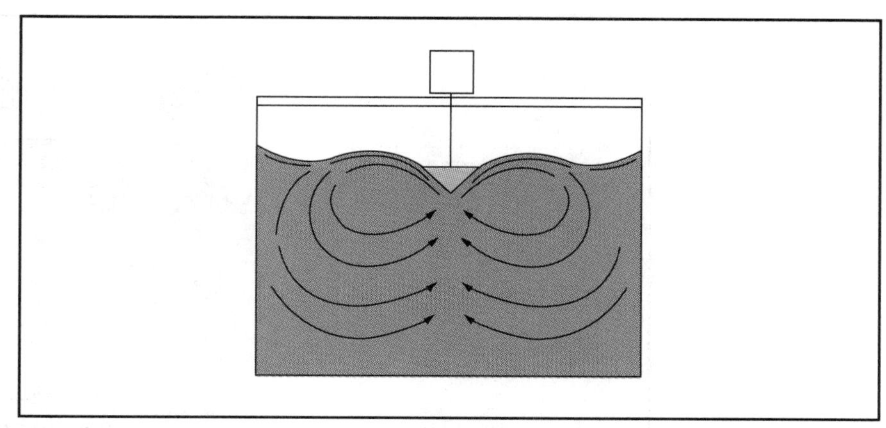

FIGURE 14-19
Mixing pattern of surface aerator

Courtesy of Eimco Process Equipment, Salt Lake City, Utah

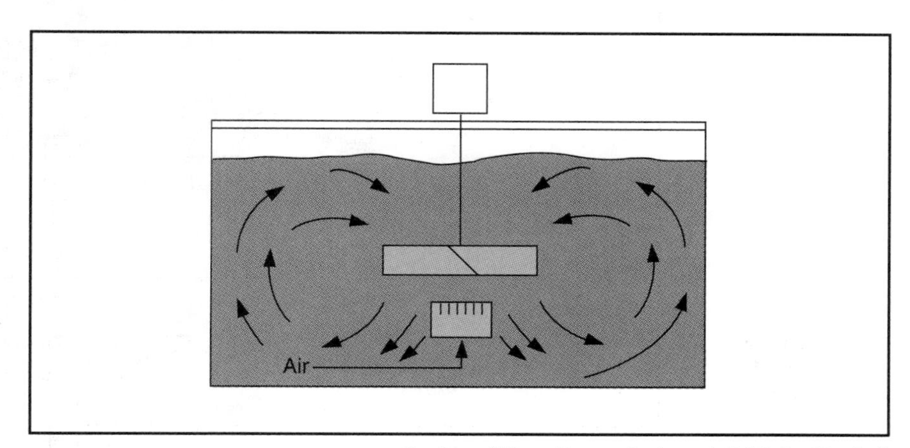

FIGURE 14-20
Mixing pattern of submerged turbine aerator

Courtesy of Eimco Process Equipment, Salt Lake City, Utah

allowing the water to pick up DO quickly. Aerated water leaves through the bottom of the tank and moves on to further treatment. Aerators of this type are used primarily to oxidize iron and manganese for later removal by filtration.

The second type of pressure aerator is diagrammed in Figure 14-23. This type has no pressure vessel; instead, air is diffused directly into a pressurized pipeline. The diffuser inside the special aeration pipe section distributes fine air bubbles into the flowing water. As in any pressure aerator, the higher the pressure, the more oxygen that will dissolve in water. The more oxygen there

FIGURE 14-21 Functioning submerged turbine aerator

Courtesy of Philadelphia Mixers Corporation

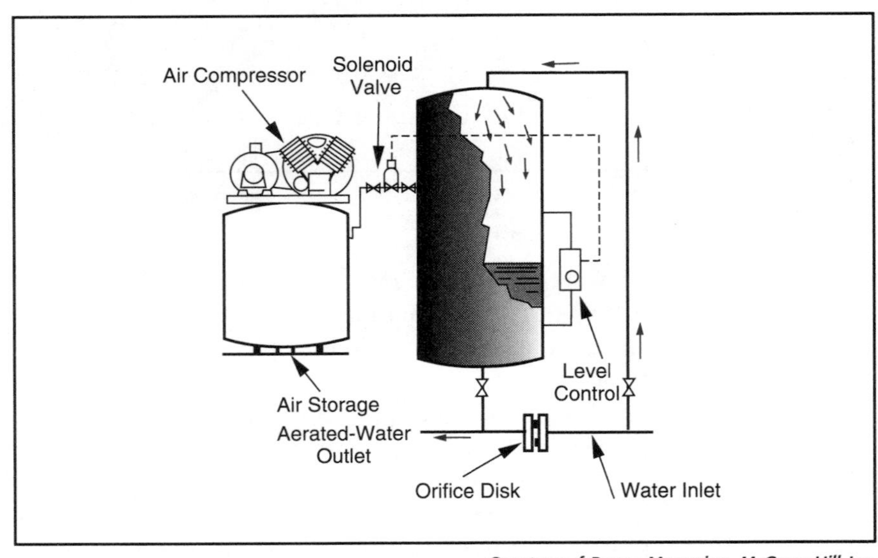

FIGURE 14-22 Pressure aerator with pressure vessel

Courtesy of Power Magazine, McGraw-Hill Inc.

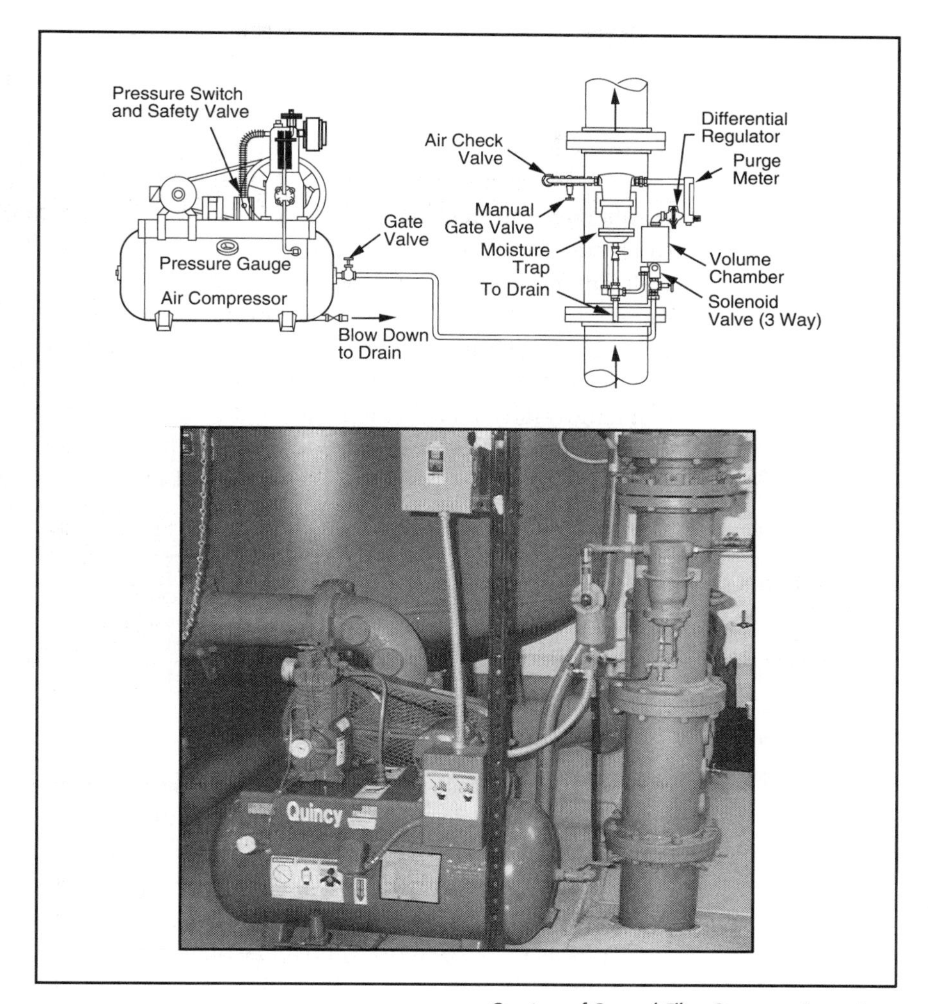

FIGURE 14-23 Pressure aerator with air diffused directly into pressure pipeline

Courtesy of General Filter Company, Ames, Iowa

is in solution, the quicker and more complete the oxidation of the iron and manganese will be.

Regulations

Regulations concerning aerators deal mostly with the removal of VOCs from contaminated groundwater and the removal of radon. In most cases, packed tower aerators will remove VOCs to an acceptable level and are generally acceptable to state authorities for potable water treatment. In a few instances, local residents opposed the proposed installation of a packed tower in a residential neighborhood because of noise, the visual impact of the relatively large tower, and/or the perceived danger from inhaling the off-gas.

Policies concerning the discharge air vary with state desires and regulations. In some areas, the towers must be provided with extra-tall discharge stacks to distribute the off-gas higher into the atmosphere. In other installations, an air scrubber using granular activated carbon has been required on the tower discharge. In any event, a permit for the aeration equipment will probably be required under the Clean Air Act.

New federal regulations restricting the concentration of radon in drinking water will affect a great number of groundwater systems. Radon is easily removed from water by aeration, so most systems will probably find that this is the most cost-effective method of meeting the requirements.

Control Tests

The three basic control tests involved in operating the aeration process are

- dissolved oxygen
- pH
- temperature

A description of each test can be found in *Water Quality*, part of this series.

The DO test is used to monitor the amount of oxygen being dissolved in the water. Knowing the needed DO concentration and monitoring the amount of DO present will prevent over- or underaeration.

The pH of the water can be used as an indicator of CO_2 removal; pH increases as CO_2 is removed. It can also be used to monitor the effective pH range for H_2S, iron, and manganese removal. The best pH range for H_2S scrubbing is 6 or less. Iron and manganese are best treated in a pH range of 8 to 9.

The amount of oxygen that will dissolve in water (the saturation concentration) varies depending on water temperature; the amount increases as the water temperature drops. As water temperature drops, the operator must adjust the aeration process to maintain the correct level of DO.

Additional control tests for the following are used to measure how effective the aeration process is in removing troublesome constituents:

- iron
- carbon dioxide
- manganese
- tastes and odors

Test frequency depends on both the variability of raw-water quality and the characteristic measured. For example, well water tends to be very constant in quality and very slow to change, so daily testing is not usually necessary.

Surface water quality is subject to more frequent change. Temperature, DO, pH, CO_2, and tastes and odors can vary daily. Therefore, sampling and testing frequencies should be set up so that all significant changes can be detected and monitored.

In addition to these specific control tests, it is advisable to make frequent visual inspections of the process. Clogged diffusers, for example, are a common problem that is most easily identified visually.

If aeration equipment is being used to remove VOCs or radon, periodic analyses of the finished water will be necessary to ensure that the process is achieving the required degree of removal. These tests are quite complicated, relatively expensive, and must be performed by a qualified laboratory. State authorities will require a minimum frequency of sampling to ensure that state requirements are being met, but more frequent sampling may be necessary for operational control.

Samples must be collected very carefully. Special containers must be used, and the laboratory will specify procedures to follow. The samples must be analyzed soon after collection, so they must be shipped by a method that will ensure delivery to the laboratory within the required holding time. If a certified laboratory is within a reasonable distance, it is often best to deliver the samples in person.

Operating Problems

The following are some common operating problems associated with aeration:

- corrosion
- floating floc in clarifiers
- false clogging of filters (air binding)

- hydrogen sulfide removal
- algae
- clogged diffusers
- energy consumption

Corrosion

Aeration easily increases the DO content of water, and excessive levels in a water distribution system can cause corrosion. Corrosion can occur whenever water and oxygen come into contact with metallic surfaces. Generally, the higher the DO concentration, the more rapid the corrosion.

The solution to the problem of excessive DO concentration is simply not to overaerate. Unfortunately, there is no definite rule as to what constitutes overaeration. The amount required will vary from plant to plant, and may vary seasonally as water quality changes. The correct amount must be determined from experience. (See Table 14-2 for the saturation levels for DO in water at various temperatures.)

Two things can be done to prevent or retard corrosion. First, adequate protective coatings can be applied and maintained on all exposed metallic surfaces. The application and reapplication of paint and other protective coatings are a vital part of any preventive maintenance program. Second, the aeration process can be operated so that it provides an adequate but not excessive level of DO. As a general rule, DO concentrations of 2 to 4 mg/L should be acceptable.

Floating Floc in Clarifiers

If clarification and filtration are performed after aeration, too much aeration may cause sedimentation problems. Small bubbles of excess air can come out of solution and attach to the particles of floc in the clarifier, causing the particles to float rather than settle. This destroys the efficiency of the sedimentation process and creates an added load on the filters.

False Clogging of Filters

In the same way that bubbles of excess air can attach to floc particles, they can also attach to the media in a filter. This will occur if the water warms as it passes through the filter, releasing the air that was in solution. It can continue until the spaces between the media particles begin to fill with air bubbles, a process called air binding. This causes the filter to behave as if it were clogged and in need of backwashing. Serious disruption of the filter media can also occur as the air bubbles burst within the bed.

Hydrogen Sulfide Removal

Hydrogen sulfide is most efficiently removed by the physical scrubbing action of aeration, not by oxidation. Which of the two removal processes actually takes place depends on the pH of the water. At a pH of 6 or less, hydrogen sulfide exists primarily as the gas H_2S. In this form, it is easily scrubbed away by aeration. However, at pH values of 8 or above, the H_2S gas ionizes and the ionized form (HS^- and HS^{-2}) cannot be removed by aeration. In fact, the oxygen provided by aeration reacts with the ionized H_2S to release the element sulfur, which occurs as a fine colloidal particle and gives water a milky-blue turbidity.

One of the better procedures for eliminating hydrogen sulfide is to lower the pH to 6 or less before beginning aeration. This will favor H_2S removal by scrubbing. The pH can be lowered by bubbling CO_2 gas into the water. After the H_2S is removed, additional aeration will remove any remaining CO_2.

Algae

In warmer climates, the slat-and-coke-tray and positive draft aerators are often located in the open air and exposed to direct sunlight. The wetted surfaces of these aerators create excellent environments for the growth of algae and slime. Although it is not practical to eliminate these problems completely, growth rates can be greatly reduced if a roof or canopy is provided to shade the aerator from direct sunlight.

Clogged Diffusers

Air diffusers can become partly clogged, either from dust in the air or from oil, debris, or chemical deposits that can collect around the diffuser opening. Operators can minimize clogging by

- maintaining clean air filters
- not overlubricating blowers
- preventing backflow of water into diffusers

Techniques for cleaning diffusers vary with the type of diffuser. Most ceramic types can be heated to burn away trapped particulates. Both fine- and large-bubble types can be cleaned with a brush and detergent. Manufacturers of ceramic diffusers should be consulted to determine the best cleaning procedure to follow.

Energy Consumption

Regardless of the method used, the aeration process will require some supply of energy. The water usually must be pumped both before and after cascade aeration; diffused aeration requires power to drive the blowers; and mechanical aeration requires power to drive the aerator and the blowers. Operators can minimize power costs by not overaerating, keeping motors and blowers properly maintained, and keeping air diffusers clean and free of debris.

Safety Precautions

Both hydrogen sulfide and methane can be produced by aeration processes, and operators must be made aware of how dangerous these gases can be.

Hydrogen sulfide is a poisonous gas and is colorless, flammable, and explosive. It kills by paralyzing a person's respiratory system; only a few minutes of exposure to a concentration of H_2S as low as 0.1 percent by volume in air is fatal. Higher concentrations, above 4.3 percent, are explosive. A very dangerous characteristic of H_2S is that its foul, rotten-egg odor is not evident at high concentrations, because these concentrations paralyze one's sense of smell (the olfactory nerves).

Methane, or natural gas, is a colorless, odorless, and tasteless gas; it is flammable and highly explosive. If the scrubbed gases containing methane are allowed to accumulate in a confined space, an explosion or fire can result. Methane can also kill by simple asphyxiation (suffocation). Because the gas has no color, taste, or odor, the victim may be completely unaware of its presence.

The safest way to deal with hydrogen sulfide and methane gases is to ensure that the aeration process is well ventilated and that there are no locations within the process where gases could accumulate. Methane is lighter than air and will accumulate at the top of an enclosure. Hydrogen sulfide is heavier than air and will collect at the bottom of confined areas. Ideally, aerators should be located in an open area with good ventilation. Where aerators must be enclosed because of climatic conditions, the enclosure must be well ventilated at all times.

Record Keeping

Following are some important aeration process records that should be maintained:

- results of periodic analysis of raw water to monitor for changes in constituents the aeration process is designed to alter
- daily quantity of water treated by the aeration process
- results of periodic analyses of finished water to determine if the aeration process is successfully altering or removing constituents from the raw water as designed
- inspections and monitoring for safety hazards, such as the release of poisonous or explosive gas from the aerators
- details of the maintenance of aeration equipment
- changes made in the equipment or process operation
- observed changes in other treatment processes that might be due to failure or improvement in the aeration process

Selected Supplementary Readings

Adams, J.Q., and R.M. Clark. 1991. Evaluating the Costs of Packed Tower Aeration and GAC for Controlling Selected Organics. ***Jour. AWWA,*** 83(1):49.

Air Stripping for Volatile Organic Contaminant Removal. 1989. Denver, Colo.: American Water Works Association.

Boyden, B.H., et al. 1992. Using Inclined Cascade Aeration to Strip Chlorinated VOCs From Drinking Water. ***Jour. AWWA,*** 84(5):62.

Controlling Radionuclides and Other Contaminants in Drinking Water Supplies: A Workbook for Small Systems. 1991. Denver, Colo.: American Water Works Association.

Dixon, K.L., et al. 1991. Evaluating Aeration Technology for Radon Removal. ***Jour. AWWA,*** 83(4):141.

Dzombak, D.A., S.B. Roy, and H-J. Fang. 1993. Air-Stripper Design and Costing Computer Program. ***Jour. AWWA,*** 85(10):63.

Jang, W., N. Nirmalakhandan, and R.E. Speece. 1989. ***Cascade Air-Stripping System for Removal of Semi-Volatile Organic Contaminants: Feasibility Study.*** Denver, Colo.: American Water Works Association Research Foundation and American Water Works Association.

Lamarche, P., and R.L. Droste. 1989. Air-Stripping Mass Transfer Correlations for Volatile Organics. ***Jour. AWWA,*** 81(1):78.

Little, J.C., and R.E. Selleck. 1991. Evaluating the Performance of Two Plastic Packings in a Crossflow Aeration Tower. ***Jour. AWWA,*** 83(6):88.

Staudinger, J., W.R. Knocke, and C.W. Randall. 1990. Evaluation of the Onda Mass Transfer Correlation for the Design of Packed Column Air Stripping. ***Jour. AWWA,*** 82(1):73.

VOCs and Unregulated Contaminants. 1990. Denver, Colo.: American Water Works Association.

Water Quality and Treatment. 4th ed. 1990. New York: McGraw-Hill and American Water Works Association (available from AWWA).

Water Treatment Plant Design. 2nd ed. 1989. Denver, Colo.: American Water Works Association.

CHAPTER 15

Membrane Processes

In 1748, the French physicist Nollet first noted that water would diffuse through a pig bladder membrane into alcohol. This was the discovery of *osmosis*, a process in which water from a dilute solution will naturally pass through a porous membrane into a concentrated solution. Over the years, scientists have attempted to develop membranes that would be useful in industrial processes, but it wasn't until the late 1950s that membranes were produced that could be used for what is known as *reverse osmosis*. In reverse osmosis, water is forced to move through a membrane from a concentrated solution to a dilute solution.

Since that time, continual improvements and new developments have been made in membrane technology, resulting in ever-increasing uses in many industries. In potable water treatment, membranes have been used for desalinization, removal of dissolved inorganic and organic chemicals, water softening, and removal of fine solids.

In particular, membrane technology enables some water systems having contaminated water sources to meet new, more stringent regulations. In some cases, it can also allow secondary sources, such as brackish groundwater, to be used. There is great potential for the continuing wider use of membrane processes in potable water treatment, especially as technology is improved and costs are reduced.

Description of Membrane Processes

In the simplest membrane processes, water is forced through a porous membrane under pressure while suspended solids, larger molecules, or ions are held back or rejected.

Types of Membrane Processes

The two general classes of membrane processes, based on the driving force used to make the process work, are

- pressure-driven processes
- electric-driven processes

Pressure-Driven Processes

The four general membrane processes that operate by applying pressure to the raw water are

- microfiltration
- ultrafiltration
- nanofiltration
- reverse osmosis

Figure 15-1 compares the sizes of substances that can be separated from water by these processes.

Microfiltration. Microfiltration (MF) is a process in which water is forced under pressure through a porous membrane. Membranes with a pore size of 0.45 μm are normally used; this size is relatively large compared with the other membrane processes. This process has not been generally applicable to drinking water treatment because it either does not remove substances that require removal from potable water, or the problem substances can be removed more economically using other processes. The current primary use of MF is by industries to remove very fine particles from process water, such as in electronics manufacturing. In addition, the process has also been used as a pretreatment for other membrane processes. In particular, RO membranes are susceptible to clogging or blinding unless the water being processed is already quite clean.

However, in recent years, microfiltration has been proposed as a filtering method for particles resulting from the direct filtration process. Traditionally, this direct filtration process has used the injection of coagulants such as alum or polymers into the raw water stream to remove turbidity such as clay or silts. The formed particles were then removed by rapid sand filters. The highly efficient filtering ability provided by microfiltration membranes has suggested their use to improve filtering efficiency, especially for small particles that could contain bacterial and protozoan life.

Ultrafiltration. Ultrafiltration (UF) is a process that uses a membrane with a pore size generally below 0.1 μm. The smaller pore size is designed to remove colloids and substances that have larger molecules, which are called

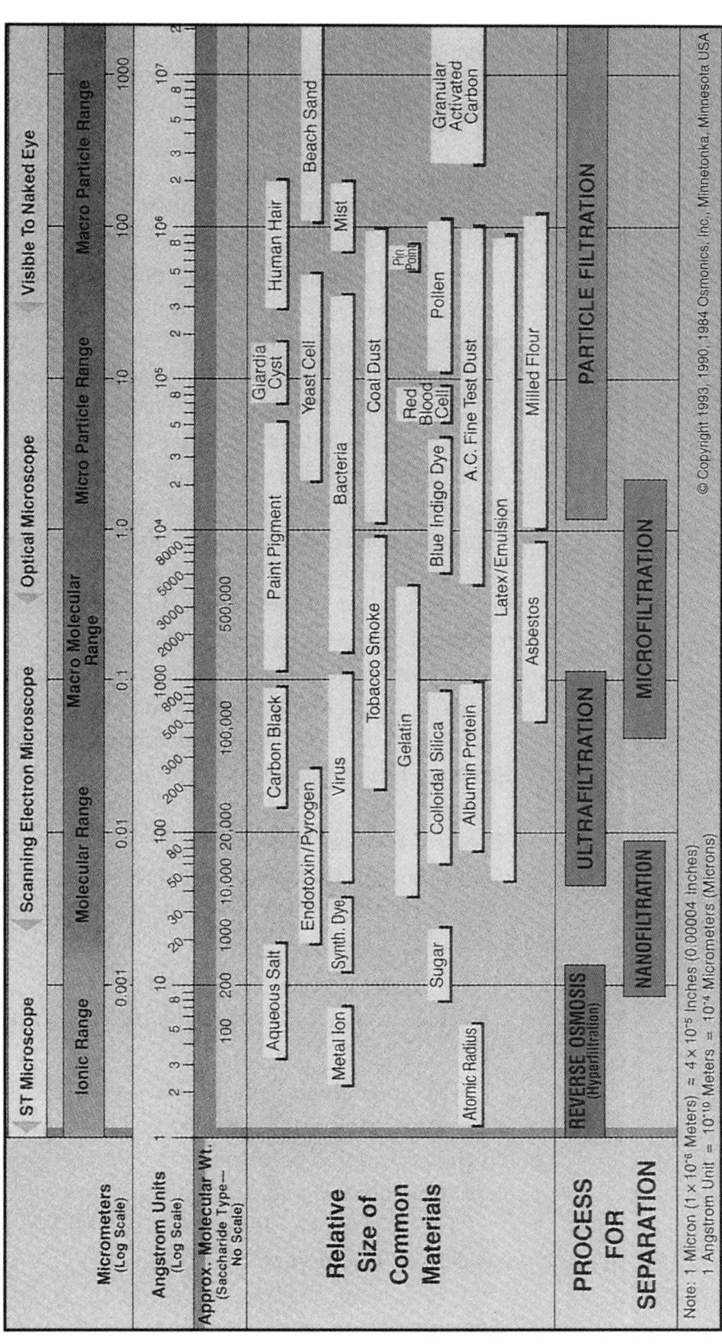

FIGURE 15-1 Comparison of sizes of particles removed by membrane processes

Courtesy of Osmonics, Minnetonka, Minn., USA

high–molecular-weight materials. UF membranes can be designed to pass materials that weigh less than or equal to a certain molecular weight. This weight is called the *molecular weight cutoff* (MWC) of the membrane. Although UF does not generally work well for removal of salt or dissolved solids, it can be used effectively for removal of most organic chemicals.

Nanofiltration. Nanofiltration (NF) is a process using membranes that will reject even smaller molecules than UF. The process has been used primarily for water softening and reduction of total dissolved solids (TDS). NF operates with less pressure than reverse osmosis and is still able to remove a significant proportion of inorganic and organic molecules. This capability will undoubtedly increase the use of NF for potable water treatment.

Reverse osmosis. Reverse osmosis (RO) is a membrane process that has the highest rejection capability of all the membrane processes. These RO membranes have very low MW8C pore sizes that can reject ions at very high rates, including chloride and sodium. Water from this process is very pure due to the high reject rates. The process has been used primarily in the water industry for desalinization of seawater because the capital and operating costs are competitive with other processes for this service. The RO also works to varying degrees of efficiency in the removal of many inorganic chemicals, most organic chemicals, and radionuclides and microorganisms. Industrial water uses such as semiconductor manufacturing is also an important RO process. RO is discussed in more detail later in this chapter.

Electric-Driven Processes

There are two membrane processes that purify a water stream by using an electric current to move ions across a membrane. These processes are

- electrodialysis
- electrodialysis reversal

These types of systems are primarily used to treat brackish water for potable use.

Electrodialysis. Electrodialysis (ED) is a process in which ions are transferred through a membrane as a result of a direct electric current applied to the solution. The current carries the ions through a membrane from the less concentrated solution to the more concentrated one.

Electrodialysis reversal. Electrodialysis reversal (EDR) is a process similar to ED, except that the polarity of the direct current is periodically reversed. The reversal in polarity reverses the flow of ions between demineralizing compartments, which provides automatic flushing of scale-forming materials from the membrane surface. As a result, EDR can often be used with little or no pretreatment of feedwater to prevent fouling. So far, ED and EDR have been used at only a few locations for drinking water treatment.

Reverse Osmosis

Because RO is the principal membrane process used in water treatment, it is described here in greater detail. However, many of these details also apply to NF. Osmosis, as applied to the water treatment industry, is a phenomenon that occurs when water can flow through a membrane but ions or molecules of dissolved or suspended substances cannot pass to a significant degree. In RO, the flow of water through the membrane is not the result of flow through definitive pores, as we commonly picture it. It is really the result of diffusion, one molecule at a time, through spaces in the molecular structure of the membrane material.

Consider saltwater as an example. When two solutions are separated by a semipermeable membrane, they tend to become equal in molecular concentration. In other words, water will naturally flow from a weaker to a stronger solution. As illustrated in Figure 15-2A, the flow of water across the membrane exerts a pressure called the *osmotic pressure*. Reverse osmosis is accomplished by applying a pressure greater than the solution osmotic pressure to the stronger solution, as shown in Figure 15-2B. The process is then reversed — relatively pure water passes through the membrane and leaves the large molecules behind.

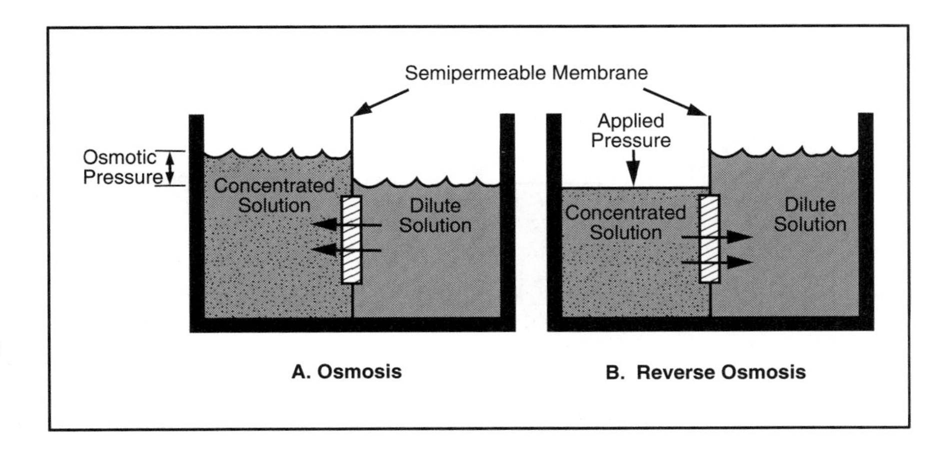

FIGURE 15-2 Illustration of the principles of osmosis and reverse osmosis

Reverse Osmosis Facilities

This section describes the equipment and chemicals associated with reverse osmosis. Figure 15-3 shows a schematic for a simple reverse osmosis installation. Figure 15-4 shows a typical package RO unit.

Membranes

The membranes used in the various membrane processes are manufactured from a variety of materials. The two most common membrane materials for water treatment are cellulose acetate and polyamide–composite. Every type of material has certain operating characteristics, such as the efficiency of salt rejection, the best pH operating range, resistance to degradation if exposed to chlorine or other oxidants, susceptibility to biological attack, and resistance to hydrolysis.

Over time, the performance of all membranes will change, mainly as a result of compaction and fouling. Compaction, which is similar to plastic or metal "creep" under compression, gradually closes the pores in the membrane. The compaction rate increases for membranes operated at higher pressures and higher temperatures. Membrane fouling occurs from substances in the feedwater, such as calcium carbonate or calcium sulfate scales, deposition of fine colloids, iron or other metal oxides, and silica. Pretreatment of water before RO is usually necessary to prevent premature fouling of the membrane.

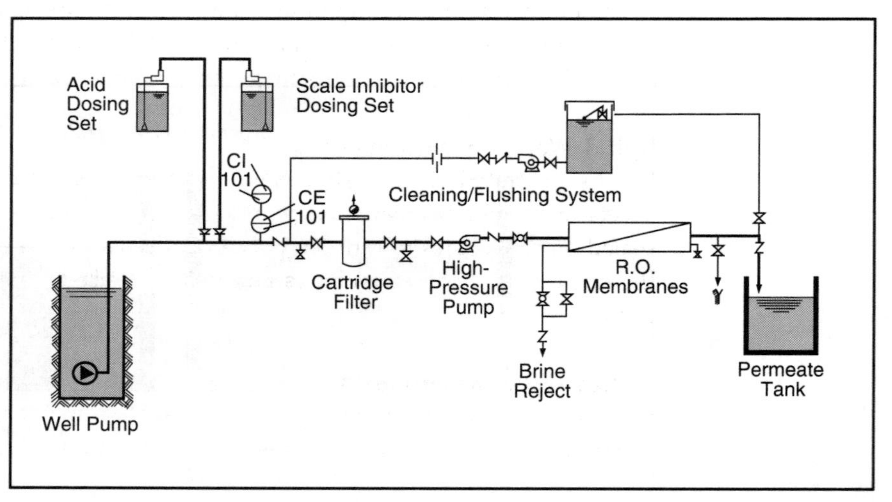

FIGURE 15-3 Typical reverse osmosis flow schematic

Courtesy of American Engineering Services, Inc., Tampa, Fla.

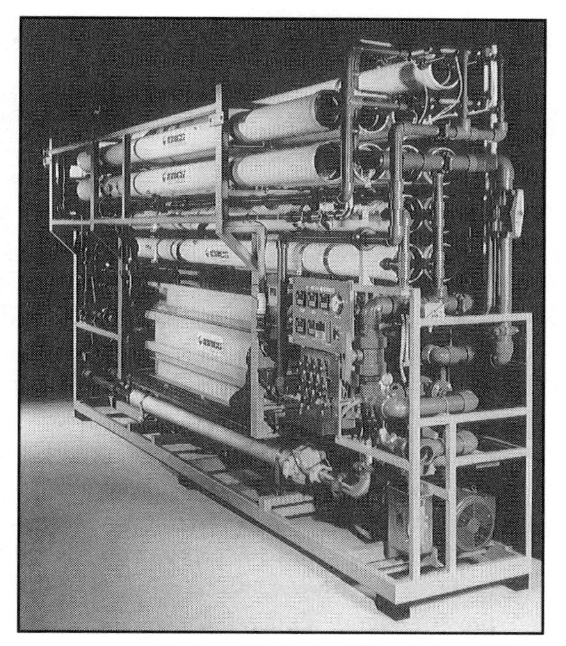

FIGURE 15-4 Typical package reverse osmosis treatment unit

Courtesy of Ionics Inc., Water Systems Division

In some cases, fouled membranes can be restored by periodic cleaning with acid. Otherwise they must be replaced when they become fouled or cease to function properly.

Four membrane configurations are currently available: spiral wound, hollow fiber, tubular, and plate and frame. Only the first two configurations are used for drinking water treatment on a commercial scale.

Spiral-Wound Membranes

As illustrated in Figure 15-5, spiral-wound membranes consist of two flat sheets of membrane material separated by porous sheets. These layers are sealed on three sides to form an envelope. Feedwater enters at one end, and the open side of the envelope is attached to a plastic tube that collects the product water.

Hollow-Fiber Membranes

As illustrated in Figure 15-6, hollow-fiber membranes consist of a compact bundle of thousands of fibers that surround the feedwater distribution

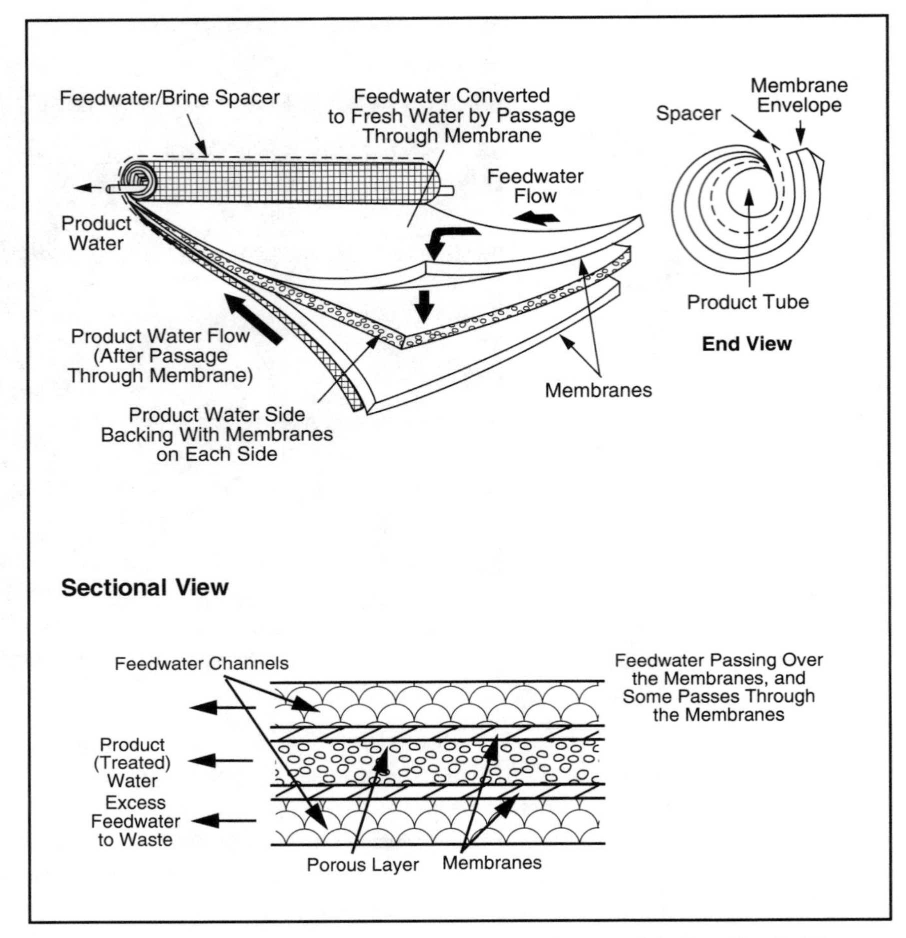

FIGURE 15-5 Details of a spiral-wound membrane

Courtesy of the Dow Chemical Company

core. Each fiber is laid in a U shape in the bundle, and both ends are encapsulated in an end sheet.

Feedwater Concerns

The term "feedwater" is used to describe raw water that has undergone pretreatment (acidification and/or scale inhibitor addition) prior to entering the membrane arrays. Feedwater quality and pressure are important issues for the reverse osmosis process. Feedwater treatment is almost always necessary.

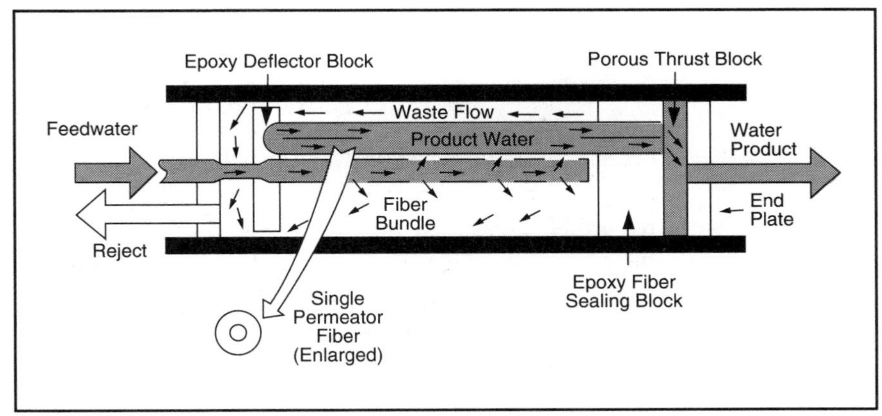

Courtesy of U.S. Filter/Permutit

FIGURE 15-6 Details of a hollow-fiber module

Feedwater Quality

The quality of feedwater passing through RO units is critical for prolonging the life of membranes. Improperly treated feedwater can rapidly cause irreversible damage to the membrane. Surface water generally requires more pretreatment and closer monitoring because its quality is not as stable. To maximize membrane life, pretreatment of RO feedwater may be needed for any one or more of the following purposes:

- turbidity reduction
- iron or manganese removal
- stabilization to prevent scale formation
- microbial control
- chlorine removal
- hardness reduction

It has also been found that the types of materials used in feedwater piping and the pressure pump can adversely affect membrane life or performance by releasing trace amounts of metals from corrosion or by allowing air to enter the system. Only noncorrosive materials such as stainless steel, polyvinyl chloride (PVC), and fiberglass should be used in piping systems that supply RO units. In addition, pumps should be either vertical-turbine types that have mechanical seals, or submersible pumps. These types of pumps ensure that no air is introduced into the system.

Feedwater Treatment

In addition to any special pretreatment for excessive levels of the items listed above, sulfuric acid is usually added to the feedwater to prevent calcium and magnesium carbonate scaling of the membrane. Cartridge filters are also commonly installed immediately ahead of RO units for final removal of any suspended matter down to the 5 or 10 μm level. This is necessary regardless of whether the supply is surface water or groundwater. Cartridge filter containers are furnished with inlet and outlet pressure gauges to monitor head loss. Cartridges should be replaced when the difference in pressure between the gauges reaches 15 psi (103 kPa).

Feedwater Pressure

The pressure differential used to make RO work depends primarily on the desired removal efficiency. Although the process can be designed to operate with a pressure as low as 50 psi (340 kPa), commercial installations are commonly operated at 300 psi (2,100 kPa) or higher.

Posttreatment

The additional treatment that must be provided for water following RO treatment typically includes the following:

- degasification for removal of carbon dioxide, hydrogen sulfide, or other undesirable gases if they are present
- pH adjustment to minimize corrosion in the distribution system
- disinfection

Membrane Cleaning

As membranes foul, water production rates and removal efficiencies decrease and higher pressure differentials are required. The following are some indications that a membrane needs cleaning:

- The passage of salt through the membrane increases by 15 percent.
- The pressure drop through the unit increases by 20 percent.
- Feed pressure requirements increase by 20 percent.
- Product water flow drops or increases by 5 percent.
- Fouling or scaling is evident.

Each membrane manufacturer has recommended procedures and chemicals that should be used to rejuvenate their membranes. These guidelines should be carefully followed, or else the membranes could be damaged.

Cleaning chemicals are usually corrosive, so all cleaning system components must be made of stainless steel or other noncorroding materials.

Reject Water

The amount of reject water created from the operation of a commercial RO plant can range from 20 to 50 percent of the feedwater to an RO unit. The amount of reject water is dependent upon the number of stages in which the membranes are configured and the feed pressure.

The reject water quantity presents two problems, which can be significant. One is that the water source must be capable of supplying up to twice the amount of water needed by the system. In areas with limited groundwater availability, consideration may have to be given to other treatment processes that do not waste as much water, even if those processes are more expensive.

The other problem is waste disposal. One acceptable method of waste disposal usually includes discharge to the local waste treatment system. However, because of the relatively large quantity of wastewater produced and the highly mineralized nature of the reject water, waste treatment authorities may find that it will overtax their system. The other common disposal method is to use deep well injection below the water supply aquifer or an evaporation pond, but care must be taken not to cause aquifer contamination.

The amount of reject water from an RO installation can be reduced to a limited extent by an increase in the feedwater pressure. However, this usually results in a shorter membrane life and increased operating costs.

A two-stage system is shown in Figure 15-7. The two first-stage modules purify 50 percent of the water fed to the system. The reject water from the first stage is then processed by the second-stage unit, which purifies it another 50 percent. The final flow ends up being 75 percent purified water and 25 percent reject water.

Operation of the Reverse Osmosis Process

Operation of most RO systems includes automatic control by programmable logic controllers (PLCs). However, operators are required to monitor PLC and system operation to ensure maximum efficiency. Some of the principal requirements in the operation of an RO treatment system are as follows:

- If surface water is being treated, the raw water quality must be routinely monitored for changes in chemical and biological quality.

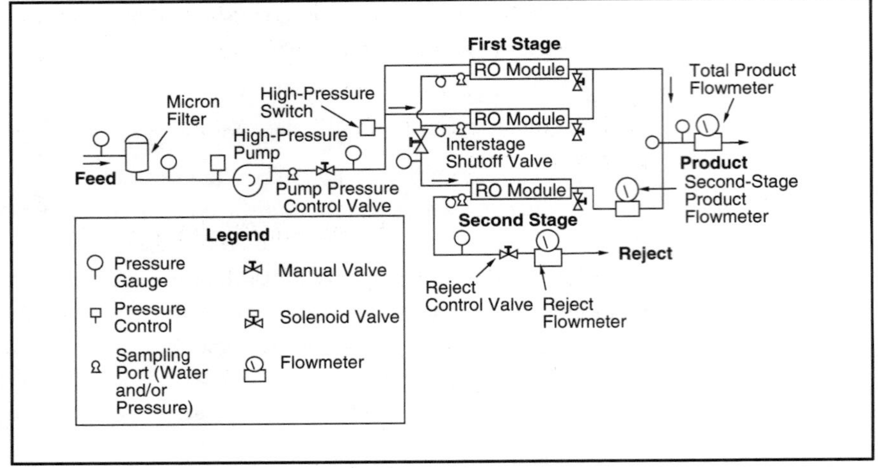

FIGURE 15-7 Schematic of a two-stage reverse osmosis treatment system

Courtesy of U.S. Filter/Permutit

- The operation of the pretreatment acid feed system must be regularly inspected and the pH level monitored to ensure that the proper proportion of acid is being fed.
- Cartridge filter drains must be flushed daily and the loss of head across the filters recorded.
- Continuous monitoring of feedwater quality and membrane product water must be done to optimize critical membrane operating parameters, including product water recovery rate (proportion of raw water recovered as potable water) and feed pressure.
- Posttreatment systems — which may include degasification, disinfection, and corrosion control treatment — must be monitored and adjusted for proper operation.

A system should have an automatic shutdown capability in the event of a problem that could damage the membranes. Problems that require shutting down the system include

- higher than normal pump pressure
- low feedwater pump suction pressure
- abnormally high or low feedwater pH
- improper feedwater flow rate
- high feedwater turbidity
- higher than normal feedwater conductivity

- abnormally high or low product flow rate
- high product water conductivity
- low antiscalant level or flow

Each of these controls should be maintained periodically and checked for proper operation.

Operating Problems

Some of the major operating problems experienced with RO installation include

- premature fouling
- flux decline
- unit shutdown
- bacteria and algae fouling
- oil and grease fouling

Premature fouling of membranes as a result of any of the factors previously mentioned generally occurs because the installation either was not designed properly or has not been operated in accordance with the design. The unit's manufacturer or membrane supplier should be consulted at the first sign of fouling to see if the problem can be corrected.

Membrane performance will gradually change as a result of compaction; this is sometimes called "flux decline." The manufacturer should be able to predict what changes can reasonably be expected and what the approximate life of the membrane should be. If the changes appear to be more drastic than predicted, the manufacturer should be consulted for advice on any changes that can be made in the operation to help prolong the membrane life.

If the plant is shut down for any period of time, membranes should be flushed with acidified feedwater or unchlorinated product water. Concentrated brine must be removed from prolonged contact with the membrane because precipitation will eventually occur, resulting in membrane scaling. If shutdown is to be for more than a few days, accumulation of biological growth must be prevented. This can be accomplished by either operating the units for a short time each day or sterilizing the membranes according to the manufacturer's recommendations.

Bacteria, bacterial slimes, and algae can cause fouling of some membranes. Some membrane materials have a limited tolerance for chlorine, and others have essentially none at all. It is safest to obtain the manufacturer's recommendation on whether controlled amounts of chlorine can be used to inhibit bacterial and algae growth, or whether biocides should be used.

Oil and grease in very low concentrations can reduce membrane performance by forming a film on the membrane surface. Only lubricants recommended by the membrane manufacturer should be used on pumps, parts, and O-rings that are installed ahead of the membrane units.

Record Keeping

The following are some types of records on the operation of an RO treatment system that should be kept:

- records of the operation of any pretreatment being provided
- records of feed rates and residual monitoring of acid and antiscalant chemicals
- daily records of the pressure drop on each cartridge filter
- daily records of the pressure drop across each membrane unit
- detailed records of each membrane-cleaning operation
- daily records of the chlorine feed rate and amount of chlorine used
- daily records of the metered flow of treated water and reject water
- records of the operation of posttreatment systems
- periodic records of the salinity of treated water as an indicator of membrane fouling

In many cases, the membrane system computer or PLC control systems can automate this data collection and provide trends in the operating data.

Selected Supplementary Readings

AWWA Membrane Technology Research Committee. 1992. Committee Report: Membrane Processes in Potable Water Treatment, ***Jour. AWWA,*** 84(1):59.

Blau, T.J., et al. 1992. DBP Control by Nanofiltration: Cost and Performance. ***Jour. AWWA,*** 84(12):104.

Conlon, W.J., and S.A. McClellan. 1989. Membrane Softening: A Treatment Process Comes of Age. ***Jour. AWWA,*** 81(11):47.

Dykes, G.M., and W.J. Conlon. 1989. Use of Membrane Technology in Florida. ***Jour. AWWA,*** 81(11):43.

El-Rehaili, A.M. 1991. Reverse Osmosis Applications in Saudi Arabia. ***Jour. AWWA,*** 83(6):72.

Jacangelo, J.G., et al. 1989. Assessing Hollow-Fiber Ultrafiltration for Particulate Removal, ***Jour. AWWA,*** 81(11):68.

———. 1991. Low-Pressure Membrane Filtration for Removing *Giardia* and Microbial Indicators. ***Jour. AWWA,*** 83(9):97.

Jacangelo, J.G., N.L. Patania, J-M Laîné, Wayne Booe, Joël Mallevialle, and General Waterworks Management and Service Company. 1992. ***Low-Pressure Membrane Filtration for Particle Removal.*** Denver, Colo.: American Water Works Association Research Foundation and American Water Works Association.

Lahoussine-Turcaus, V., et al. 1990. Coagulation Pretreatment for Ultrafiltration of a Surface Water. ***Jour. AWWA,*** 82(12):76.

Lainé, J-M., et al. 1989. Effects of Ultrafiltration Membrane Composition. ***Jour. AWWA,*** 81(11):61.

Manual M38, Electrodialysis and Electrodialysis Reversal. 1995. Denver, Colo.: American Water Works Association.

Mickley, Mike, Robert Hamilton, Lana Gallegos, and Jeffrey Truesdall. 1993. ***Membrane Concentrate Disposal.*** Denver, Colo.: American Water Works Association Research Foundation and American Water Works Association.

Pirbazari, M., et al. 1992. MF-PAC for Treating Waters Contaminated With Natural and Synthetic Organics. ***Jour. AWWA,*** 84(12):95.

Semmens, M.J., R. Qin, and A. Zander. 1989. Using a Microporous Hollow-Fiber Membrane to Separate VOCs from Water. ***Jour. AWWA,*** 81(4):162.

Taylor, J.S., S.J. Duranceau, W.M. Barrett, and J.F. Goigel. 1989. ***Assessment of Potable Water Membrane Applications and Research Needs.*** Denver, Colo.: American Water Works Association Research Foundation and American Water Works Association.

Water Quality and Treatment. 4th ed. 1990. New York: McGraw-Hill and American Water Works Association (available from AWWA).

Water Treatment Plant Design. 2nd ed., 1990. New York: McGraw-Hill and American Water Works Association (available from AWWA).

Zander, A.K., M.J. Semmens, and R.M. Narbaitz. 1989. Removing VOCs by Membrane Stripping. ***Jour. AWWA,*** 81(11):76.

CHAPTER 16

Treatment Plant Instrumentation and Control

Meters, recorders, alarms, and automatic control systems are installed in a water plant primarily to

- provide information for the operation of equipment
- meet state and federal requirements
- improve the efficiency of operation
- provide historical records
- provide more precise control of equipment
- improve the safety of working conditions
- reduce the workload on operators

The most basic instrumentation is that required for the operation of plant equipment. Instruments such as the loss-of-head gauges on filters are absolutely necessary for plant operation.

State regulations generally require data on plant pumpage, chemical quantities used, and details of equipment and process information (such as the plant effluent chlorine residual required by the Surface Water Treatment Rule). Even the smallest treatment systems must provide a number of flowmeters and analyzing tools in order to meet these requirements.

Many new instruments are available to improve the efficiency of treatment plant operations. The reliability, simplicity of operation, and cost of many types of instruments have gradually improved to where the instruments are now affordable for even relatively small systems. Continuous turbidity monitors are now common in treatment plants. Streaming current detectors and other instruments designed to assist in the efficient use of coagulants are being installed in many treatment plants.

All water systems should maintain good records of plant operations. In addition to documenting a system's operation in the event the system is questioned by the public or authorities, historical records are important to engineers who design plant improvements. Recorder charts from flowmeters, pressure indicators, and automatic analysis instruments provide important records for these purposes.

A great deal of instrumentation is available to replace manual control or visual inspection. Automatic control might be used to relieve the operator of some duties to make time available for other work, or because it is more accurate or will eliminate the chance of human error. For example, a chlorinator might be automated so that a uniform residual is maintained regardless of chlorine demand or pumping rate. This is not only more accurate than manual control, but it also eliminates the possibility of an incorrect feed caused by human error.

For these reasons, automation is finding increased use in water treatment plants every day. As a result, there is a greater number of meters and control systems that the operator must understand and use properly.

Within reason, the operator also must learn to maintain and repair this new equipment. Many analyzers, for instance, require regular cleaning, replacement of elements, and recalibration. Most electronic equipment cannot be repaired locally, but sometimes key elements can simply be replaced or returned to the factory.

This chapter is intended as a general overview of principal metering and automation equipment. For up-to-date information on the available automation and control equipment, refer to the publications listed in appendix C. Another excellent way to learn about available equipment is to visit the display area of a state or national American Water Works Association (AWWA) conference. Vendors of many products are available at these conferences to furnish information, display their products, and answer technical questions.

Flow, Pressure, and Level Measurement

Measurements of flow, pressure, and fluid levels are essential to the operation of a treatment system.

Flow Measurement

Details of metering devices are covered in more detail in *Water Distribution and Transmission* (part of this series), but the following is an overview of the needs and types of flow-measuring devices commonly used in treatment plants.

Reasons to Measure Treatment Plant Flow

Some of the primary reasons to measure flow in a treatment plant are as follows:

- The flow rate through the treatment processes needs to be controlled so that it matches distribution system use.
- It is important to determine the proper feed rate of chemicals added in the processes.
- The detention times through the treatment processes must be calculated. This is particularly applicable to surface water plants that must meet $C \times T$ values required by the Surface Water Treatment Rule.
- Flow measurement allows operators to maintain a record of water furnished to the distribution system for periodic comparison with the total water metered to customers. This provides a measure of "water accounted for," or conversely, the amount of water wasted, leaked, or otherwise not paid for.
- Flow measurement allows operators to determine the efficiency of pumps. Pumps that are not delivering their designed flow rate are probably not operating at maximum efficiency, and so power is being wasted. Pumps that do not produce at their designed rate should be checked to determine if they are worn or are not operating at their design head.
- For well systems, it is very important to maintain records of the volume of water pumped and the hours of operation for each well. The periodic computation of well pumping rates can identify problems such as worn pump impellers and blocked well screens.
- Reports that must be furnished to the state by most water systems must include records of raw- and finished-water pumpage. This may be required monthly for very small systems and daily for larger systems. The reports also usually need to include a record of the rate of chemical application, which must be computed based on the treated-water pumpage figures.

- Wastewater generated by a treatment system must also be measured and recorded. This can be done by a direct measurement of the waste stream, or indirectly by measuring the raw-water flow and subtracting the treated-water flow.
- Individual meters are often required for the proper operation of individual pieces of equipment. For instance, the makeup water to a fluoride saturator is always metered to assist in tracking the fluoride feed rate.

Flow-Measuring Devices

All of the uses just discussed create the need for a number of meters, often with different capabilities. Meters that can handle large rates of flow are required for measuring raw and finished water. Smaller meters are used on individual pieces of equipment, and special metering devices must be used for measuring corrosive or viscous liquids. Flow-measuring devices can be roughly divided into the following categories:

- pressure-differential meters
- velocity meters
- magnetic flowmeters
- ultrasonic flowmeters
- positive-displacement meters
- weirs and flumes

Pressure-differential meters. Pressure-differential meters operate on the principle of measuring pressure at two points in the flow, which provides an indication of the rate of flow that is passing by. There is a set relationship between the flow rate and volume, so the meter instrumentation automatically translates the differential pressure into a volume of flow.

The type of pressure-differential meter that has been widely used for many years to meter potable water is the venturi meter (Figure 16-1). These meters are reliable because there are no moving parts, and they are reasonably accurate over a rather wide range of flow. They have been used for measuring large volumes of water, such as the effluent of a larger treatment plant.

Velocity meters. Velocity meters use a propeller or turbine to measure the velocity of the flow passing the device (Figure 16-2). The velocity is then translated into a volumetric amount by the meter register. Velocity meters are particularly well suited for measuring intermediate flow rates on clean water.

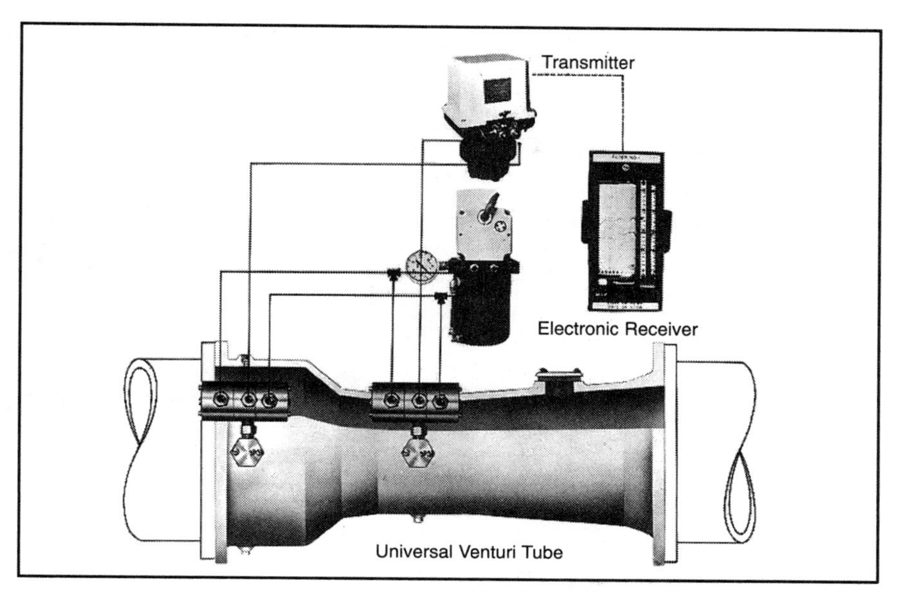

FIGURE 16-1
Venturi meter

Courtesy of Leeds & Northrup, A Division of General Signal

Magnetic flowmeters. Magnetic flowmeters are relatively new to the water industry. They measure the flow rate based on the voltage created between two electrodes as the water passes through an electromagnetic field. As illustrated in Figure 16-3, they have no internal obstructions or moving parts, so there is no head loss. They can be designed for use with corrosive or dirty liquids. Magnetic meters range from very small units that are used for measuring chemical flow to units that are several feet (meters) in diameter for mainline metering.

Ultrasonic flowmeters. Ultrasonic flowmeters use an electronic transducer to send a beam of ultrasonic sound waves through the water to another transducer on the opposite side of the unit (Figure 16-4). The velocity of the sound beam varies with the liquid flow rate, so the beam can be electronically translated to indicate flow volume. Advantages of ultrasonic meters include the absence of obstructions that create head loss and their usefulness for corrosive or dirty liquids, such as treatment plant sludge. They are available in very small to very large units.

Positive-displacement meters. Positive-displacement meters, such as the nutating-disk meter illustrated in Figure 16-5, are most commonly used for customer metering. These meters are very reliable and accurate for low flow rates because they measure the exact quantity of water passing through

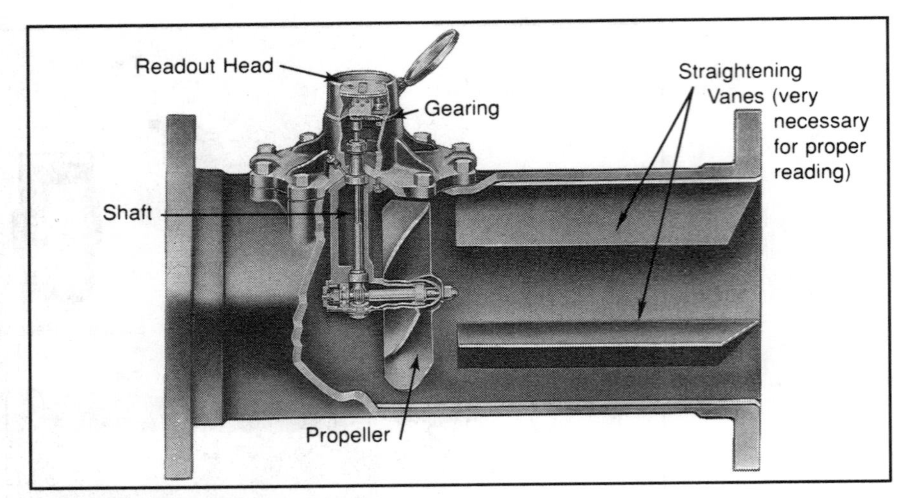

A. Propeller Meter

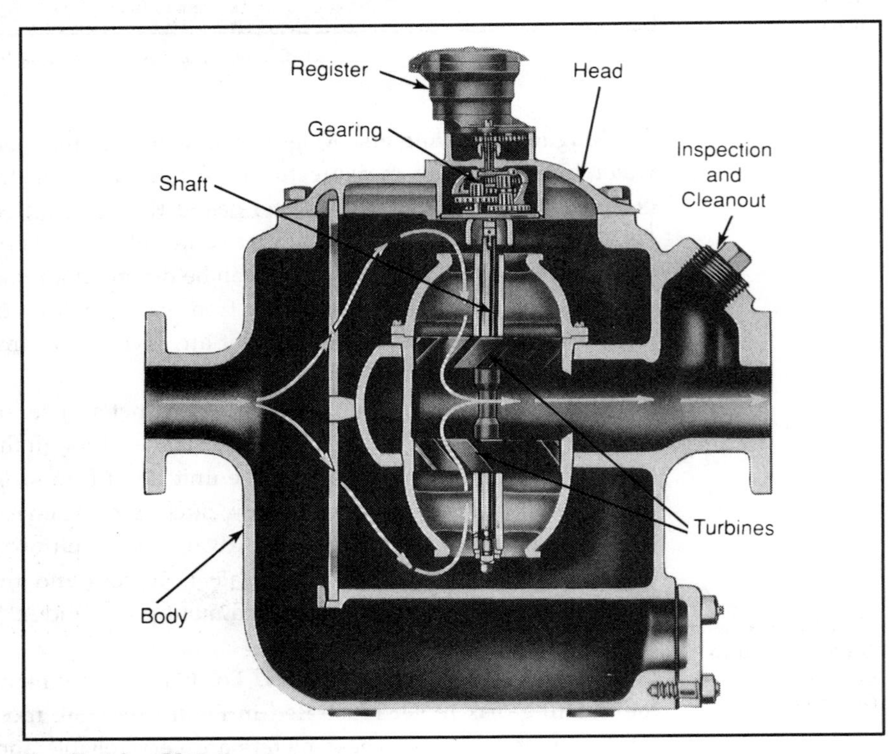

B. Turbine Meter

FIGURE 16-2
Velocity flowmeters

FIGURE 16-3
Magnetic flowmeter

Courtesy of Fischer & Porter Co.

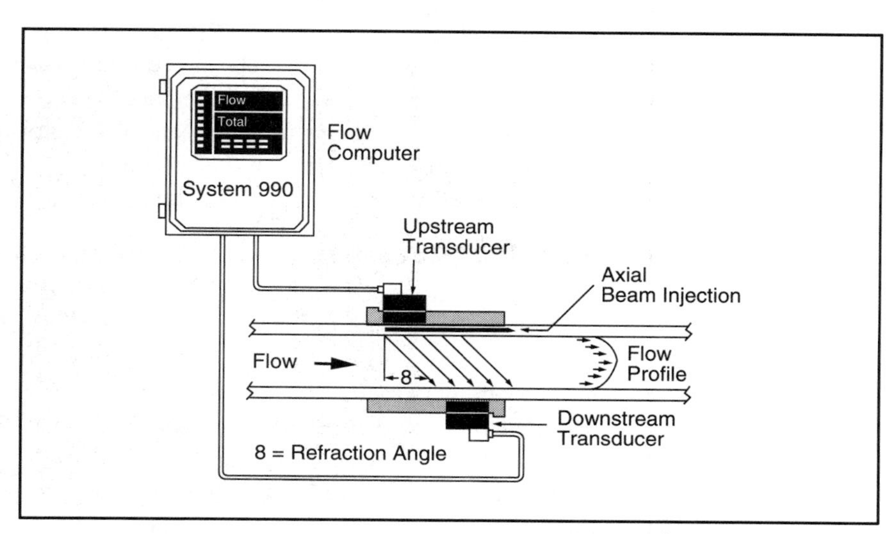

FIGURE 16-4
Transducer on an ultrasonic flowmeter

Courtesy of Controlotron Corporation

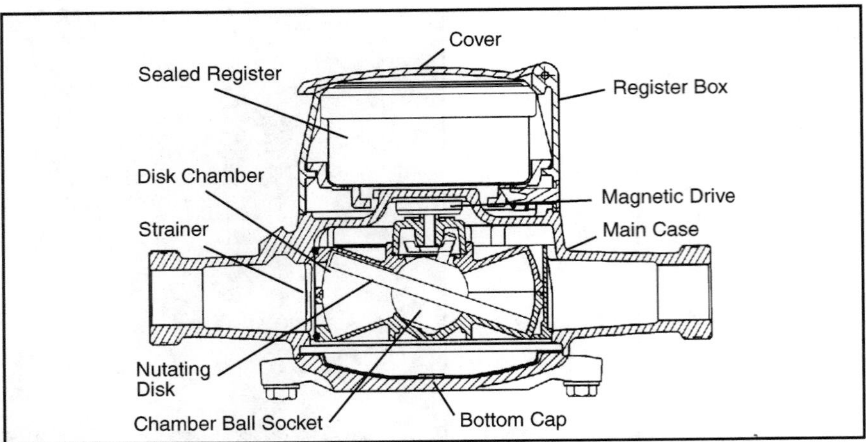

Courtesy of Schlumberger Industries Water Division

FIGURE 16-5 Nutating-disk meter

them. Positive-displacement meters are frequently used for measuring small flows in a treatment plant because of their accuracy. Repair or replacement is easy because they are so common in the distribution system.

Weirs and flumes. Weirs and flumes are used extensively for measuring flow in waste treatment plants because they are not affected by dirty water or floating solids. Two common types of weirs are shown in Figure 16-6, and a typical flume is shown in Figure 16-7. The quantity of water passing over the weir or through the flume can be determined by measuring the depth of the water and then referring to tables specific to the particular unit size. However, they can be used only for open conduits. Weirs and flumes are only occasionally used in water treatment because most of the places where flow measurement is desired are in closed pipelines.

Pressure Measurement

It is always necessary to measure pressure at several points in a treatment plant. The most important pressure is generally the plant discharge pressure, which governs the pressure being maintained on the distribution system. Pressure gauges are usually also provided at intermediate points in the treatment process, on special treatment units, and on the discharge of each pump so that their operation can be monitored.

Pressure gauges usually operate with a bellows, diaphragm, or Bourdon tube that is linked to an indicator, as illustrated in Figure 16-8. A cutaway of a typical Bourdon tube pressure gauge is shown in Figure 16-9. Circular or strip-chart recorders, illustrated in Figure 16-10, are used for important

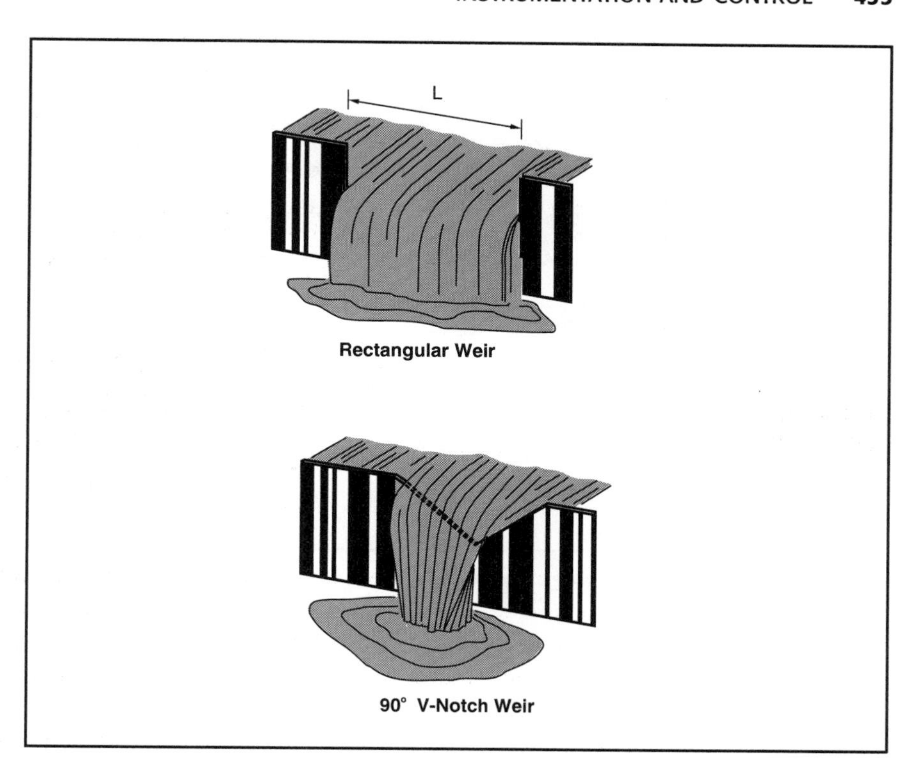

FIGURE 16-6
Two types of weirs

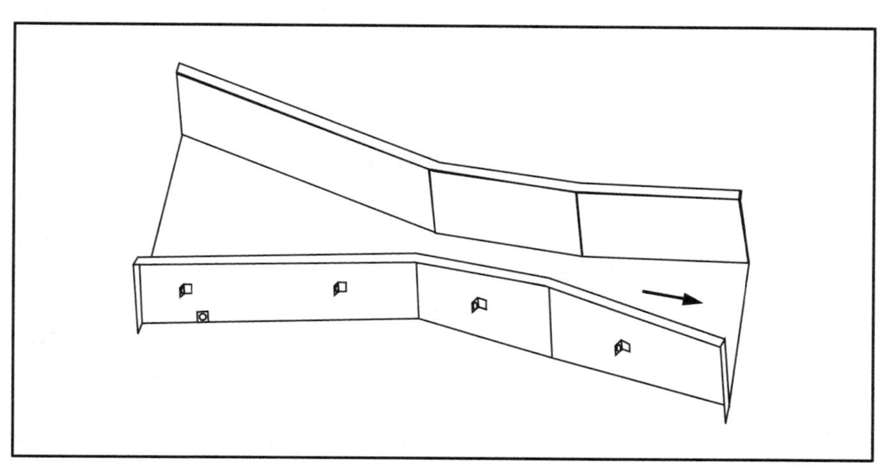

FIGURE 16-7
Fiberglass parshall flume

Courtesy of Leeds & Northrup, A Division of General Signal

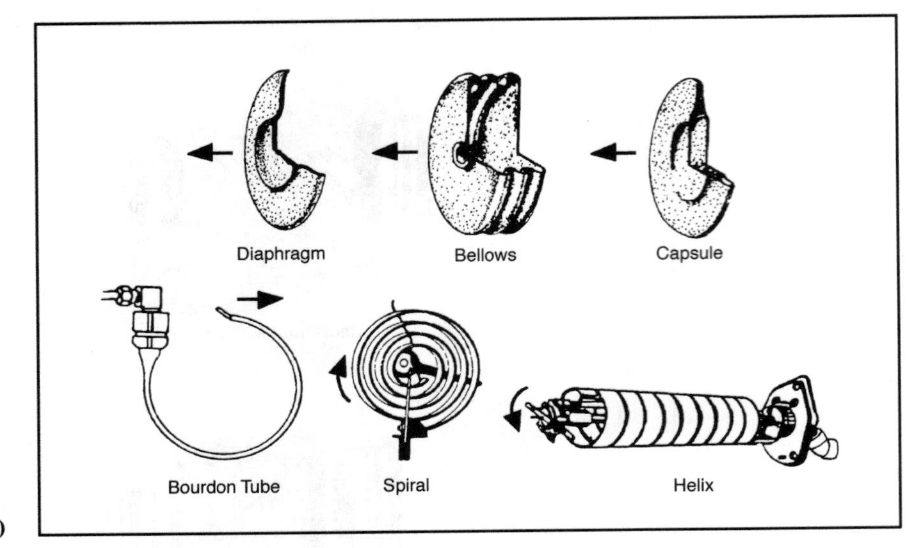

FIGURE 16-8 Typical pressure elements used to operate pressure gauges (arrows indicate movement with increased pressure)

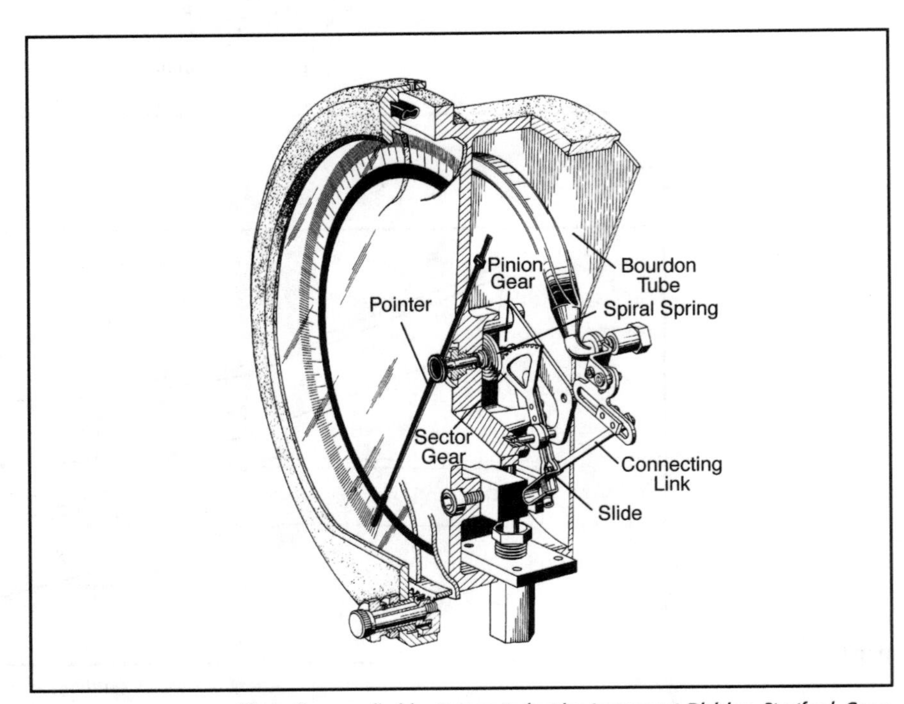

FIGURE 16-9 Cutaway view of a Bourdon tube pressure gauge

Illustration supplied by Dresser Industries Instrument Division, Stratford, Conn.

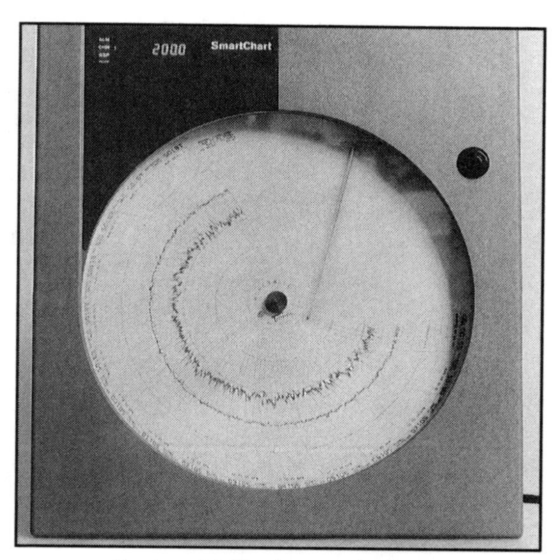

FIGURE 16-10
Circular chart recorder

Courtesy of Schlumberger Industries Water Division

pressure functions to provide a record of operations and a visual indication of operating trends. For example, when an operator sees on the distribution system pressure chart that the pressure has been slowly falling for the past hour, he or she can anticipate when additional pumps will have to be turned on to boost system pressure.

Distribution system pressure is also often transmitted to the treatment plant from one or more points on the distribution system. Common transmission points are at the base of elevated tanks and public buildings. The larger the system, the more important it is to have information on system pressure at remote locations.

Level Measurement

It is also usually necessary to measure the level of liquids at several points in the treatment process. The level-measuring devices that have been used for years are mechanical floats, bubbler tubes, and pressure gauges. There are now a number of new instruments that operate based on electrodes that sense the conductivity of the water or ultrasonic waves that are bounced off the water surface. Figure 16-11 illustrates several types of devices that measure liquid levels.

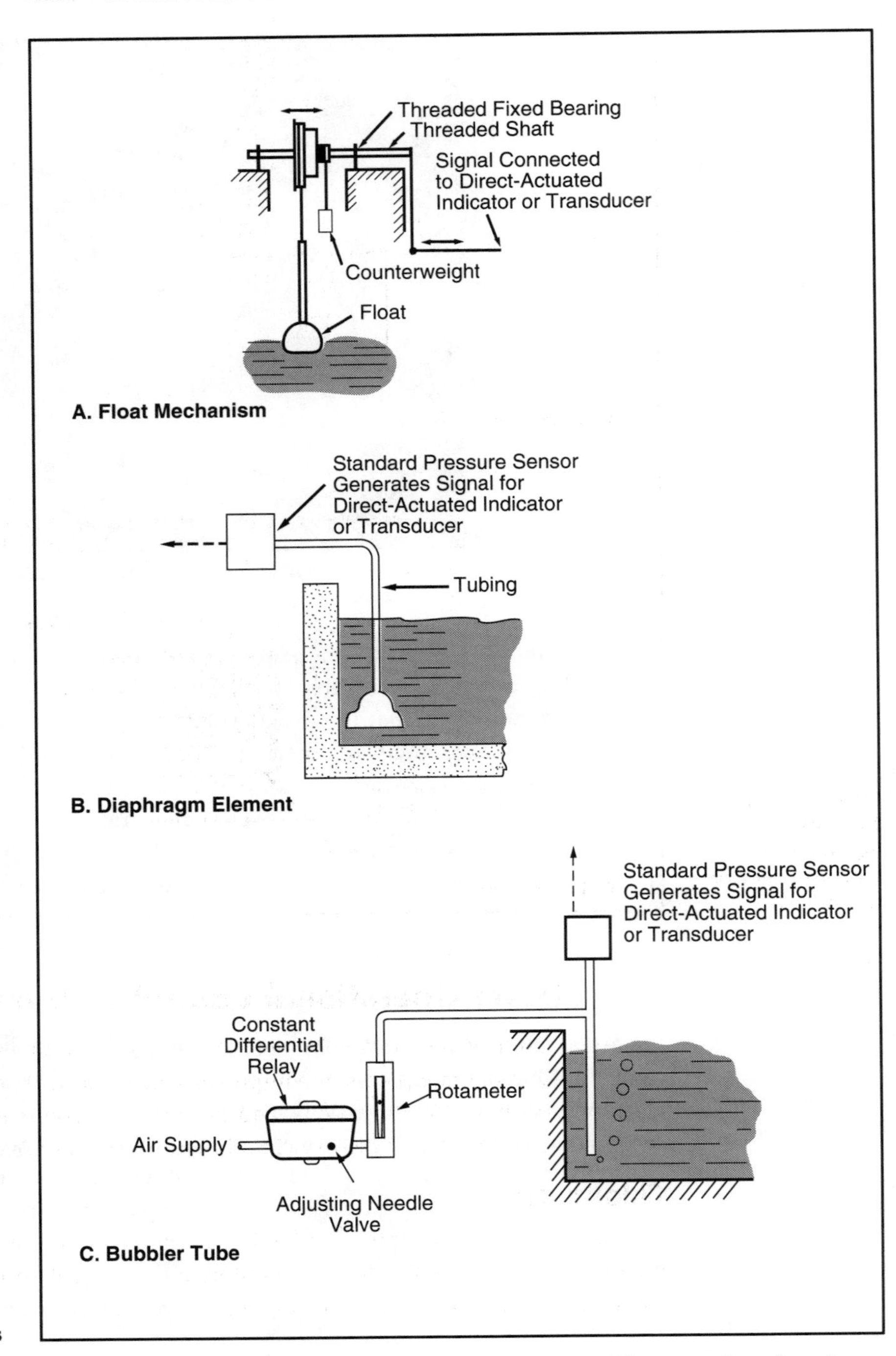

FIGURE 16-11 Types of liquid-level measuring devices

Figure continued next page

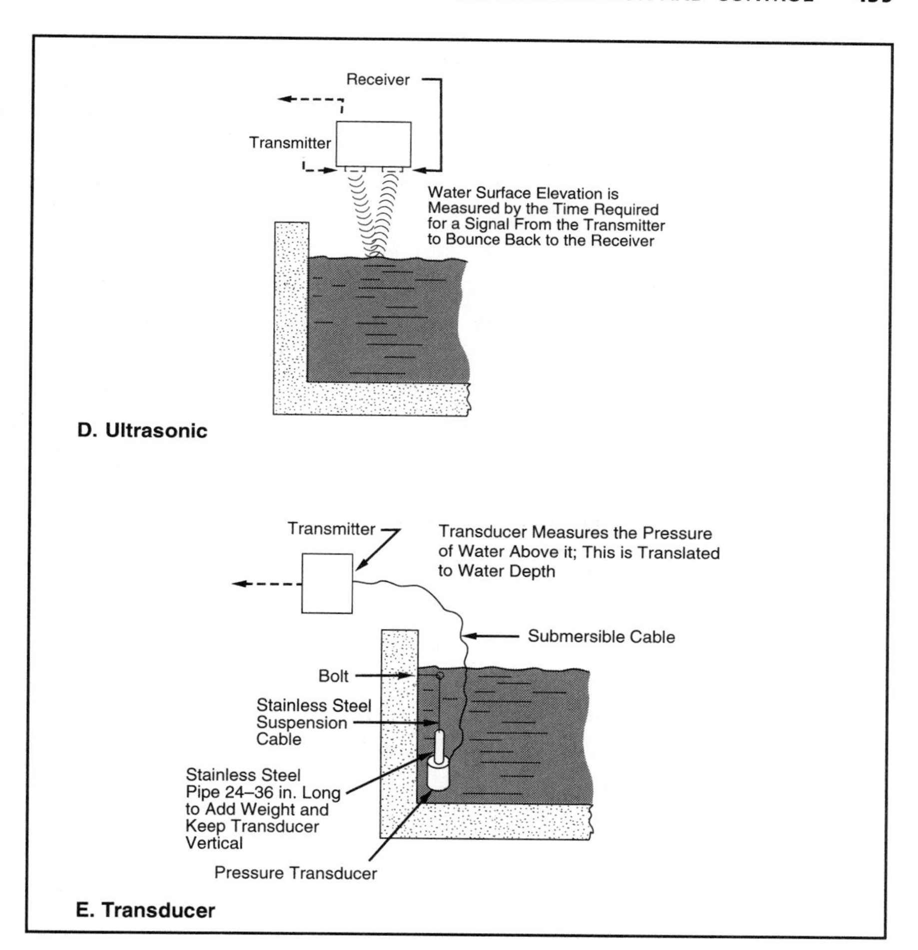

FIGURE 16-11 Types of liquid-level measuring devices (continued)

Other Operational Control Instruments

As a result of new developments in electronics, numerous instruments can be used to operate pieces of equipment automatically, monitor processes, alert the operator of malfunctions, and perform analyses automatically. This section discusses some of the principal instruments that are commonly used.

Chlorination

As explained in chapter 7, reliable equipment is now available that will monitor the chlorine residual and automatically regulate chlorine feed to compensate for changes in both flow rate and chlorine demand. Equipment is also available to provide an alarm if the residual falls outside preset limits, if the chlorine gas pressure should fail, or if chlorine gas is detected in the air.

Filtration

As discussed in chapter 4, equipment now available for controlling coagulation includes zeta potential meters, streaming current detectors, pH monitors, particle counters, and filterability test equipment. These devices either can alert the operator when the process is not operating within preset limits or can directly compensate for changes by regulating chemical feed rates or making other process adjustments.

As discussed in chapter 6, automatic turbidity monitoring is not only a good operational tool, but is also essential for meeting the effluent turbidity requirements of the Surface Water Treatment Rule. Monitors can be furnished with recorders to ensure continuous compliance with requirements. They will also alert the operator of any turbidity breakthrough that occurs.

Corrosion Control

Water systems required to install special corrosion control to meet the federal Lead and Copper Rule requirements will generally find that automation of the process is necessary to ensure continuous compliance. Continuous monitoring and recording of the pH of water entering the distribution system should be provided. In addition, if the system has a computer, a program is available to compute the Langelier saturation index quickly and easily. Chapter 9 discusses corrosion control in detail.

Automation

"Automation" of a water treatment system can be minimal or extensive. In its simplest form, automation means that individual pieces of equipment operate and adjust to changes on their own. In more elaborate forms, a whole treatment plant and distribution system can be computer controlled to operate with little or no human supervision.

Example of Automated Equipment Operation

Figures 16-12 through 16-15 demonstrate the possible degrees of automation for chlorinator control. The same basic principles can be applied to some degree to most other treatment plant processes. A reliable automatic analyzer must be available to provide the necessary control. If the process being controlled is essential to the chemical or microbiological safety of the water, the automation should also include a monitor that will sound an alarm or shut down the system if dangerous under- or overfeed occurs.

Start–Stop Control

The simplest form of automation, start–stop control, is intended only to turn the chlorinator on and off at the same time as a pressure pump (see Figure 16-12). The chlorinator feed rate must be set manually. This type of arrangement would be adequate for a single-well system where the pump rate is fixed and the chlorine demand does not vary significantly.

Proportional Pacing

Proportional pacing represents the next degree of automation (Figure 16-13). This type of arrangement may be used where the pipeline flow is variable, for example, where several different pumps discharge varying amounts but the chlorine demand is always relatively constant. The feed rate must be set manually, but the chlorinator will modulate to feed the proper concentration in relation to the total flow.

Residual Control

In cases where the chlorine demand varies but the flow is constant, an automatic residual analyzer can be set up to sample the chlorinated water some distance downstream from the chlorine application point (Figure 16-14). In these cases, the operator sets the desired chlorine residual, and the analyzer will automatically adjust the feed to provide the correct amount

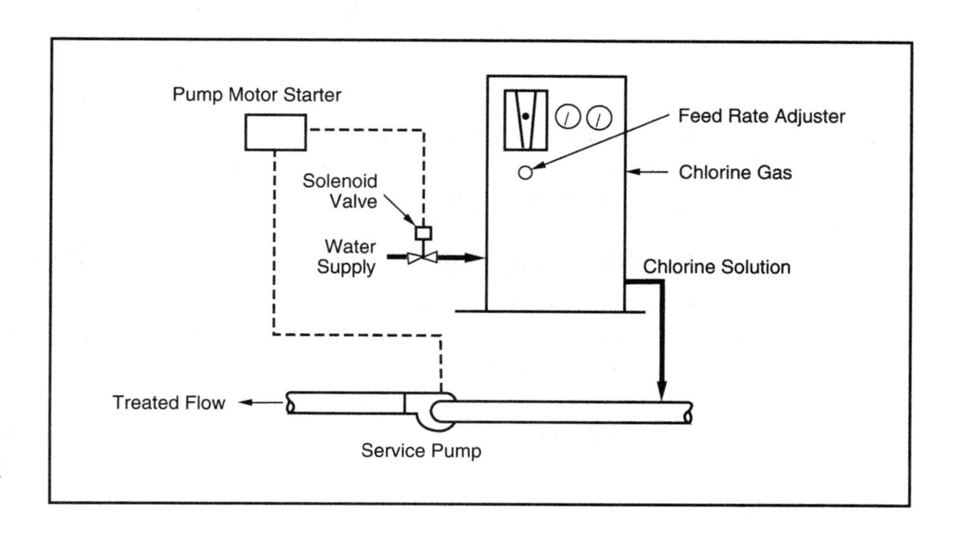

FIGURE 16-12 Start–stop control of chlorine feed

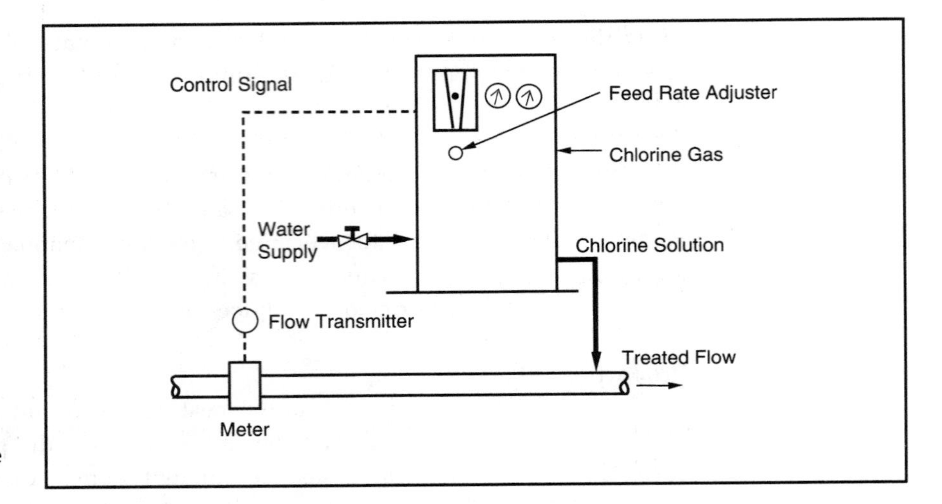

FIGURE 16-13 Proportional pacing of chlorine feed

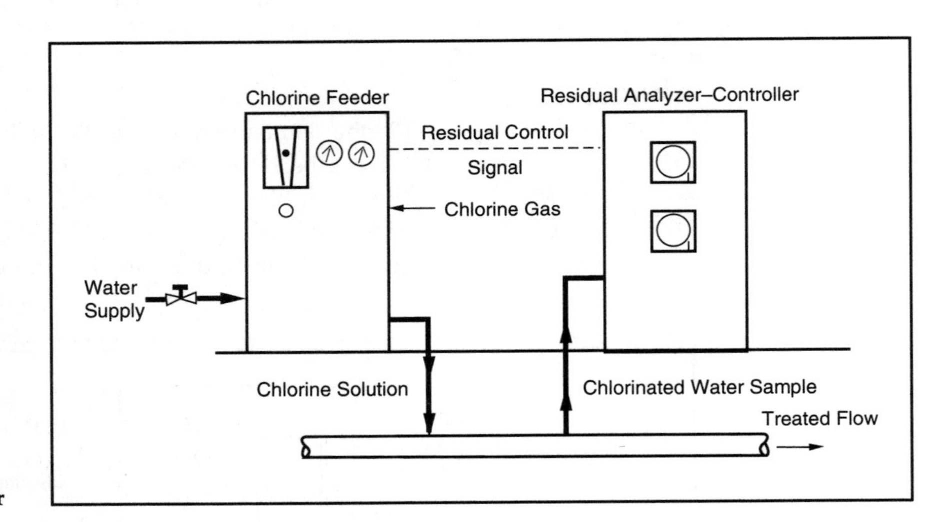

FIGURE 16-14 Residual control of chlorine feeder

(even though the flow rate and/or chlorine demand will vary). The only problem with this system is that it tends to overreact when there are abrupt changes in flow rate, causing wide swings in the residual until it eventually stabilizes.

Combined Flow and Residual Control

When both the flow rate and the chlorine demand vary, the best type of system is one that uses combined flow and residual control (Figure 16-15). In this case, the chlorinator is designed primarily to vary feed based on a signal from the flowmeter. If the flow rate is doubled, the amount of chlorine fed is immediately doubled. The analyzer then monitors the residual and varies the chlorinator feed rate, mostly in relation to the chlorine demand of the water.

Computerization

As computers and programs become more powerful, less expensive, and more "user friendly," they are finding increased use in water systems (Figure 16-16). It is becoming easier to find plant personnel who are familiar with different types of computer equipment and will make full use of their capability.

One of the first uses of computers in water systems was "data logging" — keeping track of system information. The intention was to eliminate most or all of the pressure, level, and other types of charts in a plant, and instead to record all operating data in one place on an automatic printer. In addition to keeping records for reports and future review, the system could sound an alarm and provide a special printout whenever certain parameters exceeded the desired limits.

Further degrees of computerization have included computer control of specific equipment or processes, and storage of plant records such as piping

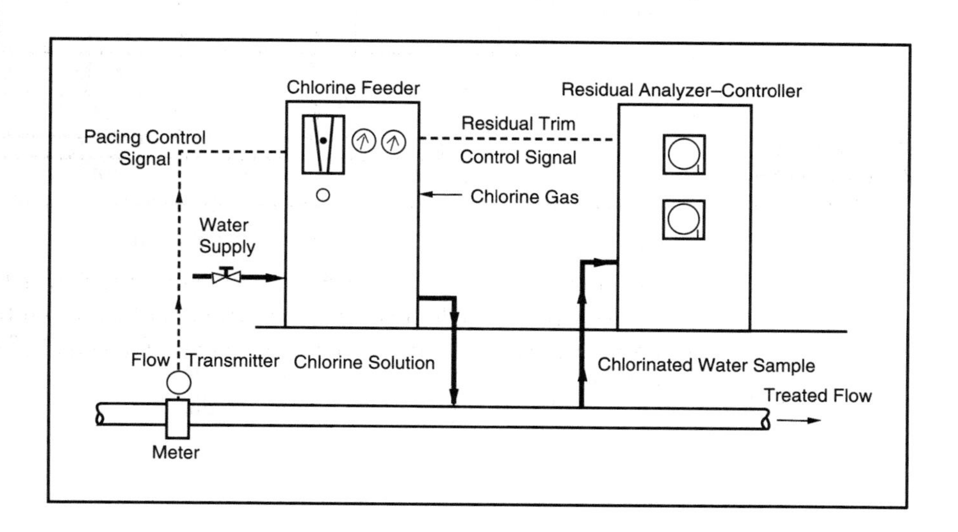

FIGURE 16-15 Combined flow and residual control of a chlorine feeder

FIGURE 16-16 A computer control system being used in a water treatment plant

diagrams, equipment operating and repair instructions, and maintenance records.

The final step is complete computer control of treatment plants and distribution system pressure. In some cases, personnel are still available to monitor the process. At other plants, the computer is allowed to operate the system without personnel during a night shift. Within limits, the computer is capable of starting and stopping pumps, adjusting chemical feed rates, and performing other functions. This, of course, requires a high degree of sophistication and reliability on the part of automatic analysis equipment, as well as a fail-safe capability. Systems operating without any on-site personnel are equipped to shut down if necessary and alert supervisors by phone when there are problems that the computer cannot correct.

State authorities are understandably cautious about allowing complicated treatment systems to function without an operator present. There are some things that only a human operator can determine about plant operation by sight, sound, or smell. State authorities will have to be thoroughly convinced that there is no public health danger that could result from the failure of an automation system.

A particularly good source of additional information on water system automation is AWWA Manual M2, *Automation and Instrumentation*, and *Water Distribution and Transmission* (part of this series).

Selected Supplementary Readings

Dentel, S.K. and K.M. Kingery. ***An Evaluation of Streaming Current Detectors.*** 1988. Denver, Colo: American Water Works Association Research Foundation and American Water Works Association.

Gotoh, K., J.K. Jacobs, S. Hosoda, and R.L. Gerstberger, eds. ***Instrumentation and Computer Integration of Water Utility Operations.*** Denver, Colo.: American Water Works Association Research Foundation, Japan Water Works Association, and American Water Works Association.

Manual M2, Automation and Instrumentation. 1994. Denver, Colo.: American Water Works Association.

Manual M33, Flowmeters in Water Supply. 1989. Denver, Colo.: American Water Works Association.

Water Treatment Plant Design. 2d ed. 1990. New York: McGraw Hill and American Water Works Association (available from AWWA).

APPENDIX A

Specifications and Approval of Treatment Chemicals and System Components

Drinking Water Additives

Many chemicals are added to potable water in the various water treatment processes. Chemical additives are commonly used in the following processes:

- coagulation and flocculation
- control of corrosion and scale
- chemical softening
- sequestering of iron and manganese
- chemical precipitation
- pH adjustment
- disinfection and oxidation
- algae and aquatic weed control

It is important that the chemicals added to water during treatment not add any objectionable tastes, odors, or color to the water; otherwise, customers will likely complain. It is of even greater importance that the chemicals themselves, or side effects that they cause, not create any danger to public health.

Coatings and Equipment in Contact With Water

In addition to chemicals that are added to water for treatment, many products come in direct contact with potable water and could, under some circumstances, cause objectionable contamination. The general categories of products that could cause an adverse effect because of their contact with water are

- pipes, faucets, and other plumbing materials
- protective materials such as paints, coatings, and linings
- joining and sealing materials
- lubricants
- mechanical devices that contact drinking water
- process media, such as filter and ion exchange media

Objectionable effects that can be caused by construction and maintenance materials include unpleasant tastes and odors, support of microbiological growth, and the liberation of toxic organic and inorganic chemicals.

NSF International Standards and Approval

For many years, the water industry relied on American Water Works Association (AWWA) standards and on approval by the US Environmental Protection Agency (USEPA) of individual products to ensure that harmful chemicals were not unknowingly added to potable water. However, there was no actual testing of the products, and the water treatment industry had to rely on the manufacturers' word that their products did not contain toxic materials.

In recent years, increasing numbers of products have been offered for public water supply use, many of them made with new manufactured chemicals that have not been thoroughly tested. In addition, recent toxicological research has revealed potential adverse health effects due to rather low levels of continuous exposure to many chemicals and substances previously considered safe.

In the early 1980s, it became evident that more exacting standards were needed, as well as more definite assurance that products positively meet safety standards. In view of the growing complexity of testing and approving water treatment chemicals and components, the USEPA awarded a grant in 1985 for the development of private-sector standards and a certification program. The grant was awarded to a consortium of partners. NSF

International was designated the responsible lead. Other cooperating organizations were the Association of State Drinking Water Administrators (ASDWA), the American Water Works Association, the American Water Works Association Research Foundation (AWWARF), and the Conference of State Health and Environmental Managers (COSHEM).

Two standards were developed by these organizations, along with the help of many volunteers from the water supply and manufacturing industries who served on development committees. These standards have now been adopted by the American National Standards Institute (ANSI), so they bear an ANSI designation in addition to the NSF reference.

ANSI/NSF Standard 60 essentially covers treatment chemicals for drinking water. The standard sets up testing procedures for each type of chemical, provides limits on the percentage of the chemical that can safely be added to potable water, and places limitations on any harmful substances that might be present as impurities in the chemical.

ANSI/NSF Standard 61 covers materials that are in contact with water, such as coatings, construction materials, and components used in processing and distributing potable water. The standard sets up testing procedures for each type of product to ensure that they do not unduly contribute to microbiological growth, leach harmful chemicals into the water, or otherwise cause problems or adverse effects on public health.

Manufacturers of chemicals and other products that are sold for the purpose of being added to water, or that will be in contact with potable water, must now submit samples of their product to NSF International or another qualifying laboratory for testing based on Standards 60 and 61. If a product qualifies, it is then "listed." There are also provisions for periodic retesting and inspection of the manufacturer's processes by the testing laboratory.

The listings of certified products provided by NSF International are used in particular by three groups in the water supply industry.

- In the design of water treatment facilities, engineers must specify that pipes, paints, caulks, liners, and other products that will be used in construction, as well as the chemicals to be used in the treatment process, are listed under one of the standards. This ensures in a very simple manner that only appropriate materials will be used. It also provides contractors with specific information on what materials qualify for use, without limiting competition.
- Individual states and local agencies have the right to impose more stringent requirements or to allow the use of products based on

other criteria, but most states have basically agreed to accept NSF International standards. When state authorities approve plans and specifications for the construction of new water systems or improvements to older systems, they will generally specify that all additives, coatings, and components must be listed as having been tested for compliance with the standards.

- Water system operators can best protect themselves and their water system from customer complaints, or possibly even lawsuits, by insisting that only listed products be used for everything that is added to, or in contact with, potable water. Whether it concerns purchasing paint for plant maintenance or taking bids for supplies of chemicals, the manufacturer or representative should be asked to provide proof that the exact product has been tested and is listed.

Copies of the current listing of products approved based on Standards 60 and 61 should be available at state drinking water program offices. A copy of the current listing can also be obtained by contacting the nearest NSF International regional office or the following address:

NSF International
3475 Plymouth Rd.
Ann Arbor, MI 48105
(313) 769-5362

AWWA Standards

Since 1908, AWWA has developed and maintained a series of voluntary consensus standards for products and procedures used in the water supply community. These standards cover products such as pipe, valves, and water treatment chemicals. They also cover procedures such as disinfection of storage tanks and design of pipe. As of 1994, AWWA had approximately 120 standards in existence and more than a dozen new standards under development. A list of standards currently available may be obtained from AWWA at any time.

AWWA offers and encourages use of its standards by anyone on a voluntary basis. AWWA has no authority to require the use of its standards by any water utility, manufacturer, or other person. Many individuals in the water supply community, however, choose to use AWWA standards. Manufacturers often produce products complying with the provisions of AWWA standards. Water utilities and consulting engineers frequently include the provisions of AWWA standards in their specifications for projects or purchase of products. Regulatory agencies require compliance with AWWA standards as part of their public water supply regulations. All of these uses of AWWA standards can establish a mandatory relationship, for example, between a buyer and a seller, but AWWA is not part of that relationship.

AWWA recognizes that others use its standards extensively in mandatory relationships and takes great care to avoid provisions in the standards that could give one party a disadvantage relative to another. Proprietary products are avoided whenever possible in favor of generic descriptions of functionality or construction. AWWA standards are not intended to describe the highest level of quality available, but rather describe minimum levels of quality and performance expected to provide long and useful service in the water supply community.

While maintaining product and procedural standards, AWWA does not endorse, test, approve, or certify any product. No product is or ever has been AWWA approved. Compliance with AWWA standards is encouraged, and demonstration of such compliance is entirely between the buyer and seller, with no involvement by AWWA.

AWWA standards are developed by balanced committees of persons from the water supply community who serve on a voluntary basis. Product users and producers, as well as those with general interest, are all involved in AWWA committees. Persons from water utilities, manufacturing companies, consulting engineering firms, regulatory agencies, universities, and others gather to provide their expertise in developing the content of the standards. Agreement of such a group is intended to provide standards, and thereby products, that serve the water supply community well.

Appendix B

Point-of-Use Treatment

In recent years, the public has shown increasing interest in installing special devices on their water service to further treat water supplied to them by a public water system. One reason is that they are not satisfied with the aesthetic qualities of the water, such as hardness, taste, and odor, and wish to improve the quality for their own use.

Another major reason is fear that the water is harmfully contaminated. This idea is fostered primarily by media stories of water contamination at specific locations and advertising by some bottled water and point-of-use device companies.

Home treatment devices are also widely used by owners of private wells to improve water quality. All of this public interest in improving water quality has led many new companies to enter the field of providing small treatment units.

Terminology

It is necessary to define the terms associated with these treatment devices. The term *home drinking water treatment* is fairly descriptive of the field. It covers the vast majority of uses, but many individual treatment devices are also installed in factories, offices, and other buildings.

There is a need to differentiate between units installed to treat all of the water at a building and those at just one water tap, so the following definitions have come into general use.

Point-of-entry (POE) devices are treatment units that are connected so they treat *all* of the water entering a building. A common example is a water softener installed to treat all of the water used in a building.

Point-of-use (POU) devices are units connected to treat water at a single location in a building. POU devices can be further subdivided into the following categories:

- *Stationary*. An example is a treatment unit mounted under a kitchen sink to treat all water to the cold-water faucet.
- *Line bypass*. An example is a treatment unit, mounted under a kitchen sink, that takes water from the cold-water pipe but delivers it to a separate, third faucet.
- *Countertop*. This generally refers to a small treatment unit that rests on the counter near a sink and is connected to the regular faucet by a hose when it is needed.
- *Faucet mounted*. This refers to a small treatment unit that is clamped on the spout of the regular faucet. These units usually have a bypass valve to select either treated or untreated water.
- *Pour-through*. These are portable, self-contained units, similar to a coffee maker. Water is poured through the unit for treatment.

Types of Treatment

This section discusses the principal types of POE and POU treatment devices.

Water Softeners

Home-type water softeners generally use the ion exchange process. In addition to reducing hardness, they remove some other impurities. But at the same time, they increase the salt level in the treated water.

Coarse salt is typically used for regeneration, and most units are furnished with a time clock to operate the regeneration cycle automatically in the middle of the night on set days of the week. Softeners are normally classified as POE devices because they are often installed to treat all water to the building. In other instances, softeners are connected to treat only the hot-water supply to a building.

Physical Filters

Physical filters are rather like strainers in that they are intended to remove suspended matter from the water. Some can remove material as small as cysts, larger bacteria, and asbestos fibers. Most units are designed to remove only larger particles such as grit, dirt, and rust. The filter unit may be made of fabric, fiber, ceramic, screening, diatomaceous earth, or other materials. Physical filters are most often used as a prefilter ahead of other treatment devices.

Activated Carbon Filters

As detailed in chapter 13, activated carbon can remove many organic chemical contaminants from water — provided there is sufficient contact time with the water. Carbon filters can remove most tastes, odors, and color as well as synthetic organic chemicals such as pesticides and volatile organic chemicals (VOCs). Carbon also effectively removes radon and chlorine from water. It is not generally effective for removing inorganic chemicals.

Carbon does not effectively remove microorganisms. As a matter of fact, if there are any microorganisms in the source water, they can actually thrive by feeding on the organic material that has previously been retained on the carbon particles.

The carbon in filters will eventually become loaded with contaminants so that the unit will no longer function. If a carbon unit is used after this point, it will periodically start to discharge quantities of organics into the effluent. As a result, the water being used from the unit may, at times, have a higher level of organics than the influent.

Unfortunately, there is no definite way of determining when carbon is exhausted. The life of the carbon in a unit depends on such factors as the concentration and types of contaminants in the water, the quantity of granular activated carbon (GAC) in the treatment unit, and the amount of water put through the unit. It is strongly recommended that when a GAC treatment unit is used, the carbon be changed frequently. A cross section of a typical small activated carbon filter is shown in Figure B-1. The inlet water is distributed throughout the container so that the flow through the carbon filter element is relatively even to the center core. The treated water then leaves the unit through the filter outlet.

Reverse Osmosis

A reverse osmosis (RO) unit cannot be used by itself for water treatment. Some membranes will be destroyed by exposure to chlorine, and the efficiency of a membrane can be fouled by suspended matter in the influent water. For this reason, an RO module is always preceded by prefilters (Figure B-2).

As discussed in chapter 15, RO units must be operated under pressure. Home-type RO units do not work very efficiently because they must operate on water system pressure, so about 75 percent of the water introduced to the unit is usually piped to waste.

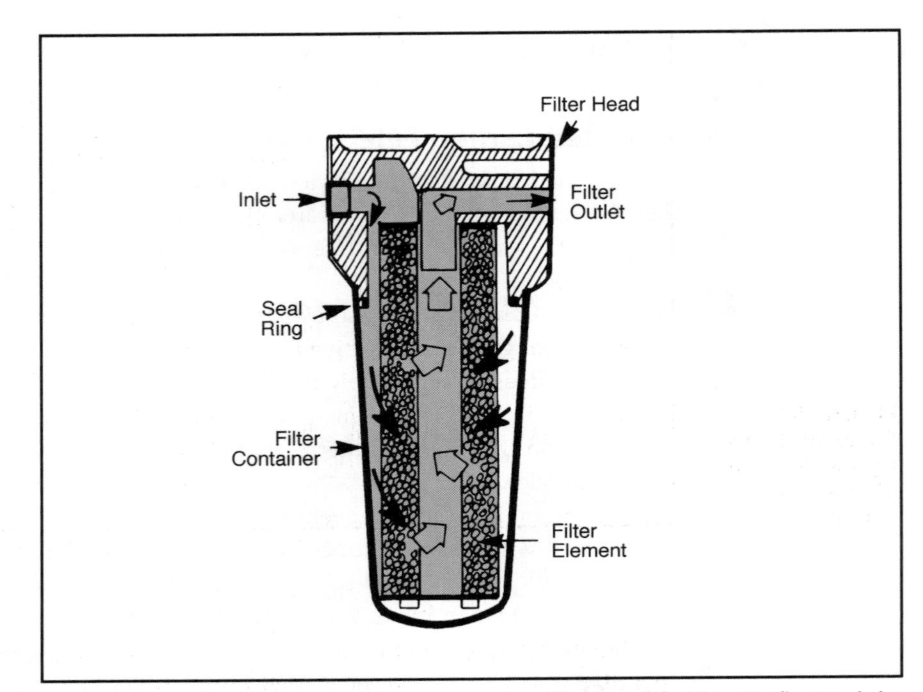

FIGURE B-1
Cross section of a typical activated carbon filter

Courtesy of the Water Quality Association

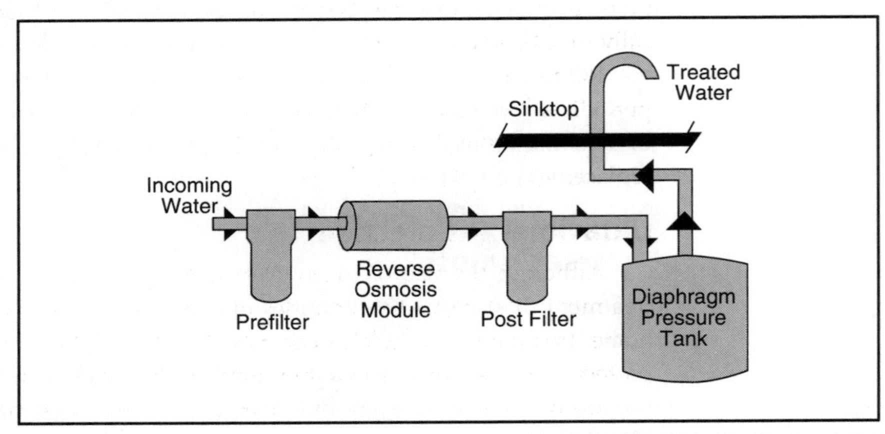

FIGURE B-2
Schematic of a reverse osmosis treatment unit

Courtesy of the Water Quality Association

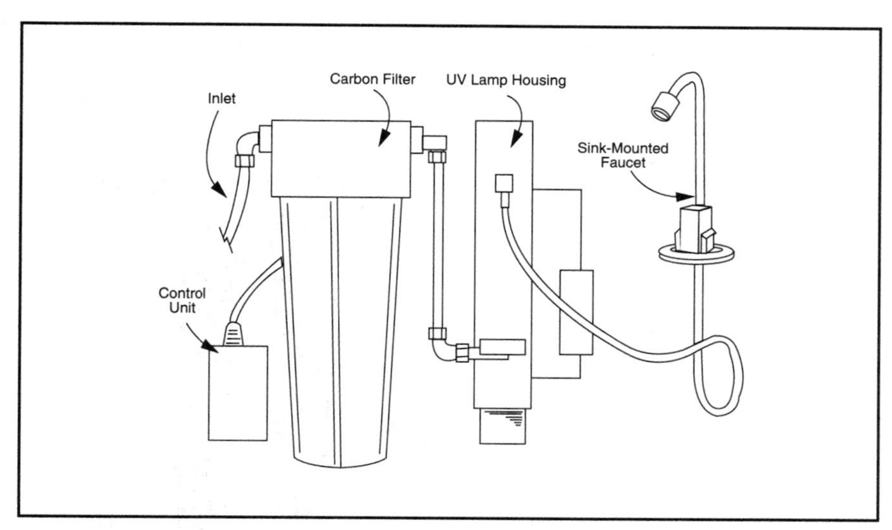

FIGURE B-3 Components of a typical ultraviolet disinfection system

Courtesy of the Water Quality Association

Small RO units produce a very small flow, so the treated water is piped to a diaphragm pressure tank, which supplies water to a separate faucet. RO is effective in removing most inorganics, salts, metals, and organics having larger molecules. Although RO is known to effectively remove *Giardia lamblia* cysts, bacteria, and viruses, it is not recommended for use on microbiologically unsafe water.

RO membrane units, as well as the pre- and postfilters, must be replaced periodically. Persons purchasing a small RO unit should have a clear understanding with the supplier on the projected life of the unit and what the replacement costs will be.

Ultraviolet Disinfection

Although the use of ultraviolet (UV) light is not cost effective for the treatment of any sizable quantities of water, it can be practically applied for home treatment use. Units consist of a UV light source enclosed in a protective transparent sleeve, mounted so that water can flow past the light. (Figure B-3) UV light destroys bacteria and inactivates viruses. However, its effectiveness against spores and cysts is questionable.

Both dissolved solids and turbidity adversely affect the performance of UV disinfection. The most common operating problem is a buildup of dirt on the transparent shield, so units should frequently be inspected and cleaned.

Distillers

Various sizes of distillers are available for home use. They all work on the principle of vaporizing water and then condensing the vapor (Figure B4 A–C). In the process, dissolved solids such as salt, metals, minerals, asbestos fibers, and other particles are removed. Some organic chemicals are also removed, but those that are more volatile are often vaporized and condensed with the product water. Distillers are effective in killing all microorganisms.

The principal problems with a distiller are that a small unit can produce only 2–3 gal (7.5–11 L) a day, and that the power cost for operation will be substantially higher than the operating cost of other types of treatment devices.

Special Treatment

Special treatment units for home use are also available for removal of methane, hydrogen sulfide, nitrate, fluoride, radon, and other objectionable contaminants. Most often, this type of treatment is used on private wells.

The Use of POE Treatment Instead of Central Treatment

Under recent US Environmental Protection Agency (USEPA) regulations, water systems with a source water exceeding a maximum contaminant level (MCL) can install POE devices on all water services (instead of providing central treatment) to satisfy the requirement of providing water to the public below the MCL. An example of where this might be a viable alternative is a system having water so hard that a large proportion of the customers have already installed home softeners. If the contaminant exceeding the MCL in the source water can be removed by ion exchange softening, the system might meet the federal requirements by ensuring that every customer has a softener installed.

At the same time, a number of problems must be overcome. It is difficult for the system to ensure that *all* drinking water furnished to the public is treated. Another problem is that the water system must have a plan for checking and maintaining all the units to ensure that they are always properly operated. It is expected that some very small systems may find this option workable under certain circumstances. The American Water Works Association has recommended that all treatment required for a public water supply be in the form of central treatment.

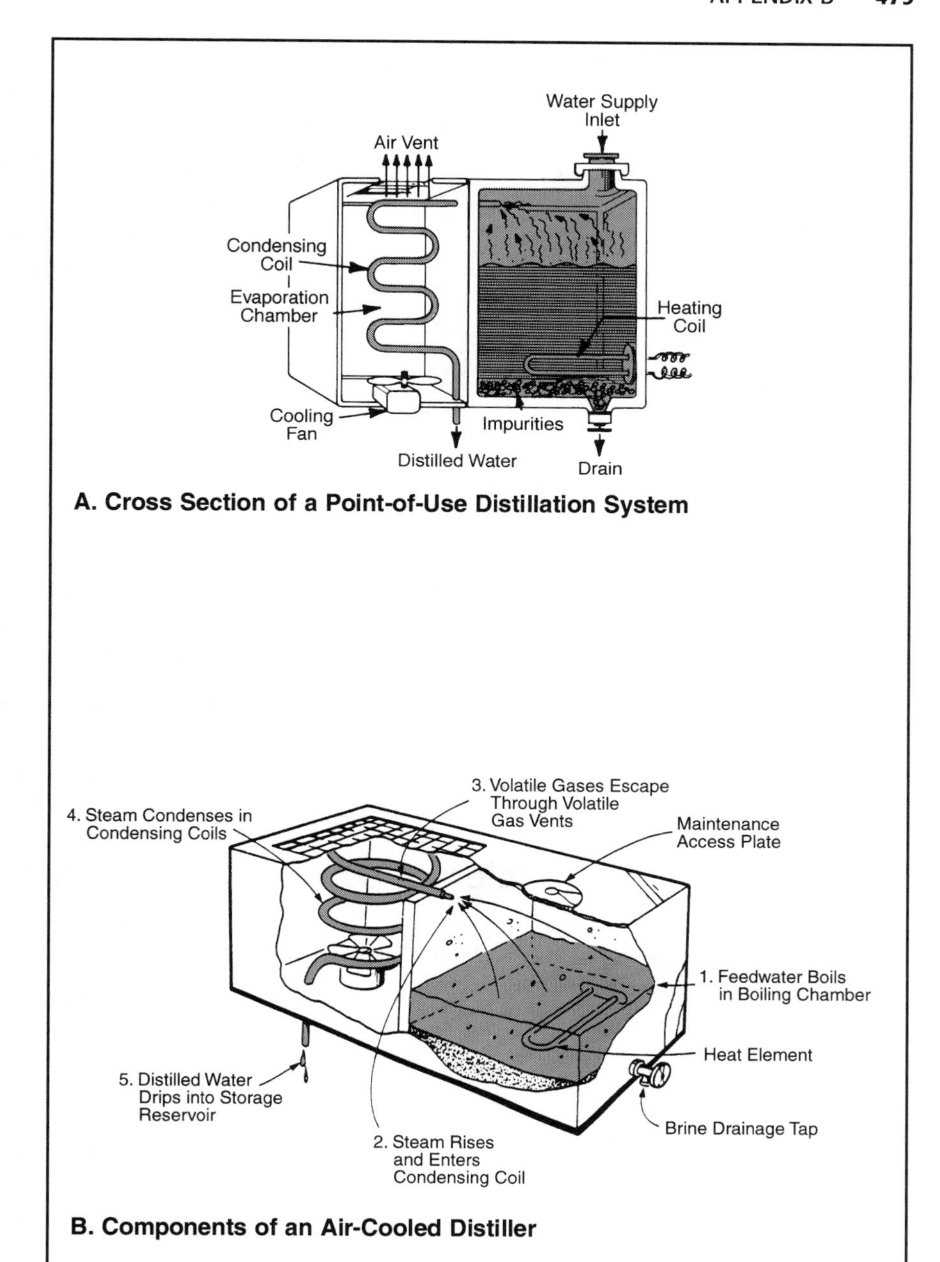

FIGURE B-4
Types of distillers

Courtesy of the Water Quality Association

Figure continued next page

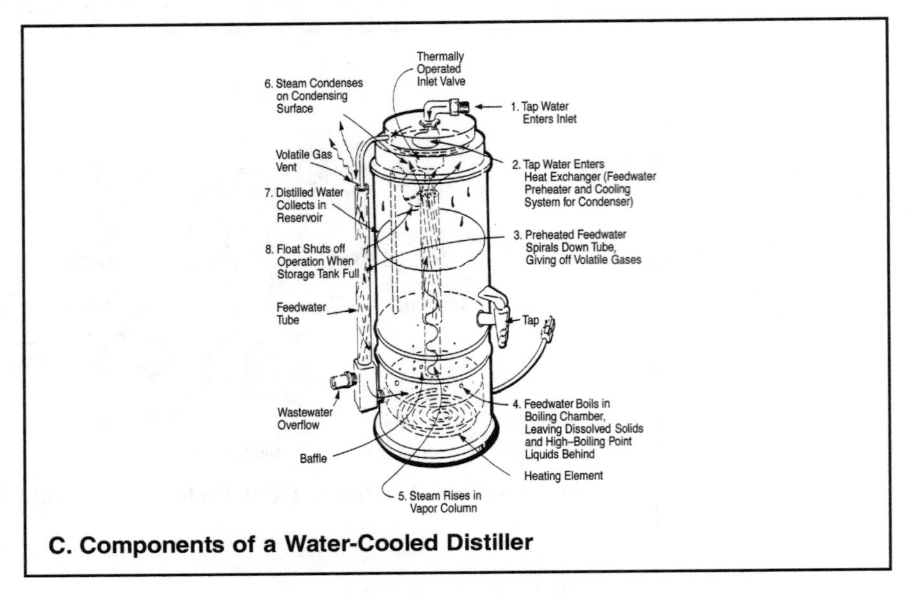

C. Components of a Water-Cooled Distiller

FIGURE B-4
Types of distillers (continued)

Further Information

County and state public health departments usually can provide additional details concerning POU and POE devices and installation. Other sources of information are the Water Quality Association, which is the trade association for the POU device industry, and NSF International, which provides testing of POE and POU units.

Selected Supplementary Readings

Bell, F.A., Jr. 1991. Review of Effects of Silver-Impregnated Carbon Filters on Microbial Quality, ***Jour. AWWA,*** 83(8):74.

Distillation for Home Water Treatment. 1990. East Lansing, Mich.: Michigan State University Cooperative Extension Service.

Fox, K.R. 1989. Field Experience With Point-of-Use Treatment Systems for Arsenic Removal. ***Jour. AWWA,*** 81(2):94.

Guide to Point-of-Use Treatment Devices for Removal of Inorganic/Organic Contaminants from Drinking Water. New Jersey Department of Environmental Protection.

Lykins, B.W., R.M. Clark, and J.A. Goodrich. ***Point-of-Use/Point-of-Entry for Drinking Water Treatment.*** Boca Raton, Fla.: Lewis Publishers.

Reverse Osmosis for Home Treatment of Drinking Water. 1990. Bulletin WQ24. East Lansing, Mich.: Michigan State University Cooperative Extension Service.

APPENDIX C

Other Sources of Information

Many trade associations, publishers, and other groups will supply specialized information to water system operators upon request. Some organizations provide product literature, association standards, installation manuals, slide presentations, and video programs. Some groups will also provide technical advice on their area of expertise.

Following are the names and addresses of some of these organizations:

American Concrete Pressure Pipe Association
8300 Boone Blvd., Suite 400
Vienna, VA 22182
(703) 893-4350

American Dental Association
211 E. Chicago Ave.
Chicago, IL 60611
(312) 440-2500

American Public Works Association
106 W. 11th St., Suite 1800
Kansas City, MO 64105-8106

Automatic Meter Reading Association
60 Revere Dr., Suite 500
Northbrook, IL 60062
(708) 480-9628

The Chlorine Institute, Inc.
2001 L Street, NW, Suite 506
Washington, DC 20036
(202) 775-2790

Ductile Iron Pipe Research Association
245 Riverchase Parkway East, Suite O
Birmingham, AL 35244
(205) 988-9870

Health Education Services
Division of Health Research, Inc.
P.O. Box 7126
Albany, NY 12224
(518) 439-7286

International Ozone Association
Pan American Committee
83 Oakwood Ave.
Norwalk, CT 06850
(203) 847-8169

International Society for Backflow Prevention
and Cross-Connection Control
P.O. Box 335
Eastlake, CO 80614
(303) 450-6651

National Ground Water Association
6375 Riverside Dr.
Dublin, OH 43017
(614) 761-1711

National Lime Association
3601 N. Fairfax Dr.
Arlington, VA 22201
(703) 243-5463

National Small Flows Clearinghouse
258 Stewart St.
P.O. Box 6064
Morgantown, WV 26506
(304) 293-4191

North American Society for Trenchless Technology
435 N. Michigan Ave., Suite 1717
Chicago, IL 60611
(312) 644-0828

NSF International
3475 Plymouth Rd.
P.O. Box 130140
Ann Arbor, MI 48113
(313) 769-8010

Plastics Pipe Institute
Wayne Interchange, Plaza II
155 Rte. 46 West
Wayne, NJ 07470
(201) 812-9076

Salt Institute
700 N. Fairfax St.
Fairfax Plaza, Suite 600
Alexandria, VA 22314
(703) 549-4648

Steel Tube Institute of North America
8500 Station St., Suite 270
Mentor, Ohio 44060
(216) 269-8447

Trench Shoring & Shielding Association
25 N. Broadway
Tarrytown, NY 10591
(914) 332-0040

Uni-Bell PVC Pipe Association
2655 Villa Creek Dr., Suite 155
Dallas, TX 75234
(214) 243-3902

Valve Manufacturers Association of America
050 17th St., NW, Suite 701
Washington, DC 20036
(202) 331-8105

Water Quality Association
4151 Naperville Rd.
Lisle, IL 60532
(708) 505-0160

The following are sources of current information on water system equipment:

American Water Works Association Buyers' Guide (covering the field of water supply and treatment). Published annually by the American Water Works Association, 6666 W. Quincy Ave., Denver, CO 80235.

Pollution Equipment News Buyer's Guide (covering the water supply and pollution fields). Published annually by Pollution Equipment News, 8650 Babcock Blvd., Pittsburgh, PA 15237-5821.

Public Works Manual (covering the fields of general operations, streets and highways, water supply and treatment, and water pollution control). Published annually by Public Works Journal Corporation, 200 S. Broad St., Ridgewood, NJ 07451.

GLOSSARY

activated alumina The chemical compound aluminum oxide, which is used to remove fluoride and arsenic from water by adsorption.

activated carbon A highly adsorptive material used to remove organic substances from water. (See *adsorption*.)

activated silica A coagulant aid used to form a denser, stronger floc.

activation The process of producing a highly porous structure in carbon by exposing the carbon to high temperatures in the presence of steam.

adhesion A condition in which particles stick together.

adsorbent Any material, such as activated carbon, used to adsorb substances from water.

adsorption The water treatment process used primarily to remove organic contaminants from water. Adsorption involves the adhesion of the contaminants to an adsorbent such as activated carbon.

aeration The process of bringing water and air into close contact to remove or modify constituents in the water.

after-precipitation The continued precipitation of a chemical compound (primarily calcium carbonate in the softening process) after leaving the sedimentation or solids-contact basin. This can cause scale formation on the filter media and in the distribution system.

agglomeration The action of microfloc particles colliding and sticking together to form larger settleable floc particles.

aggressive See *corrosive*.

air binding A condition that occurs in filters when air comes out of solution as a result of pressure decreases and temperature increases. The air clogs the voids between the media grains, which causes increased head loss through the filter and shorter filter runs.

air gap A method to prevent backflow by physically disconnecting the water supply and source of contamination.

air scouring The practice of admitting air through the underdrain system to ensure complete cleaning of media during filter backwash. Normally an alternative to using a surface wash system.

air stripper A packed-tower aerator consisting of a cylindrical tank filled with a packing material made of plastic or other material. Water is usually distributed over the packing at the top of the tank and air is forced in at the bottom using a blower.

alizarin-visual test A laboratory procedure for determining the fluoride concentration in water.

alum The most common chemical used for coagulation. It is also called aluminum sulfate.

aluminum sulfate See *alum*.

anaerobic Having no air or free oxygen.

anionic Having a negative ionic charge.

anionic polyelectrolyte A polyelectrolyte that forms negatively charged ions when dissolved in water.

anode Positive end (pole) of an electrolytic system.

anti-siphon device See *vacuum breaker*.

arching A condition that occurs when dry chemicals bridge the opening from the hopper to the dry feeder, clogging the hopper.

Asiatic clam A freshwater clam (*Corbicula fluminea*) that was introduced to US waters from southeast Asia in 1938. It is now present in almost all rivers south of 40° latitude. It presents a particular problem to surface water systems by clogging intakes and mechanical systems.

aspirator See *eductor*.

auxiliary scour See *filter agitation*.

auxiliary tank valve In a chlorination system, a union or yoke-type valve connected to the chlorine container or cylinder. It acts as a shutoff valve in case the container valve is defective.

backflow A hydraulic condition, caused by a difference in pressures, that causes nonpotable water or other fluid to flow into a potable water system.

backwash The reversal of flow through a filter to remove the material trapped on and between the grains of filter media.

baffle A metal, wooden, or plastic plate installed in a flow of water to slow the water velocity and provide a uniform distribution of flow.

bar screen A series of straight steel bars welded at their ends to horizontal steel beams, forming a grid. Bar screens are placed on intakes or in waterways to remove large debris.

Baumé The Baumé scale is a means of expressing the strength of a solution based on the solution's specific gravity.

bed life The time it takes for a bed of adsorbent to lose its adsorptive capacity. When this occurs, the bed must be replaced with fresh adsorbent.

bivalent ion An ion that has a valence charge of two. The charge can be positive or negative.

body feed In diatomaceous earth filters, the continuous addition of diatomaceous earth during the filtering cycle to provide a fresh filtering surface as the suspended material clogs the precoat.

breakpoint The point at which the chlorine dosage has satisfied the chlorine demand.

breakthrough The point in a filtering cycle at which turbidity-causing material starts to pass through the filter.

calcium carbonate ($CaCO_3$) The principal hardness- and scale-causing compound in water.

calcium hardness The portion of total hardness caused by calcium compounds such as calcium carbonate and calcium sulfate.

carbon dioxide (CO_2) A common gas in the atmosphere that is very soluble in water. High concentrations in water can cause the water to be corrosive. Carbon dioxide is added to water after the lime-softening process to lower the pH in order to reduce calcium carbonate scale formation. This process is known as recarbonation.

carbonate hardness Hardness caused primarily by compounds containing carbonate (CO_3), such as calcium carbonate and magnesium carbonate.

carcinogen A chemical compound that can cause cancer in animals or humans.

cathode The negative end (pole) of an electrolytic system.

cation A positive ion.

cation exchange Ion exchange involving ions that have positive charges, such as calcium and sodium.

cationic Having a positive ionic charge.

cationic polyelectrolyte Polyelectrolyte that forms positively charged ions when dissolved in water.

centrate The water that is separated from sludge and discharged from a centrifuge.

centrifugation In water treatment, a method of dewatering sludge by using a mechanical device (centrifuge) that spins the sludge at a high speed.

chlorination The process of adding chlorine to water to kill disease-causing organisms or to act as an oxidizing agent.

chlorinator Any device that is used to add chlorine to water.

clarification Any process or combination of processes that reduces the amount of suspended matter in water.

clarifier See *sedimentation basin*.

coagulant A chemical used in water treatment for coagulation. Common examples are aluminum sulfate and ferric sulfate.

coagulant aid A chemical added during coagulation to improve the process by stimulating floc formation or by strengthening the floc so it holds together better.

coagulation The water treatment process that causes very small suspended particles to attract one another and form larger particles. This is accomplished by the addition of a chemical, called a coagulant, that neutralizes the electrostatic charges on the particles that cause them to repel each other.

coagulation–flocculation The water treatment process that converts small particles of suspended solids into larger, more settleable clumps.

colloidal solid Finely divided solid that will not settle out of water for very long periods of time unless the coagulation–flocculation process is used.

combined chlorine residual The chlorine residual produced by the reaction of chlorine with substances in the water. Because the chlorine is "combined," it is not as effective a disinfectant as free chlorine residual.

concentration cell corrosion A form of localized corrosion that can form deep pits and tubercules.

contactor A vertical, steel cylindrical pressure vessel used to hold the activated carbon bed.

container valve The valve mounted on a chlorine container or cylinder.

conventional filtration A term that describes the treatment process used by most US surface water systems, consisting of the steps of coagulation, flocculation, sedimentation, and filtration.

corrosion The gradual deterioration or destruction of a substance or material by chemical action. The action proceeds inward from the surface.

corrosive Tending to deteriorate material, such as pipe, through electrochemical processes.

coupon test A method of determining the rate of corrosion or scale formation by placing metal strips (coupons) of a known weight in the pipe and examining them for corrosion after a period of time.

cross connection Any connection between safe drinking water and a nonpotable water or fluid.

$C \times T$ **value** The product of the residual disinfectant concentration C, in milligrams per liter, and the corresponding disinfectant contact time T, in minutes. Minimum $C \times T$ values are specified by the Surface Water Treatment Rule as a means of ensuring adequate kill or inactivation of pathogenic microorganisms in water.

cyclone degritter A centrifugal sand-and-grit removal device.

cylinder valve See *container valve.*

DE filter See *diatomaceous earth filter.*

debris rack See *bar screen.*

decant To draw off the liquid from a basin or tank without stirring up the sediment in the bottom.

density current A flow of water that moves through a larger body of water, such as a reservoir or sedimentation basin, and does not become mixed with the other water because of a density difference. This difference usually occurs because the incoming water has a different temperature or suspended solids content than the water body.

destratification Use of a method to prevent a lake or reservoir from becoming stratified. Typically consists of releasing diffused compressed air at a low point on the lake bottom.

detention time The average length of time a drop of water or a suspended particle remains in a tank or chamber. Mathematically, it is the volume of water in the tank divided by the flow rate through the tank.

dewatering (of reservoirs) A physical method for controlling aquatic plants in which a water body is completely or partially drained and the plants allowed to die.

dewatering (of sludge) A process to remove a portion of water from sludge.

diaphragm pump See *diaphragm-type metering pump.*

diaphragm-type metering pump A pump in which a flexible rubber, plastic, or metal diaphragm is fastened at the edges in a vertical cylinder. As the diaphragm is pulled back, suction is exerted and the liquid is drawn into the pump. When it is pushed forward, the liquid is discharged.

diatom A type of algae characterized by the presence of silica in its cell walls. Diatoms in source water can clog filters.

diatomaceous earth filter A pressure filter using a medium made from diatoms. The water is forced through the diatomaceous earth by pumping.

diffuser (1) Section of a perforated pipe or porous plates used to inject a gas, such as carbon dioxide or air, under pressure into water. (2) A type of pump.

direct filtration A filtration method that includes coagulation, flocculation, and filtration but excludes sedimentation. Only applicable to raw water relatively low in turbidity because all suspended matter must be trapped by the filters.

disinfectant residual An excess of chlorine left in water after treatment. The presence of residuals indicates that an adequate amount of chlorine has been added at the treatment stage to ensure completion of all reactions with some chlorine remaining.

disinfection The water treatment process that kills disease-causing organisms in water, usually by the addition of chlorine.

disinfection by-products (DBPs) New chemical compounds that are formed by the reaction of disinfectants with organic compounds in water. At high concentrations, many disinfection by-products are considered a danger to human health.

dissolved air flotation A clarification process in which gas bubbles are generated in a basin so that they will attach to solid particles to cause them to rise to the surface. The sludge that accumulates on the surface is then periodically removed by flooding or mechanical scraping.

dissolved solid Any material that is dissolved in water and can be recovered by evaporating the water after filtering the suspended material.

divalent See *bivalent ion*.

dredging A physical method for controlling aquatic plants in a lake or reservoir in which a dragline or similar mechanical equipment is used to remove plants and the bottom mud in which they are rooted.

drip leg A small piece of pipe installed on a chlorine cylinder or container that prevents collected moisture from draining back into the container.

dual-media filtration A filtration method designed to operate at a higher rate by using two different types of filter media, usually sand and finely granulated anthracite.

eductor A device used to mix a chemical with water. The water is forced through a constricted section of pipe (venturi) to create a low pressure, which allows the chemical to be drawn into the stream of water.

effluent launder A trough that collects the water flowing from a basin (effluent) and transports it to the effluent piping system.

ejector The portion of a chlorination system that feeds the chlorine solution into a pipe under pressure.

emergent weed An aquatic plant, such as cattails, that is rooted in the bottom mud of a water body but projects above the water surface.

empty bed contact time (EBCT) The volume of the tank holding an *activated carbon* bed, divided by the flow rate of water. The EBCT is expressed in minutes and corresponds to the detention time in a sedimentation basin.

epilimnion The upper, warmer layer of water in a stratified lake.

erosion The wearing away of a material by physical means.

evaporator A device used to increase release of chlorine gas from a container by heating the liquid chlorine.

excess-lime treatment A modification of the lime–soda ash method that uses additional lime to remove magnesium compounds.

ferric sulfate A chemical commonly used for coagulation.

filter agitation A method used to achieve more effective cleaning of a filter bed. The system typically uses nozzles attached to a fixed or rotating pipe installed just above the filter media. Water or an air–water mixture is fed through the nozzles at high pressure to help agitate the media and break loose accumulated suspended matter. It can also be called auxiliary scour or surface washing.

filter sand Sand that is prepared according to detailed specifications for use in filters.

filter tank The concrete or steel basin that contains the filter media, gravel support bed, underdrain, and wash-water troughs.

filtration The water treatment process involving the removal of suspended matter by passing the water through a porous medium such as sand.

flash mixing See *rapid mixing*.

floating weed An aquatic plant, such as water lilies, that floats partly or entirely on the surface of the water.

floc Collections of smaller particles (such as silt, organic matter, and microorganism) that have come together (agglomerated) into larger, more settleable particles as a result of the coagulation– flocculation process.

flocculation The water treatment process, following coagulation, that uses gentle stirring to bring suspended particles together so that they will form larger, more settleable clumps called floc.

flow measurement A measurement of the volume of water flowing through a given point in a given amount of time.

flow proportional control A method of controlling chemical feed rates by having the feed rate increase or decrease as the flow increases or decreases.

flow tube One type of primary element used in a pressure-differential meter. It measures flow velocity based on the amount of pressure drop through the tube. It is similar to a venturi tube.

fluoridation The water treatment process in which a chemical is added to the water to increase the concentration of fluoride ions to an optimal level. The purpose of fluoridation is to reduce the incidence of dental cavities in children.

fluorosis Staining or pitting of the teeth due to excessive amounts of fluoride in the water.

fluosilicic acid A strongly acidic liquid used to fluoridate drinking water.

free chlorine residual The residual formed once all the chlorine demand has been satisfied. The chlorine is not combined with other constituents in the water and is free to kill microorganisms.

galvanic corrosion A form of localized corrosion caused by the connection of dissimilar metals in an electrolyte such as water.

galvanic series A listing of metals and alloys according to their corrosion potential.

granular activated carbon (GAC) Activated carbon in a granular form, which is used in a bed, much like a conventional filter, to adsorb organic substances from water.

gravel bed Layers of gravel of specific sizes that support the filter media and help distribute the backwash water uniformly.

gravimetric dry feeder See *gravimetric feeder*.

gravimetric feeder Chemical feeder that adds specific weights of dry chemical.

hardness A characteristic of water, caused primarily by the salts of calcium and magnesium. Hardness causes deposition of scale in boilers, damage in some industrial processes, and sometimes objectionable taste.

harvesting A physical method for controlling aquatic plants in which the plants are pulled or cut and raked from the water body.

head loss The amount of energy used by water in moving from one point to another.

herbicide A compound, usually a synthetic organic substance, used to stop or retard plant growth.

humic substance Material resulting from the decay of leaves and other plant matter.

hydrogen sulfide (H_2S) A toxic gas produced by the anaerobic decomposition of organic matter and by sulfate-reducing bacteria. Hydrogen sulfide has a very noticeable rotten-egg odor.

hypochlorination Chlorination using solutions of calcium hypochlorite or sodium hypochlorite.

hypolimnion The lower layer of water in a stratified lake. The water temperature is near 39.2°F (4°C), at which water attains its maximum density.

injector See *ejector*.

inlet zone The initial zone in a sedimentation basin. It decreases the velocity of the incoming water and distributes it evenly across the basin.

insecticide A compound, usually a synthetic organic substance, used to kill insects.

interference substances All of the substances with which chlorine reacts before a chlorine residual can be available.

ion An atom that is electrically unstable because it has a different number of electrons than protons. A positive ion (one with more protons than electrons) is called a cation. A negative ion (one with fewer protons than electrons) is called an anion.

ion exchange process A process used to remove hardness from water that depends on special materials known as resins. The resins trade nonhardness-causing ions (usually sodium) for the hardness-causing ions calcium and magnesium. The process removes practically all the hardness from water.

ion exchange water softener A treatment unit that removes calcium and magnesium from water using ion exchange resins.

ionize To change or be changed into ions.

iron An abundant element found naturally in the earth. As a result, dissolved iron is found in most water supplies. When the concentration of iron exceeds 0.3 mg/L, it causes red stains on plumbing fixtures and other items in contact with the water. Dissolved iron can also be present in water as a result of corrosion of cast-iron or steel pipes. This is usually the cause of red-water problems.

iron bacteria Bacteria that use dissolved iron as an energy source. They can create serious problems in a water system because they form large, slimy masses that clog well screens, pumps, and other equipment.

jar test A laboratory procedure for evaluating coagulation, flocculation, and sedimentation processes. Used to estimate the proper coagulant dosage.

lamellar plates A series of thin, parallel plates installed at a 45° angle for shallow-depth sedimentation.

Langelier saturation index (LI) A numerical index that indicates whether calcium carbonate will be deposited or dissolved in a distribution system. The index is also used to indicate the corrosivity of water.

Leopold filter bottom A patented filter underdrain system using a series of perforated vitrified clay blocks with channels to carry the water.

lime–soda ash method A process used to remove carbonate and noncarbonate hardness from water.

limestone contactor A treatment device consisting of a bed of limestone through which water is passed to dissolve calcium carbonate. The addition of calcium carbonate to the water decreases corrosivity by increasing the pH, calcium concentration, and alkalinity of the water.

lining A physical method for controlling aquatic plants by placing a permanent lining, such as synthetic rubber, on the bottom of a water body.

loading rate The flow rate per unit area at which the water is passed through a filter or ion exchange unit.

localized corrosion A form of corrosion that attacks a small area.

magnesium hardness The portion of total hardness caused by magnesium compounds such as magnesium carbonate and magnesium sulfate.

magnetic flowmeter A flow-measuring device in which the movement of water induces an electrical current proportional to the rate of flow.

manganese An abundant element found naturally in the earth. Dissolved manganese is found in many water supplies. At concentrations above 0.05 mg/L, it causes black stains on plumbing fixtures, laundry, and other items in contact with the water.

manifold A pipe with several branches or fittings to allow water or gas to be discharged at several points. In aeration, manifolds are used to spray water through several nozzles.

manual solution feed A method of feeding a chemical solution for small water systems. The chemical is dissolved in a small plastic tank, transferred to another tank, and fed to the water system by a positive-displacement pump.

maximum contaminant level (MCL) The maximum allowable concentration of a contaminant in drinking water, as established by state and/or federal regulations. Primary MCLs are health related and mandatory. Secondary MCLs are related to the aesthetics of the water and are highly recommended but not required.

membrane processes Water treatment processes in which relatively pure water passes through a porous membrane while particles, molecules, or ions of unwanted matter are excluded. The membrane process used primarily for potable water treatment is reverse osmosis.

metering pump A chemical solution feed pump that adds a measured volume of solution with each stroke or rotation of the pump.

methane (CH_4) A colorless, odorless, flammable gas formed by the anaerobic decomposition of organic matter. When dissolved in water, methane causes a garlic-like taste. It is also called swamp gas.

microfloc The initial floc formed immediately after coagulation, composed of small clumps of solids.

microstrainer A rotating drum lined with a finely woven material such as stainless steel. Microstrainers are used to remove algae and small debris before they enter the treatment plant.

milk of lime The lime slurry formed when water is mixed with calcium hydroxide.

monovalent ion An ion having a valence charge of one. The charge can be either positive or negative.

mottling The staining of teeth due to excessive amounts of fluoride in the water.

mudball An accumulation of media grains and suspended material that creates clogging problems in filters.

multimedia filter A filtration method designed to operate at a high rate by utilizing three or more different types of filter media. The media types typically used are silica sand, anthracite, and garnet sand.

muriatic acid Another name for hydrochloric acid (HCl).

negative head A condition that can develop in a filter bed when the head loss gets too high. When this occurs, the pressure in the bed can drop to less than atmospheric.

nephelometric turbidimeter An instrument that measures turbidity by measuring the amount of light scattered by turbidity in a water sample. It is the only instrument approved by the US Environmental Protection Agency to measure turbidity in treated drinking water.

nephelometric turbidity unit (ntu) The amount of turbidity in a water sample as measured by a nephelometric turbidimeter.

noncarbonate hardness Hardness caused by the salts of calcium and magnesium.

nonionic Not having an ionic charge.

nonionic polyelectrolyte Polyelectrolyte that forms both positively and negatively charged ions when dissolved in water.

nonsettleable solids Finely divided solids, such as bacteria and fine clay particles, that will stay suspended in water for long periods of time.

ntu See *nephelometric turbidity unit.*

nucleus (plural: nuclei) The center of an atom, made up of positively charged particles called protons and uncharged particles called neutrons.

on-line turbidimeter A turbidimeter that continuously samples, monitors, and records turbidity levels in water.

organic substance (organic) A chemical substance of animal or vegetable origin, having carbon in its molecular structure.

organobromine compound The chemical compound formed when chlorine reacts with bromine.

orifice plate A type of primary element used in a pressure-differential meter, consisting of a thin plate with a precise hole through the center. Pressure drops as the water passes through the hole.

outlet zone The final zone in a sedimentation basin. It provides a smooth transition from the settling zone to the effluent piping.

overflow weir A steel or fiberglass plate designed to distribute flow evenly. In a sedimentation basin, the weir is attached to the effluent launder.

oxidation (1) The chemical reaction in which the valence of an element increases because of the loss of electrons from that element. (2) The conversion of organic substances to simpler, more stable forms by either chemical or biological means.

oxidize To chemically combine with oxygen.

ozone contactor A tank used to transfer ozone to water. A common type applies ozone under pressure through a porous stone at the bottom of the tank.

ozone generator A device that produces ozone by passing an electrical current through air or oxygen.

packed tower A cylindrical tank containing packing material, with water distributed at the top and airflow introduced from the bottom by a blower. Commonly referred to as an air-stripper.

packing material The material placed in a packed tower to provide a very large surface area over which water must pass to attain a high liquid–gas transfer.

pathogen A disease-causing organism.

pellet reactor A conical tank, filled about halfway with calcium carbonate granules, in which softening takes place quite rapidly as water passes up through the unit.

percolation The movement or flow of water through the pores of soil, usually downward.

permanent hardness Another term for noncarbonate hardness, derived from the fact that the hardness-causing noncarbonate compounds do not precipitate when the water is boiled.

pesticide Any substance or chemical used to kill or control troublesome organisms including insects, weeds, and bacteria.

photochemically Referring to chemical reactions that depend on light.

photosynthesis The process by which plants, using the chemical chlorophyll, convert the energy of the sun into food energy. Through photosynthesis, all plants, and ultimately all animals that feed on plants or on other plant-eating animals, obtain the energy of life from sunlight.

piezometer An instrument that measures pressure head in a conduit, tank, or soil by determining the location of the free water surface.

pilot filter A small tube, containing the same media as treatment plant filters, through which flocculated plant water is continuously passed, with a recording turbidimeter continuously monitoring the effluent. The amount of water passing through the pilot filter before turbidity breakthrough can be correlated to the operation of the plant filters under the same coagulant dosage.

pipe lateral system A filter underdrain system using a main pipe (header) with several smaller perforated pipes (laterals) branching from it on both sides.

piston pump A positive-displacement pump that uses a piston moving back and forth in a cylinder to deliver a specific volume of liquid being pumped.

plain sedimentation The sedimentatlon of suspended matter without the use of chemicals or other special means.

point-of-entry (POE) treatment A water treatment device, installed on a water customer's service, intended to treat all of the water entering a building.

point-of-use (POU) treatment A water treatment device used by a water customer to treat water at only one point, such as at a kitchen sink. The term is also sometimes used interchangeably with POE to cover all treatment installed on customer services.

polyelectrolyte High–molecular-weight, synthetic organic compound that forms ions when dissolved in water. It is also called a polymer.

polymer See *polyelectrolyte.*

polystyrene resin The most common resin used in the ion exchange process.

porous plate A filter underdrain system using ceramic plates supported above the bottom of the filter tank. This system is often used without a gravel layer so that the plates are directly beneath the filter media.

positive-displacement pump A pump that delivers a specific volume of liquid for each stroke of the piston or rotation of the impeller.

powdered activated carbon (PAC) Activated carbon in a fine powder form. It is added to water in a slurry form primarily for removing those organic compounds causing tastes and odors.

precipitate (1) A substance separated from a solution or suspension by a chemical reaction. (2) To form such a substance.

precoating The initial step in DE filtration, in which a thin coat of diatomaceous earth is applied to a support surface called a septum. This provides an initial layer of media for the water to pass through.

precursor compound Any of the organic substances that react with chlorine to form trihalomethanes.

preliminary treatment Any physical, chemical, or mechanical process used before the main water treatment processes. It can include screening, presedimentation, and chemical addition. Also called pretreatment.

presedimentation A preliminary treatment process used to remove gravel, sand, and other gritty material from the raw water before it enters the main treatment plant. This is usually done without the use of coagulating chemicals.

presedimentation impoundment A large earthen or concrete basin used for presedimentation of raw water. It is also useful for storage and for reducing the impact of raw-water quality changes on water treatment processes.

pressure-differential meter Any flow-measuring device that creates and measures a difference in pressure proportionate to the rate of flow. Examples include the venturi meter, orifice meter, and flow nozzle.

pressure-sand filter A sand filter placed in a cylindrical steel pressure vessel. The water is forced through the media under pressure.

pretreatment See *preliminary treatment*.

primary drinking water regulations Regulations on drinking water quality that are considered essential for preservation of public health.

primary element The part of a pressure-differential meter that creates a signal proportional to the water velocity through the meter.

progressive cavity pump A type of positive-displacement pump.

propeller meter A meter for measuring (1) flow rate by measuring the speed at which a propeller spins, and hence (2) the velocity at which the water is moving through a conduit of known cross-sectional area.

proportional meter Any flowmeter that diverts a small portion of the main flow and measures the flow rate of that portion as an indication of the rate of the main flow. The rate of the diverted flow is proportional to the rate of the main flow.

quicklime Another name for calcium oxide (CaO), which is used in water softening and stabilization.

radial flow Flow that moves across a basin from the center to the outside edges or vice versa.

rapid mixing The process of quickly mixing a chemical solution uniformly through the water.

rate-of-flow controller A control valve used to maintain a fairly constant flow through the filter.

reactivate To remove the adsorbed materials from spent activated carbon and restore the carbon's porous structure so that it can be used again. The reactivation process is similar to that used to activate carbon.

recarbonation The reintroduction of carbon dioxide into the water, either during or after lime–soda ash softening, to lower the pH of the water.

receiver The part of a pressure-differential meter that converts the signal from the receiver into a flow rate that can be read by the operator.

rectilinear flow Uniform flow in a horizontal direction.

red water Rust-colored water resulting from the formation of ferric hydroxide from iron naturally dissolved in the water or from the action of iron bacteria.

reducing agent Any chemical that decreases the positive valence of an ion.

regeneration The process of reversing the ion exchange softening reaction of ion exchange materials. Hardness ions are removed from the used materials and replaced with nontroublesome ions, thus rendering the materials fit for reuse in the softening process.

regeneration rate The flow rate per unit area of an ion exchange resin at which the regeneration solution is passed through the resin.

reject water The water that does not pass through a membrane, carries away the rejected matter, and must be disposed of.

residual See *disinfectant residual*.

residual flow control A method of controlling the chlorine feed rate based on the residual chlorine after the chlorine feed point.

resin In water treatment, the synthetic, bead-like material used in the ion exchange process.

respiration The process by which a living organism takes in oxygen from the air or water, uses it in oxidation, and gives off the products of oxidation, especially carbon dioxide. Breathing is an example of respiration.

reverse osmosis A pressure-drive process in which almost-pure water is passed through a semipermeable membrane. Water is forced through the membrane and most ions (salts) are left behind. The process is principally used for desalination of sea water.

rotameter A flow measurement device used for gases.

sand boil The violent washing action in a filter caused by uneven distribution of backwash water.

sand trap An enlargement of a conduit carrying raw water that allows the water velocity to slow down so that sand and other grit can settle.

saturation point The point at which a solution can no longer dissolve any more of a particular chemical. Precipitation of the chemical will occur beyond this point.

saturator A piece of equipment that feeds a sodium fluoride solution into water for fluoridation. A layer of sodium fluoride is placed in a plastic tank and water is allowed to trickle through the layer, forming a solution of constant concentration that is fed to the water system.

schmutzdecke The layer of solids and biological growth that forms on top of a slow sand filter, allowing the filter to remove turbidity effectively without chemical coagulation.

screening A pretreatment method that uses coarse screens to remove large debris from the water to prevent clogging of pipes or channels to the treatment plant.

secondary drinking water regulations Regulations developed under the Safe Drinking Water Act that establish maximum levels for substances affecting the aesthetic characteristics (taste, odor, or color) of drinking water.

sedimentation The water treatment process that involves reducing the velocity of water in basins so that the suspended material can settle out by gravity.

sedimentation basin A basin or tank in which water is retained to allow settleable matter, such as floc, to settle by gravity. Also called a settling basin, settling tank, or sedimentation tank.

sedimentation tank See *sedimentation basin*.

sequestering agent A chemical compound such as EDTA or certain polymers that chemically tie up (sequester) other compounds or ions so that they cannot be involved in chemical reactions.

settleability test A determination of the settleability of solids in a suspension by measuring the volume of solids settled out of a measured volume of sample in a specified interval of time, usually reported in milliliters per liter.

settling basin See *sedimentation basin.*

settling tank See *sedimentation basin.*

settling zone The zone in a sedimentation basin that provides a calm area so that the suspended matter can settle.

shading A physical method for controlling aquatic plants by limiting the amount of sunlight reaching the bottom of the water body.

shallow-depth sedimentation A modification of the traditional sedimentation process using inclined tubes or plates to reduce the distance the settling particles must travel to be removed.

short-circuiting A hydraulic condition in a basin in which the actual flow time of water through the basin is less than the design flow time (detention time).

slake The addition of water to quicklime (calcium oxide) to form calcium hydroxide, which can then be used in the softening or stabilization processes.

slaker The part of a quicklime feeder that mixes the quicklime with water to form hydrated lime (calcium hydroxide).

slow sand filtration A filtration process that involves passing raw water through a bed of sand at low velocity, resulting in particulate removal by physical and biological mechanisms.

sludge The accumulated solids separated from water during treatment.

sludge-blanket clarifier See *solids-contact basin.*

sludge blowdown The controlled withdrawal of sludge from a solids-contact basin to maintain the proper level of settled solids in the basin.

sludge zone The bottom zone of a sedimentation basin. It receives and stores the settled particles.

slurry A thin mixture of water and any insoluble material, such as activated carbon.

sodium fluoride A dry chemical used in the fluoridation of drinking water. It is commonly used in saturators.

sodium silicofluoride A dry chemical used in the fluoridation of drinking water. It is derived from fluosilicic acid.

softening The water treatment process that removes calcium and magnesium, the hardness-causing constituents in water.

solids-contact basin A basin in which the coagulation, flocculation, and sedimentation processes are combined. The water flows upward through the basin. It is used primarily in the lime softening of water. It can also be called an upflow clarifier or sludge-blanket clarifier.

solids-contact process A process combining coagulation, flocculation, and sedimentation in one treatment unit in which the flow of water is vertical.

SPADNS method A procedure used to determine the concentration of fluoride ion in water; a color change takes place following addition of a chemical reagent. SPADNS is the chemical reagent used in the test.

spray tower A tower built around a spray aerator to keep the wind from blowing the spray and to prevent the water from freezing during cold temperatures.

stabilization The water treatment process intended to reduce the corrosive or scale-forming tendencies of water.

static mixer A device designed to produce turbulence and mixing of chemicals with water, by means of fixed sloping vanes within the unit, without the need for any application of power.

sterilization The destruction of all organisms in water.

streaming current monitor An instrument that passes a continuous sample of coagulated water past a streaming current detector (SCD). The measurement is similar in theory to zeta potential determination and provides a reading that can be used to optimize chemical application.

submerged weed An aquatic plant, such as pondweed, that grows entirely beneath the surface of the water.

supersaturation A condition in which water contains very high concentrations of dissolved oxygen.

surface overflow rate A measurement of the amount of water leaving a sedimentation tank per square foot of tank surface area.

surface washing See *filter agitation.*

surface weed See *floating weed.*

suspended solid Solid organic and inorganic particle that is held in suspension by the action of flowing water.

synthetic organic chemical A carbon-containing chemical that has been manufactured, as opposed to occurring in nature.

synthetic resin See *resin*.

tare weight The initial weight of an item.

temporary hardness Another term for carbonate hardness, derived from the fact that the hardness-causing carbonate compounds precipitate when water is heated.

terminal head loss The head loss in a filter at which water can no longer be filtered at the desired rate because the suspended matter fills the voids in the filter and greatly increases the resistance to flow (head loss).

thermocline The temperature transition zone in a stratified lake, located between the epilimnion and the hypolimnion.

THM See *trihalomethane*.

total organic carbon (TOC) The amount of carbon bound in organic compounds in a water sample as determined by a standard laboratory test.

tracer study A study using a substance that can readily be identified in water (such as a dye) to determine the distribution and rate of flow in a basin, pipe, or channel.

transmitter The part of a pressure-differential meter that measures the signal from the primary element and sends another signal to the receiver.

trash rack See *bar screen*.

trihalomethane (THM) A compound formed when natural organic substances from decaying vegetation and soil (such as humic and fulvic acids) react with chlorine.

trivalent ion An ion having three valence charges. The charges can be positive or negative.

trunnion A roller device placed under ton containers of chlorine to hold them in place.

tube settlers A series of plastic tubes about 2 in. (50 mm) square, used for shallow-depth sedimentation.

tube-settling A shallow-depth sedimentation process that uses a series of inclined tubes.

tubercules Knobs of rust formed on the interior of cast-iron pipes as a result of corrosion.

turbidity A physical characteristic of water making the water appear cloudy. The condition is caused by the presence of suspended matter.

turbine meter A meter that measures flow rates by measuring the speed at which a turbine spins in water, indicating the velocity at which the water is moving through a conduit of known cross-sectional area.

turbulence A flow of water in which there are constant changes in flow velocity and direction resulting in agitation.

ultrasonic flowmeter A water meter that measures flow rate by measuring the difference in the velocity of sound beams directed through the water.

underdrain The bottom part of a filter that collects the filtered water and uniformly distributes the backwash water.

uniform corrosion A form of corrosion that attacks a material at the same rate over the entire area of its surface.

unstable Corrosive or scale forming.

upflow clarifier See *solids-contact basin.*

UV disinfection Disinfection using an ultraviolet light.

vacuum breaker A mechanical device that prevents backflow. It uses a siphoning action created by a partial vacuum to allow air into the piping system to break the vacuum.

Van der Waals force The attractive force existing between colloidal particles that allows the coagulation process to take place.

velocity meter A meter that measures water velocity by using a rotor with vanes (such as a propeller). It operates on the principle that the vanes move at about the same velocity as the flowing water.

venturi tube A type of primary element used in a pressure-differential meter that measures flow velocity based on the amount of pressure drop through the tube. Also used in a filter rate-of-flow controller.

viscosity The resistance of a fluid to flowing due to internal molecular forces.

viscous Having a sticky quality.

volatile Capable of turning to vapor (evaporating) easily.

volatile organic chemicals (VOCs) A class of manufactured, synthetic chemicals that are generally used as industrial solvents. They are classified as known or suspected carcinogens or as causing other adverse health effects. They are of particular concern to the water supply industry because they have widely been found as contaminants in groundwater sources.

volumetric dry feeder See *volumetric feeder*.

volumetric feeder A chemical feeder that adds specific volumes of dry chemical.

wash-water trough A trough placed above the filter media to collect the backwash water and carry it to the drainage system.

waterborne disease A disease caused by a waterborne organism or toxic substance.

weighting agent A material, such as bentonite, added to low-turbidity waters to provide additional particles for good floc formation.

weir overflow rate A measurement of the number of gallons per day of water flowing over each foot of weir in a sedimentation tank or circular clarifier. Mathematically, it is the gallons-per-day flow over the weir divided by the total length of the weir in feet.

wet scrubber A device installed to remove dust from a dry chemical feeder by means of a continuous water spray.

wheeler bottom A patented filter underdrain system using small porcelain spheres of various sizes in conical depressions.

wire-mesh screen A screen made of a wire fabric attached to a metal frame. The screen is usually equipped with a motor so that it can move continuously through the water and be automatically cleaned with a water spray. It is used to remove finer debris from the water than the bar screen is able to remove.

zebra mussel A freshwater mussel (*Dreissena polymorpha*) that was first found in the United States in Lake St. Clair in 1988. It multiplies extremely rapidly, and an infestation can quickly block a water intake and reduce the capacity of an intake pipeline.

zeta potential A measurement (in millivolts) of the particle charge strength surrounding colloidal solids. The more negative the number, the stronger the particle charge and the repelling force between particles.

INDEX

Note: an *f.* following a page number refers to a figure; a *t.* refers to a table.

A

B

C

D

E

F

G

H

I

J

L

M

N

O

Q

R

S

T

U

V

W

Z